Intimate Relationships

Marriage, Family, and Lifestyles through Literature

Edited by

Rose M. Somerville

San Diego State University

PRENTICE-HALL, INC., *Englewood Cliffs, New Jersey*

Library of Congress Cataloging in Publication Data

SOMERVILLE, ROSE M   comp.
  Intimate relationships.

  1. Family—Literary collections. 2. Interpersonal
relations—Literary collections. I. Title.

PN6071.F2S6   301.42   74-11366

ISBN 0-13-476861-2
ISBN 0-13-476879-5 (pbk.)

To my husband, John Somerville,
and our sons, Greg and Kent,
and to Judy, Susan, and Tara

Printed in the United States of America

10 9 8 7 6 5 4 3 2 1

PRENTICE-HALL INTERNATIONAL, INC., *London*
PRENTICE-HALL OF AUSTRALIA, PTY. LTD., *Sydney*
PRENTICE-HALL OF CANADA, LTD., *Toronto*
PRENTICE-HALL OF INDIA PRIVATE LIMITED, *New Delhi*
PRENTICE-HALL OF JAPAN, INC., *Tokyo*

ACKNOWLEDGMENTS

We gratefully acknowledge permission to reprint the following:

Roger Angell, "Summer in the Mountains." From *The Stone Arbor;* originally published, in somewhat different form, in *The New Yorker.* Copyright © 1960 by Roger Angell. Reprinted by permission of the author and Little, Brown and Company.

James Baldwin, "Sonny's Blues." Excerpt from "Sonny's Blues," from *Going to Meet the Man.* Copyright © 1957 by James Baldwin. Originally appeared in *Partisan Review.* Reprinted courtesy of the author and by permission of The Dial Press.

Rudolf Besier, *The Barretts of Wimpole Street.* Copyright 1930, 1958 by Rudolf Besier. Reprinted by permission of Little, Brown and Company and Victor Gollancz Ltd.

Kay Boyle, "Winter Night." From *Thirty Stories.* Copyright 1946 by Kay Boyle; first published in *The New Yorker.* Reprinted by permission of New Directions Publishing Corporation.

Jesus Colón, "Little Things Are Big." Excerpted from *A Puerto Rican in New York,* copyright © 1961. Reprinted by permission of New Outlook Publishers & Distributors, Inc.

Gregory Corso, "Marriage." From *The Happy Birthday of Death,* copyright © 1960 by New Directions Publishing Corporation and reprinted by their permission.

Harry Dolan, "I Remember Papa." From Budd Schulberg, ed., *From the Ashes,* 1967. Reprinted by permission of Ad Schulberg, agent.

Euripides, *Medea.* From the translation by Coleridge, in Oates and O'Neill, *7 Famous Greek Plays,* abridged and reprinted by permission of G. Bell & Sons, Ltd.

Edna Ferber, "Old Man Minick." From *Gigolo.* Copyright 1922 by Crowell Publishing Company, copyright renewed 1949 by Edna Ferber. All rights reserved. Reprinted by permission of the Proprietors of the Edna Ferber Estate.

Merrill Joan Gerber, "A Daughter of My Own." From *Stop Here, My Friend.* Copyright © 1964 by Merrill Joan Gerber. Reprinted by permission of the author and Cyrilly Abels, Literary Agent.

Herbert Gold, "Love and Like." Copyright © 1958 by Herbert Gold. Originally published in *The Hudson Review.* Reprinted by permission of the author and his agent, James Brown Associates, Inc.

Maxim Gorky, "Birth of a Man." Excerpted from "Birth of a Man," from *A Book of Short Stories by Maxim Gorky,* edited by Avrahm Yarmolinsky and Baroness Moura Budberg. Copyright 1939 by Holt, Rinehart and Winston, Inc. Copyright © 1967 by Avrahm Yarmolinsky and Moura Budberg. Reprinted by permission of Holt, Rinehart and Winston, Inc., and Jonathan Cape Ltd., on behalf of the Estate of Maxim Gorky.

Lorraine Hansberry, *A Raisin in the Sun.* Copyright © 1958, 1959 by Robert Nemiroff as Executor of the Estate of Lorraine Hansberry. Scene 1 of Act I reprinted by permission of Random House, Inc., and William Morris Agency, Inc.

Henrik Ibsen, *A Doll's House.* From the book *A Doll's House, The Wild Duck and The Lady from the Sea,* by Henrik Ibsen. Trans. by R. Farquharson Sharp, Eleanor Marx-Aveling; revised by Linda Hannas. Everyman's Library Edition. Reprinted by permission of the publishers, E. P. Dutton & Co., Inc., and J. M. Dent & Sons Ltd.

Christopher La Farge, "Mary Mulcahy." Reprinted by permission of Coward, McCann and Geoghegan, Inc., and Marie Rodell from *All Sorts and Kinds* by Christopher La Farge. Copyright © 1949 by Christopher La Farge. Originally published in *The New Yorker.*

Sinclair Lewis, *Main Street.* Excerpts reprinted by permission of Harcourt Brace Jovanovich, Inc., and Jonathan Cape Ltd.

Victoria Lincoln, "Down in the Reeds by the River." Originally published in *The New Yorker.* Copyright 1946 by the author. Reprinted by permission of Harold Ober Associates.

Michael Mason, "Saul." Originally published in *New Campus Writing No. 2,* edited by Nolan Miller. Copyright © 1957 by Bantam Books, Inc., and reprinted by their permission.

Mary McCarthy, "Yonder Peasant, Who Is He?" From *Memories of a Catholic Girlhood,* copyright 1948 by the author. First published in *The New Yorker.* Reprinted by permission of Harcourt Brace Jovanovich, Inc., A. M. Heath & Company Ltd., and William Heinemann Ltd.

Edna St. Vincent Millay, "Love Is Not All." From *Collected Poems,* Harper & Row, Publishers. Copyright 1931, 1958 by Edna St. Vincent Millay and Norma Millay Ellis. Reprinted by permission of Norma Millay Ellis.

Gabriela Mistral, "Madre mexicana." From *Lecturas para mujeres,* Mexico, 1967. Reprinted by permission of Libreria de Porrua Unos, Mexico.

Gerhard Neubeck, "Jim/Ann/Victoria and Fran/Greg/Phil." From *The Family in Search of a Future,* edited by Herbert A. Otto. Copyright © 1970 by Meredith Corporation. Reprinted by permission of Appleton-Century-Crofts, Educational Division, Meredith Corporation.

Joyce Carol Oates, "Normal Love." From *Marriages and Infidelities,* first published in *The Atlantic Monthly.* Reprinted by permission of the author.

Tillie Olsen, "I Stand Here Ironing." Copyright © 1956 by Tillie Olsen. Reprinted from the book *Tell Me a Riddle* by Tillie Olsen. Delacorte Press/Seymour Lawrence, Inc. Reprinted by permission of the author and the publisher. Originally appeared in *Pacific Spectator.*

Anne Petry, *The Street.* Copyright 1946 by Anne Petry. Excerpts reprinted by permission of Houghton Mifflin Company.

Carol Reilley, "The Dog." Reprinted by permission from *Practical English* (now *Scholastic Voice*), copyright 1947 by Scholastic Magazines, Inc.

Robert Rimmer, *Proposition 31.* Copyright © 1968 by Robert H. Rimmer. Excerpts reprinted by courtesy and permission of Robert H. Rimmer.

Anne Sexton, "For My Lover, Returning to His Wife." From *Love Poems,* copyright ©
1966, 1967, 1968, 1969 by Anne Sexton. Reprinted by permission of Houghton Mifflin
Company and The Sterling Lord Agency.

George Bernard Shaw, *Mrs. Warren's Profession.* Excerpt reprinted by permission of The
Society of Authors, on behalf of the Bernard Shaw Estate.

August Strindberg, *The Father.* From *The Plays of Strindberg,* Vol. I, translated by
Michael Meyer. Copyright © 1964 by Michael Meyer. Act I, Scene 5 of Act II, and
Scene 7 of Act III reprinted by permission of Random House, Inc., and David Higham
Associates, Ltd.

Ruth Suckow, "A Start in Life." From *Carryover.* Reprinted by permission of Ferner Nuhn.

Jeanne A. Taylor, "A House Divided." Copyright © 1967 by The Antioch Review, Inc.
First published in *The Antioch Review,* Vol. 27, No. 3; reprinted by permission of the
editors.

Piri Thomas, "Puerto Rican Paradise" and "If You Ain't Got Heart, You Ain't Got Nada."
From *Down These Mean Streets.* Copyright © 1967 by Piri Thomas. Reprinted by
permission of Alfred A. Knopf, Inc., and International Famous Agency.

Leo N. Tolstoy, "The Death of Ivan Ilyich." Reprinted in abridged form from *A Treasury
of Russian Literature,* edited by Bernard Guilbert Guerney, by permission of Vanguard
Press, Inc. Copyright 1943, renewed 1970, by Vanguard Press, Inc.

John Updike, "The Family Meadow." Copyright © 1965 by John Updike. Reprinted from
The Music School, by John Updike, by permission of Alfred A. Knopf, Inc., and Andre
Deutsch Limited. Originally appeared in *The New Yorker.*

Shirley Williams, "Tell Martha Not to Moan." Reprinted from *The Massachusetts Review,*
copyright © 1968 by The Massachusetts Review, Inc., by permission of the editors.

Sarellen M. Wuest, "A Sense of Tribe." Originally published in *New Campus Writing
No. 2,* edited by Nolan Miller. Copyright © 1957 by Bantam Books, Inc., and reprinted
by their permission.

Richard Yates, "The Best of Everything." From *Eleven Kinds of Loneliness,* copyright ©
1954 by the author. Reprinted by permission of Monica McCall, International Famous
Agency.

We also gratefully acknowledge permission to reproduce paintings as follows:

The Brooklyn Museum, for William Hogarth, *Gin Lane,* 1751. Bequest of Samuel E.
Haslett.

The Butler Institute of American Art, Youngstown, Ohio, for Winslow Homer, *Snap the
Whip.*

The Cleveland Museum of Art, for John Sloan, *Woman's Work,* gift of Amelia Elizabeth
White.

The Columbus Gallery of Fine Arts, for Jerome Myers, *An Interlude,* 1916, Ferdinand
Howald Collection.

The Hispanic Society of America, for Plate 14, *Que Sacrificio!* from Goya's *Los Caprichos.*

Instituto Nacional de Bellas Artes y Literatura—Mexico, for Diego Rivera, *Festival of
All Souls in the City and Festival of All Souls in the Country.*

Museo Frida Kahlo, for Frida Kahlo de Rivera, *My Pre-Natal Life,* 1936, and *Child
with Skeleton's Mask,* 1938.

Musée du Louvre, cliché des Musées Nationaux, for Degas, *The Bellelilli Family,* and
Veronese, *The Wedding at Cana.*

Museum of Fine Arts, Boston, for Jerome Thompson, *A 'Pic Nick,' Camden, Maine.*

National Gallery of Art, Washington, D.C., for William Hogarth, *Marriage à la Mode,*
Plate II, *The Unhappy Pair at Home,* Rosenfeld Collection; Amedeo Modigliani, *Gypsy
Woman with Baby,* Chester Dale Collection; Pablo Picasso, *The Lovers,* Chester Dale
Collection.

The Rose and John Somerville Collection, for Judith Somerville, *Aged Woman,* 1970.

Contents

part 2   *Parenting . . .*

part 3   *Extending . . .*

part 4   Losing . . .

part 5   **Innovating . . .**

Introduction

Brief works of fiction can be likened to tiny gems which enchant by their very concentration of quality. It is rarely surprising to find a novel or a very large jewel impressive. There is always, however, an element of pleasant surprise in the smaller work.

The stories and poems as well as the excerpts from plays, biographies, and novels that are offered in this collection have much to tell that is significant in human relationships generally and in family relationships specifically. In a world that moves at a rapid pace, readers who seek insights into their own values and behaviors and broadened understanding of the people around them are likely to be grateful for the shorter works of fiction that offer this opportunity. Brief works can be read between classes, on planes and buses, in waiting rooms. Cultural enrichment can come, then, in the interstices of busy lives, in leisure moments that must be snatched between chores or guests or exams. Even as leisure moments increase—and for some this does occur with age—the attention span may not broaden proportionately, and there is still appreciation for beauty and meaning packed into small spaces.

Given the facts of frenetic existence, it is indeed fortunate that the labor of the creative writer lightens our own as readers. It is a rare novelist who has not tried shorter forms of fiction, some successfully, some not. The conciseness of the short story is a challenge to the writer; many tellers of tales may shine brightly once or twice and then not be able to recapture the magic. But for all the shooting stars that descend into darkness and oblivion, there are hundreds more that continue to illuminate and inspire. Fictional art is flourishing. For the anthologist who wishes to bring together meaningful reflections of the complex realities in the relationships between generations, between men and women, between the family and the larger community, the problem is not scarcity but abundance.

No matter how many stories, plays, poems, and biographies the anthologist may read in this quest, there is the tantalizing thought that the next book on the shelf and the next among the thousands published each year will bring a gem more worthy of selection. An anthology, unlike a work of fiction, is never finished or complete. It can only represent an approximation of what was sought. In this collection a lifetime of reading, with particular intensity in the past two decades, lies back of the selection, together with years of teaching in the behavioral sciences and getting feedback on fiction from students of all ages.

Among the many possible, why these works of fiction and not others? The anthologist tends to weed out those that did not strike a spark, did not call out to be taken seriously, did not compel re-reading. All this, of course, is relative to time and circumstances. The discarded story might have carried a more significant message if read at another moment in the life of the anthologist. The inclusion of some works cannot be taken, then, as a negative judgment on those omitted.

Indeed, there is a rich storehouse awaiting the reader whose appetite has grown with tasting from the offerings in this collection. Hopefully the reader who is not accustomed to reading fiction will become "addicted," and the reader who has bypassed some form of fiction or has put off reading in full some of the longer works excerpted here, will find a broader world opening up. The greatest tribute that can be paid this collection is moving on from it to other works by the authors presented or to works necessarily omitted for lack of space in a single volume.

The fiction selected for this collection is focused on various stages of relationship: choosing or refusing or losing an intimate partner; becoming parents or foregoing this complex role; visiting kinfolk or being denied access; enjoying one's sexuality or being troubled by it; growing or shriveling in transactions with significant others in our lives or with the social institutions that affect us. Prejudice of various kinds becomes evident in these pages. The writers' sensitive perceptions alert us to our own ways of functioning, and nobody who has grown up in a family or plans to create or ignore one can fail to see the sometimes frightening images reflected in the mirror of literature.

Hopefully the vicarious experiences offered in this collection will send the reader not only to more fiction but to readings in the behavioral sciences for explanations of the acts and the attitudes which are feelingly depicted but rarely explained by the creative artist. Anyone can enjoy a story, and even the saddest tale can offer enjoyment. To go beyond enjoyment to full understanding may require the aid of sociologists, psychologists, and others who have studied larger samples of the relationships which the imaginative writer succeeds in spotlighting in a single brilliant instance. An instructor's guide is available to teachers at various academic levels and in all the many disciplines (sociology, psychology, home economics, counseling, health education, and, of course, literature) that can make use of the fiction offered in this collection.

The reader may, of course, follow any rhythm of reading preferred. The anthologist can only suggest that consideration be given to reading one story or poem or excerpt from a novel, play, or biography at a time, savoring its impact, re-reading it, and thinking about it before rushing on to the next. Perhaps only the reader who has dieted will recognize the pleasure that comes from the one sweet permitted from a full, tempting box. Just as rushing through the box can create feelings of regret and guilt, too many works of fiction in rapid succession can stir feelings of despair and unease, for the positive elements in intimate relationships are more rarely explored than the problems. Time lapses between stories may allow the reader to reflect upon some of these positive elements and, so fortified, be emotionally ready for the next challenge to complacency.

In this collection the choice of fiction from several cultures and historical periods is deliberate. The goal is not to suggest the universality of the human condition, although some of this does come through, but to encourage thought about the different kinds of expectations that each culture and subculture creates for its families. What kinds of deceit are practiced with the tyrannical father, or with the permissive father? What kinds of feelings are appropriate for the discarded spouse or lover? What kinds of behaviors are expected of a mother, a father, an in-law, a young child, an older child, a brother and sister? The names of these roles remain the same in the various cultures; what varying content is to be found in the role deriving from the norms of the given society or subsociety and relatedly from the personality needs of the individual? Is childlessness always considered a tragedy? Is old age always seen as a problem, and for whom? Is emotional distance between husband and wife prescribed or proscribed? Are male-female roles changing? Is adultery acceptable and in what circumstances?

For any reader fiction is full of "alternative lifestyles," as glimpses are provided of diverse attitudes and living patterns. However, the more recent variations, in the form of communal living, cohabitation, one-parent families by choice, triadic or group marriages, have yet to find full literary expression. Another collection a decade from now may offer more choices in this regard. Meanwhile the fiction available shows sufficient diversity of ways of relating to give readers much to ponder over. Awareness of the values implicit in the choices made by fictional characters will hopefully help readers to clarify their own values and to see not only their own lives but their own societies in clearer perspective. These psychosocial dividends will perhaps be seen by readers as fully repaying their investment of time and may be regarded ultimately as more precious than the immediate enjoyment offered by the works of art.

The paintings and etchings offered in this volume were selected with a view to focusing attention on the intimate relationships of men and women, of children and parents, of kin and non-kin, and on the feelings of joy, sadness, rivalry, boredom, degradation and elevation that mark the broad array of their interactions. Just as the literature is intended to stir discussion of social norms and personal choices, rather than issues of aesthetics, so too the art will hopefully serve not for exploration of art history but as kinds of projective tests of diverse feelings and values, and will thereby enrich the consideration of our human, and inevitably because human, our intimate relationships. In the literature as in the art, the selection has been based on considerations both of quality and of pertinence to various aspects of intimacy.

ROSE M. SOMERVILLE

part 1

Choosing...

Pablo Picasso. *The Lovers.*

The Best of Everything

Nobody expected Grace to do any work the Friday before her wedding. In fact, nobody would let her, whether she wanted to or not.

A gardenia corsage lay in a cellophane box beside her typewriter—from Mr. Atwood, her boss—and tucked inside the envelope that came with it was a ten-dollar gift certificate from Bloomingdale's. Mr. Atwood had treated her with a special shy courtliness ever since the time she necked with him at the office Christmas party, and now when she went in to thank him he was all hunched over, rattling desk drawers, blushing and grinning and barely meeting her eyes.

"Aw, now, don't mention it, Grace," he said. "Pleasure's all mine. Here, you need a pin to put that gadget on with?"

"There's a pin that came with it,"

she said, holding up the corsage. "See? A nice white one."

Beaming, he watched her pin the flowers high on the lapel of her suit. Then he cleared his throat importantly and pulled out the writing panel of his desk, ready to give the morning's dictation. But it turned out there were only two short letters, and it wasn't until an hour later, when she caught him handing over a pile of Dictaphone cylinders to Central Typing, that she realized he had done her a favor.

"That's very sweet of you, Mr. Atwood," she said, "but I do think you ought to give me all your work today, just like any oth—"

"Aw, now, Grace," he said, "you only get married once."

The girls all made a fuss over her too, crowding around her desk and giggling, asking again and again to

4

see Ralph's photograph ("Oh, he's *cute!*"), while the office manager looked on nervously, reluctant to be a spoilsport, but anxious to point out that it was, after all, a working day.

Then at lunch there was the traditional little party at Schrafft's—nine women and girls, giddy on their unfamiliar cocktails, letting their chicken à la king grow cold while they pummeled her with old times and good wishes. There were more flowers and another gift—a silver candy dish for which all the girls had whisperingly chipped in.

Grace said "Thank you" and "I certainly do appreciate it" and "I don't know what to say" until her head rang with the words and the corners of her mouth ached from smiling, and she thought the afternoon would never end.

Ralph called up about four o'clock, exuberant. "How ya doin', honey?" he asked, and before she could answer he said, "Listen. Guess what I got?"

"I don't know. A present or something? What?" She tried to sound excited but it wasn't easy.

"A bonus. Fifty dollars." She could almost see the flattening of his lips as he said "fifty dollars" with the particular earnestness he reserved for pronouncing sums of money.

"Why, that's lovely, Ralph," she said, and if there was any tiredness in her voice he didn't notice it.

"Lovely, huh?" he said with a laugh, mocking the girlishness of the word. "Ya *like* that, huh Gracie? No, but I mean I was really surprised, ya know it? The boss siz, 'Here, Ralph,' and he hands me this envelope. He don't even crack a smile or nothin', and I'm wonderin', what's

the deal here? I'm getting fired here, or what? He siz, 'G'ahead, Ralph, open it.' So I open it, and then I look at the boss and he's grinning a mile wide." He chuckled and sighed. "Well, so listen honey. What time ya want me to come over tonight?"

"Oh, I don't know. Soon as you can, I guess."

"Well listen. I gotta go over to Eddie's house and pick up that bag he's gonna loan me, so I might as well do that, go on home and eat, and then come over to your place around eight-thirty, nine o'clock. Okay?"

"All right," she said. "I'll see you then, darling." She had been calling him "darling" for only a short time —since it had become irrevocably clear that she was, after all, going to marry him—and the word still had an alien sound. As she straightened the stacks of stationery in her desk (because there was nothing else to do) a familiar little panic gripped her: She couldn't marry him—she hardly even *knew* him. Sometimes it occurred to her differently, that she couldn't marry him because she knew him too well, and either way it left her badly shaken, vulnerable to all the things that Martha, her roommate, had said from the very beginning.

"Isn't he funny?" Martha had said after their first date. "He says 'terlet.' I didn't know people really said 'terlet.'" And Grace had giggled, ready enough to agree that it *was* funny. That was a time when she had been ready to agree with Martha on practically anything—when it often seemed, in fact, that finding a girl like Martha from an ad in the *Times* was just about the luckiest thing that ever happened to her.

But Ralph had persisted all through the summer, and by fall she had begun standing up for him. "But what don't you *like* about him, Martha? He's perfectly nice."

"Oh, everybody's perfectly nice, Grace," Martha would say in her college voice, making perfectly nice a faintly absurd thing to be, and then she'd look up crossly from the careful painting of her fingernails. "It's just that he's such a little—a little *white worm.* Can't you see that?"

"Well, I certainly don't see what his *complexion* has to do with—"

"Oh, God, *you* know what I mean. Can't you see what I *mean?* Oh, and all those friends of his, his Eddie and his Marty and his George with their mean, ratty little clerks' lives and their mean, ratty little . . . It's just that they're all *alike,* those people. All they ever say is 'Hey, wha' happen t'ya Giants?' and 'Hey, wha' happen t'ya Yankees?' and they all live way out in some crowded little community, and their mothers have those damn little china elephants on the mantelpiece." And Martha would frown over her nail polish again, making it clear that the subject was closed.

All that fall and winter she was confused. For a while she tried going out only with Martha's kind of men —the kind that used words like "amusing" all the time and wore small-shouldered flannel suits like a uniform; and for a while she tried going out with no men at all. She even tried that crazy business with Mr. Atwood at the office Christmas party. And all the time Ralph kept calling up, hanging around, waiting for her to make up her mind. Once

she took him home to meet her parents in Pennsylvania (where she never would have dreamed of taking Martha), but it wasn't until Easter time that she finally gave in.

They had gone to a dance somewhere in Queens, one of those big American Legion dances that Ralph's crowd was always going to, and when the band played "Easter Parade" he held her very close, hardly moving, and sang to her in a faint, whispering tenor. It was the kind of thing she'd never have expected Ralph to do—a sweet, gentle thing—and although it probably wasn't just then that she decided to marry him, it always seemed so afterwards.

That night she had told Martha, and she could still see the look on Martha's face. "Oh, Grace, you're not —surely you're not *serious.* I mean, I mean, I thought he was more or less of a *joke*—you can't really mean you want to—"

"Shut up! You just shut up, Martha!" And she'd cried all night. Even now she hated Martha for it; even as she stared blindly at a row of filing cabinets along the office wall, half sick with fear that Martha was right.

The noise of giggles swept over her, and she saw with a start that two of the girls—Irene and Rose— were grinning over their typewriters and pointing at her. "*We* saw ya!" Irene sang. "*We* saw ya! Mooning again, huh Grace?" Then Rose did a burlesque of mooning, heaving her meager breasts and batting her eyes, and they both collapsed in laughter.

With an effort of will Grace resumed the guileless, open smile of a bride. The thing to do was concentrate on plans.

Tomorrow morning, "bright and early" as her mother would say, she would meet Ralph at Penn Station for the trip home. They'd arrive about one, and her parents would meet the train. "Good t'see ya, Ralph!" her father would say, and her mother would probably kiss him. A warm, homely love filled her; *they* wouldn't call him a white worm; *they* didn't have any ideas about Princeton men and "interesting" men and all the other kinds of men Martha was so stuck-up about. Then her father would probably take Ralph out for a beer and show him the paper mill where he worked (and at least Ralph wouldn't be snobby about a person working in a paper mill, either), and then Ralph's family and friends would come down from New York in the evening.

She'd have time for a long talk with her mother that night, and the next morning, "bright and early" (her eyes stung at the thought of her mother's plain, happy face), they would start getting dressed for the wedding. Then the church and the ceremony, and then the reception (Would her father get drunk? Would Muriel Ketchel sulk about not being a bridesmaid?), and finally the train to Atlantic City, and the hotel. But from the hotel on she couldn't plan any more. A door would lock behind her and there would be a wild, fantastic silence, and nobody else in all the world but Ralph to lead the way.

"Well, Grace," Mr. Atwood was saying, "I want to wish you every happiness." He was standing at her desk with his hat and coat on, and all around her was the chattering and scraping-back of chairs that meant it was five o'clock.

"Thank you, Mr. Atwood." She got to her feet, suddenly surrounded by all the girls in a bedlam of farewell.

"All the luck in the world, Grace."

"Drop us a card, huh, Grace? From Atlantic City?"

"So long, Grace."

"G'night, Grace, and listen: the best of everything."

Finally she was free of them all, out at the elevator, out of the building, hurrying through the crowds to the subway.

When she got home Martha was standing in the door of the kitchenette, looking very svelte in a crisp, new dress.

"Hi, Grace, I bet they ate you alive today, didn't they?"

"Oh, no," Grace said. "Everybody was—real nice." She sat down, exhausted, and dropped the flowers and the wrapped candy dish on a table. Then she noticed that the whole apartment was swept and dusted, and the dinner was cooking in the kitchenette. "Gee, everything looks wonderful," she said. "What'd you do all this for?"

"Oh, well, I got home early anyway," Martha said. Then she smiled, and it was one of the few times Grace had ever seen her look shy. "I just thought it might be nice to have the place looking decent for a change, when Ralph comes over."

"Well," Grace said, "it certainly was nice of you."

The way Martha looked now was even more surprising: she looked awkward. She was turning a greasy spatula in her fingers, holding it delicately away from her dress and examining it, as if she had something difficult to say. "Look, Grace," she began. "You do understand why I can't come to the wedding, don't you?"

"Oh, sure," Grace said, although in fact she didn't, exactly. It was something about having to go up to Harvard to see her brother before he went into the Army, but it had sounded like a lie from the beginning.

"It's just that I'd hate you to think I—well, anyway, I'm glad if you do understand. And the other thing I wanted to say is more important."

"What?"

"Well, just that I'm sorry for all the awful things that I used to say about Ralph. I never had a right to talk to you that way. He's a very sweet boy and I—well, I'm sorry, that's all."

It wasn't easy for Grace to hide a rush of gratitude and relief when she said, "Why, that's all right, Martha. I—"

"The chops are on fire!" Martha bolted for the kitchenette. "It's all right," she called back. "They're edible." And when she came out to serve dinner all her old composure was restored. "I'll have to eat and run," she said as they sat down. "My train leaves in forty minutes."

"But I thought you didn't have to go until *tomorrow!*"

"Well, I don't, actually," Martha said, "but I decided to go tonight. Because you see, Grace, another thing —if you can stand one more apology —another thing I'm sorry for is that I've hardly ever given you and Ralph a chance to be alone here. So tonight I'm going to clear out." She hesitated. "It'll be a sort of wedding gift from me, okay?" And then she smiled, not shyly this time but in a way that was more in character

—the eyes subtly averted after a flicker of special meaning. It was a smile that Grace—through stages of suspicion, bewilderment, awe, and practiced imitation—had long ago come to associate with the word "sophisticated."

"Well, that's very sweet of you," Grace said, but she really didn't get the point just then. It wasn't until long after the meal was over and the dishes washed, until Martha had left for her train in a whirl of cosmetics and luggage and quick good-bys, that she began to understand.

She took a deep, voluptuous bath and spent a long time drying herself, posing in the mirror, filled with a strange, slow excitement. In her bedroom, from the rustling tissues of an expensive white box, she drew the prizes of her trousseau—a sheer nightgown of white nylon and a matching negligee—put them on, and went to the mirror again. She had never worn anything like this before, or felt like this, and the thought of letting Ralph see her like this sent her into the kitchenette for a glass of the special dry sherry Martha kept for cocktail parties. Then she turned out all the lights but one and, carrying her glass, went to the sofa and arranged herself there to wait for him. After a while she got up and brought the sherry bottle over to the coffee table, where she set it on a tray with another glass.

When Ralph left the office he felt vaguely let down. Somehow, he's expected more of the Friday before his wedding. The bonus check had been all right (though secretly he'd been counting on twice that amount), and the boys had bought him a drink at lunch and kidded around in the appropriate way ("Ah, don't feel too bad, Ralph—worse things could happen"), but still, there ought to have been a real party. Not just the boys in the office but Eddie, and *all* his friends. Instead there would only be meeting Eddie at the White Rose like every other night of the year, and riding home to borrow Eddie's suitcase and to eat, and then having to ride all the way back to Manhattan just to see Grace for an hour or two. Eddie wasn't in the bar when he arrived, which sharpened the edge of his loneliness. Morosely he drank a beer, waiting.

Eddie was his best friend and an ideal Best Man because he'd been in on the courtship of Grace from the start. It was in this very bar, in fact, that Ralph had told him about their first date last summer: "Ooh, Eddie —what a figger!"

And Eddie had grinned. "Yeah? So what's the roommate like?"

"Ah, you don't want the roommate, Eddie. The roommate's a dog. A snob, too, I think. No, but this *other* one, this little *Gracie*—boy, I mean, she is *stacked*."

Half the fun of every date—even more than half—had been telling Eddie about it afterwards, exaggerating a little here and there, asking Eddie's advice on tactics. But after today, like so many other pleasures, it would all be left behind. Gracie had promised him at least one night off a week to spend with the boys, after they were married, but even so it would never be the same. Girls never understood a thing like friendship.

There was a ball game on the bar's television screen and he watched it idly, his throat swelling in a sentimental pain of loss. Nearly all his

life had been devoted to the friendship of boys and men, to trying to be a good guy, and now the best of it was over.

Finally Eddie's stiff finger jabbed the seat of his pants in greeting. "Whaddya say, sport?"

Ralph narrowed his eyes to indolent contempt, and slowly turned around. "Wha' happen ta you, wise guy? Get lost?"

"Whaddya—in a hurry a somethin'?" Eddie barely moved his lips when he spoke. "Can't wait two minutes?" He slouched on a stool and slid a quarter at the bartender. "Draw one, there, Jack."

They drank in silence for a while, staring at the television. "Got a little bonus today," Ralph said. "Fifty dollars."

"Yeah?" Eddie said. "Good."

A batter struck out; the inning was over and the commercial came on. "So?" Eddie said, rocking the beer around in his glass. "Still gonna get married?"

"Why not?" Ralph said with a shrug. "Listen, finish that, willya? I wanna get a move on."

"Wait awhile, wait awhile. What's ya hurry?"

"C'mon, willya?" Ralph stepped impatiently away from the bar. "I wanna go pick up ya bag."

"Ah, bag schmagg."

Ralph moved up close again and glowered at him. "Look, wise guy. Nobody's gonna *make* ya loan me the goddam bag, ya know. I don't wanna break ya *heart* or nothin'—"

"Arright, arright, arright. You'll getcha bag. Don't worry so much." He finished the beer and wiped his mouth. "Let's go."

Having to borrow a bag for his wedding trip was a sore point with

Ralph; he'd much rather have bought one of his own. There was a fine one displayed in the window of a luggage shop they passed every night on their way to the subway— a big, tawny Gladstone with a zippered compartment on the side, at thirty-nine ninety-five—and Ralph had had his eye on it ever since Easter time. "Think I'll buy that," he'd told Eddie, in the same offhand way that a day or so before he had announced his engagement ("Think I'll marry the girl"). Eddie's response to both remarks had been the same: "Whaddya—crazy?" Both times Ralph had said, "Why not?" and in defense of the bag he had added, "Gonna get married, I'll *need* somethin' like that." From then on it was as if the bag, almost as much as Gracie herself, had become the symbol of the new and richer life he sought. But after the ring and the new clothes and all the other expenses, he'd found at last that he couldn't afford it; he had settled for the loan of Eddie's which was similar but cheaper and worn, and without the zippered compartment.

Now as they passed the luggage shop he stopped, caught in the grip of a reckless idea. "Hey wait awhile, Eddie. Know what I think I'll do with that fifty-dollar bonus? I think I'll buy that bag right now." He felt breathless.

"Whaddya—*crazy?* Forty bucks for a bag you'll use maybe one time a year? Ya crazy, Ralph. C'mon."

"Ah—I dunno. Ya think so?"

"Listen, you better *keep* ya money, boy. You're gonna *need* it."

"Ah—yeah," Ralph said at last. "I guess ya right." And he fell in step with Eddie again, heading for the subway. This was the way things usually turned out in his life; he could never own a bag like that until he made a better salary, and he accepted it—just as he'd accepted without question, after the first thin sigh, the knowledge that he'd never possess his bride until after the wedding.

The subway swallowed them, rattled and banged them along in a rocking, mindless trance for half an hour, and disgorged them at last into the cool early evening of Queens.

Removing their coats and loosening their ties, they let the breeze dry their sweated shirts as they walked. "So what's the deal?" Eddie asked. "What time we supposed to show up in this Pennsylvania burg tomorra?"

"Ah, suit yourself," Ralph said. "Any time in the evening's okay."

"So whadda we do then? What the hell can ya *do* in a hillbilly town like that, anyway?"

"Ah, I dunno," Ralph said defensively. "Sit around and talk, I guess; drink beer with Gracie's old man or somethin'."

"Some week end," Eddie said. "Big, big deal."

Ralph stopped on the sidewalk, suddenly enraged, his damp coat wadded in his fist. "Look, you bastid. Nobody's gonna *make* ya come, ya know—you or Marty or George or any a the rest of 'em. Get that straight. You're not doin' *me* no favors, unnastand?"

"Whatsa matta?" Eddie inquired. "Whatsa matta? Can'tcha take a joke?"

"Joke," Ralph said. "You're fulla jokes." And plodding sullenly in Eddie's wake he felt close to tears.

They turned off into the block where they both lived, a double row of neat, identical houses, bordering the street where they'd fought and loafed and played stick-ball all their lives. Eddie pushed open the front door of his house and ushered Ralph into the vestibule, with its homely smell of cauliflower and overshoes. "G'wan in," he said, jerking a thumb at the closed living-room door, and he hung back to let Ralph go first.

Ralph opened the door and took three steps inside it before it hit him like a sock on the jaw. The room, dead silent, was packed deep with grinning, red-faced men— Marty, George, the boys from the block, the boys from the office— everybody, all his friends, all on their feet and poised motionless in a solid mass. Skinny Maguire was crouched at the upright piano, his spread fingers high over the keys, and when he struck the first rollicking chords they all roared into song, beating time with their fists, their enormous grins distorting the words:

"Fa he's a jally guh fella
Fa he's a jally guh fella
Fa he's a jally guh fell-ah
That nobody can deny!"

Weakly Ralph retreated a step on the carpet and stood there wide-eyed, swallowing, holding his coat. "That nobody can deny!" they sang, "That nobody can deny!" And as they swung into the second chorus Eddie's father appeared through the dining-room curtains, bald and beaming, in full song, with a great glass pitcher of beer in either hand. At last Skinny hammered out the final line:

"That—no—bod—dee—
can—dee—nye!"

And they all surged forward cheering, grabbing Ralph's hand, pound-ing his arms and his back while he stood trembling, his own voice lost under the noise. "Gee, fellas— thanks. I—I don't know what to— thanks, fellas. . . ."

Then the crowd cleaved in half, and Eddie made his way slowly down the middle. His eyes gleamed in a smile of love, and from his bashful hand the suitcase—not his own, but a new one: the big, tawny Gladstone with the zippered compartment on the side.

"Speech!" they were yelling. *"Speech! Speech!"*

But Ralph couldn't speak and couldn't smile. He could hardly even see.

At ten o'clock Grace began walking around the apartment and biting her lip. What if he wasn't coming? But of course he was coming. She sat down again and carefully smoothed the billows of nylon around her thighs, forcing herself to be calm. The whole thing would be ruined if she was nervous.

The noise of the doorbell was like an electric shock. She was halfway to the door before she stopped, breathing hard, and composed herself again. Then she pressed the buzzer and opened the door a crack to watch for him on the stairs.

When she saw he was carrying a suitcase, and saw the pale serious-ness of his face as he mounted the stairs, she thought at first that he knew; he had come prepared to lock the door and take her in his arms. "Hello, darling," she said softly, and opened the door wider.

"Hi, baby." He brushed past her and walked inside. "Guess I'm late, huh? You in bed?"

"No." She closed the door and

leaned against it with both hands holding the doorknob at the small of her back, the way heroines close doors in the movies. "I was just— waiting for you."

He wasn't looking at her. He went to the sofa and sat down, holding the suitcase on his lap and running his fingers over its surface. "Gracie," he said, barely above a whisper. "Look at this."

She looked at it, and then into his tragic eyes.

"Remember," he said. "I told you about that bag I wanted to buy? Forty dollars?" He stopped and looked around. "Hey, where's Martha? She in bed?"

"She's gone, darling," Grace said, moving slowly toward the sofa. "She's gone for the whole week end." She sat down beside him, leaned close and gave him Martha's special smile.

"Oh yeah?" he said. "Well anyway, listen. I said I was gonna borrow Eddie's bag instead, remember?"

"Yes."

"Well, so tonight at the White Rose I siz, 'C'mon, Eddie, let's go home pick up ya bag.' He siz, 'Ah, bag schmagg.' I siz, 'Whatsa matta?' but he don't say nothin', see? So we go home to his place and the living room door's shut, see?"

She squirmed closer and put her head on his chest. Automatically he raised an arm and dropped it around her shoulders, still talking. "He siz, 'G'head, Ralph, open the door.' I siz, 'Whatsa deal?' He siz, 'Never mind, Ralph, open the door.' So I open the door, and oh Jesus." His fingers gripped her shoulder with such intensity that she looked up at him in alarm.

"They was all there, Gracie," he said. "All the fellas. Playin' the piana, singin', cheerin'—" His voice wavered and his eyelids fluttered shut, their lashes wet. "A big surprise party," he said, trying to smile. "Fa me. Can ya beat that, Gracie? And then—and then Eddie comes out and —Eddie comes out and hands me this. The very same bag I been looking at all this time. He bought it with his own money and he didn't say nothin', just to give me a surprise. 'Here, Ralph,' he siz. 'Just to let ya know you're the greatest guy in the world.'" His fingers tightened again, trembling. "I cried, Gracie," he whispered. "I couldn't help it. I don't think the fellas saw it or anything, but I was cryin'." He turned his face away and worked his lips in a tremendous effort to hold back the tears.

"Would you like a drink, darling?" she asked tenderly.

"Nah, that's all right, Gracie. I'm all right." Gently he set the suitcase on the carpet. "Only, gimme a cigarette, huh?"

She got one from the coffee table, put it in his lips and lit it. "Let me get you a drink," she urged.

He frowned through the smoke. "Whaddya got, that sherry wine? Nah, I don't like that stuff. Anyway, I'm fulla beer." He leaned back and closed his eyes. "And then Eddie's mother feeds us this terrific meal," he went on, and his voice was almost normal now. "We had *steaks;* we had French-fried *potatas,*" his head rolled on the sofa-back with each item, "lettuce-and-tomato *salad, pickles, bread, butter*—everything. The works."

"Well," she said. "Wasn't that nice?"

"And afterwards we had ice cream

and coffee," he said, "and all the beer we could drink. It was a real spread."

Grace ran her hands over her lap, partly to smooth the nylon and partly to dry the moisture on her palms. "Well, that was certainly nice of them," she said. They sat there silent for what seemed a long time.

"I can only stay a minute, Gracie," Ralph said at last. "I promised 'em I'd be back."

Her heart thumped under the nylon. "Ralph, do you—do you like this?"

"What, honey?"

"My negligee. You weren't supposed to see it until—after the wedding, but I thought I'd—"

"Nice," he said, feeling the flimsy material between thumb and index finger, like a merchant. "Very nice. Wudga pay fa this, honey?"

"Oh—I don't know. But do you like it?"

He kissed her and began, at last, to stroke her with his hands. "Nice," he kept saying. "Nice. Hey, I like this." His hand hesitated at the low neckline, then slipped inside.

"I do love you, Ralph," she whispered. "You know that, don't you?"

His fingers pinched her breast, once, and slid quickly out again. The policy of restraint, the habit of months was too strong to break. "Sure," he said. "And I love you, baby. Now, you be a good girl and get ya beauty sleep, and I'll see ya in the morning. Okay?"

"Oh, Ralph. Don't go. Stay."

"Ah, I promised the fellas, Gracie." He stood up and straightened his clothes. "They're waitin' fa me, out to Eddie's."

She blazed to her feet, but the cry that was meant for a woman's appeal came out, through her tightening lips, as the whine of a wife. "Can't they wait?"

"Whaddya—*crazy?*" He backed away, eyes round with righteousness. She would *have* to understand. If this was the way she acted *before* the wedding, how the hell was it going to be *afterwards?* "Have a *heart,* willya? Keep the fellas waitin' *tonight?* After all they done *fa me?*"

After a second or two, during which her face became less pretty than he had ever seen it before, she was able to smile. "Of course not, darling. You're right."

He came forward again and gently brushed the tip of her chin with his fist, a husband reassured. " 'At's more like it," he said. "So I'll see ya, Penn Station, nine o'clock tomorra. Right, Gracie? Only, before I go—" He winked and slapped his belly. "I'm fulla beer. Mind if I use ya terlet?"

When he came out of the bathroom she was waiting to say goodnight, standing with her arms folded across her chest, as if for warmth. Lovingly he hefted the new suitcase and joined her at the door. "Okay, then, baby," he said, and kissed her. "Nine o'clock. Don't forget, now."

She smiled tiredly and opened the door for him. "Don't worry, Ralph," she said. "I'll be there."

Marriage

GREGORY CORSO

Should I get married? Should I be good?
Astound the girl next door with my velvet suit and faustus hood?
Don't take her to movies but to cemeteries
tell all about werewolf bathtubs and forked clarinets
then desire her and kiss her and all the preliminaries
and she going just so far and I understanding why
not getting angry saying You must feel! It's beautiful to feel!
Instead take her in my arms lean against an old crooked tombstone
and woo her the entire night the constellations in the sky—

When she introduces me to her parents
back straightened, hair finally combed, strangled by a tie,
should I sit knees together on their 3rd degree sofa
and not ask Where's the bathroom?
How else to feel other than I am,
often thinking Flash Gordon soap—
O how terrible it must be for a young man
seated before a family and the family thinking
We never saw him before! He wants our Mary Lou!
After tea and homemade cookies they ask What do you do for a living?
Should I tell them? Would they like me then?
Say All right get married, we're losing a daughter
but we're gaining a son—
And should I then ask Where's the bathroom?

O God, and the wedding! All her family and her friends
and only a handful of mine all scroungy and bearded
just wait to get at the drinks and food—
And the priest! he looking at me as if I masturbated
asking me Do you take this woman for your lawful wedded wife?
And I trembling what to say say Pie Glue!
I kiss the bride all those corny men slapping me on the back
She's all yours, boy! Ha-ha-ha!
And in their eyes you could see some obscene honeymoon going on—
Then all that absurd rice and clanky cans and shoes
Niagara Falls! Hordes of us! Husbands! Wives! Flowers! Chocolates!
All streaming into cozy hotels
All going to do the same thing tonight
The indifferent clerk he knowing what was going to happen
The lobby zombies they knowing what
The whistling elevator man he knowing
The winking bellboy knowing
Everybody knowing! I'd be almost inclined not to do anything!
Stay up all night! Stare that hotel clerk in the eye!
Screaming: I deny honeymoon! I deny honeymoon!
running rampant into those almost climactic suites
yelling Radio belly! Cat shovel!
O I'd live in Niagara forever! in a dark cave beneath the Falls
I'd sit there the Mad Honeymooner
devising ways to break marriages, a scourge of bigamy
a saint of divorce—

But I should get married I should be good
How nice it'd be to come home to her
and sit by the fireplace and she in the kitchen
aproned young and lovely wanting my baby
and so happy about me she burns the roast beef
and comes crying to me and I get up from my big papa chair
saying Christmas teeth! Radiant brains! Apple deaf!
God what a husband I'd make! Yes, I should get married!
So much to do! like sneaking into Mr Jones' house late at night
and cover his golf clubs with 1920 Norwegian books
Like hanging a picture of Rimbaud on the lawnmower
like pasting Tannu Tuva postage stamps all over the picket fence
like when Mrs Kindhead comes to collect for the Community Chest
grab her and tell her There are unfavorable omens in the sky!
And when the mayor comes to get my vote tell him
When are you going to stop people killing whales!
And when the milkman comes leave him a note in the bottle
Penguin dust, bring me penguin dust, I want penguin dust—

Yet if I should get married and it's Connecticut and snow

and she gives birth to a child and I am sleepless, worn,
up for nights, head bowed against a quiet window, the past behind me,
finding myself in the most common of situations a trembling man
knowledged with responsibility not twig-smear nor Roman coin soup—
O what would that be like!
Surely I'd give it for a nipple a rubber Tacitus
For a rattle a bag of broken Bach records
Tack Della Francesca all over its crib
Sew the Greek alphabet on its bib
and build for its playpen a roofless Parthenon

No, I doubt I'd be that kind of father
not rural not snow no quiet window
but hot smelly tight New York City
seven flights up, roaches and rats in the walls
a fat Reichian wife screeching over potatoes Get a job
And five nose running brats in love with Batman
And the neighbors all toothless and dry haired
like those hag masses of the 18th century
all wanting to come in and watch TV
The landlord wants his rent
Grocery store Blue Cross Gas & Electric Knights of Columbus
Impossible to lie back and dream Telephone snow, ghost parking—
No! I should not get married I should never get married!
But—imagine If I were married to a beautiful sophisticated woman
tall and pale wearing an elegant black dress and long black gloves
holding a cigarette holder in one hand and a highball in the other
and we lived high up in a penthouse with a huge window
from which we could see all of New York and ever farther on clearer days
No, can't imagine myself married to that pleasant prison dream—

O but what about love? I forget love
not that I am incapable of love
it's just that I see love as odd as wearing shoes—
I never wanted to marry a girl who was like my mother
And Ingrid Bergman was always impossible
And there's maybe a girl now but she's already married
And I don't like men and—
but there's got to be somebody!
Because what if I'm 60 years old and not married,
all alone in a furnished room with pee stains on my underwear
and everybody else is married! All the universe married but me!

Ah, yet well I know that were a woman possible as I am possible
then marriage would be possible—
Like SHE in her lonely alien gaud waiting her Egyptian lover
so I wait—bereft of 2,000 years and the bath of life.

from

Main Street

Sinclair Lewis

Carol was extravagant, but at least she did not try to clear herself of blame by going about whimpering, "I know I'm terribly extravagant but I don't seem to be able to help it."

Kennicott had never thought of giving her an allowance. His mother had never had one! As a wage-earning spinster Carol had asserted to her fellow librarians that when she was married, she was going to have an allowance and be businesslike and modern. But it was too much trouble to explain to Kennicott's kindly stubbornness that she was a practical housekeeper as well as a flighty playmate. She bought a budget-plan account book and made her budgets as exact as budgets are likely to be when they lack budgets.

For the first month it was a honeymoon jest to beg prettily, to confess, "I haven't a cent in the house, dear," and to be told, "You're an extravagant little rabbit." But the budget book made her realize how inexact were her finances. She became self-conscious; occasionally she was indignant that she should always have to petition him for the money with which to buy his food. She caught herself criticizing his belief that, since his joke about trying to keep her out of the poorhouse had once been accepted as admirable humor, it should continue to be his daily *bon mot*. It was a nuisance to have to run down the street after him because she had forgotten to ask him for money at breakfast.

But she couldn't "hurt his feelings," she reflected. He liked the lordliness of giving largess.

She tried to reduce the frequency of begging by opening accounts and having the bills sent to him. She had found that staple groceries, sugar, flour, could be most cheaply purchased at Axel Egge's rustic general store. She said sweetly to Axel:

"I think I'd better open a charge account here."

"I don't do no business except for cash," grunted Axel.

She flared, "Do you know who I am?"

"Yuh, sure, I know. The doc is good for it. But that's yoost a rule I made. I make low prices. I do business for cash."

She stared at his red impassive face, and her fingers had the undignified desire to slap him, but her reason agreed with him. "You're quite right. You shouldn't break your rule for me."

Her rage had not been lost. It had been transferred to her husband. She wanted ten pounds of sugar in a hurry, but she had no money. She ran up the stairs to Kennicott's office. On the door was a sign advertising a headache cure and stating, "The doctor is out, back at——" Naturally, the blank space was not filled out. She stamped her foot. She ran down to the drug store—the doctor's club.

As she entered she heard Mrs. Dyer demanding, "Dave, I've got to have some money."

Carol saw that her husband was there, and two other men, all listening in amusement.

Dave Dyer snapped, "How much do you want? Dollar be enough?"

"No, it won't! I've got to get some underclothes for the kids."

"Why, good Lord, they got enough now to fill the closet so I couldn't find my hunting boots, last time I wanted them."

"I don't care. They're all in rags. You got to give me ten dollars——"

Carol perceived that Mrs. Dyer was accustomed to this indignity. She perceived that the men, particularly Dave, regarded it as an excellent jest. She waited—she knew what would come—it did. Dave yelped, "Where's that ten dollars I gave you last year?" and he looked to the other men to laugh. They laughed.

Cold and still, Carol walked up to Kennicott and commanded, "I want to see you upstairs."

"Why—something the matter?"

"Yes!"

He clumped after her, up the stairs, into his barren office. Before he could get out a query she stated:

"Yesterday, in front of a saloon, I heard a German farmwife beg her husband for a quarter, to get a toy for the baby—and he refused. Just now I've heard Mrs. Dyer going through the same humiliation. And I—I'm in the same position! I have to beg you for money. Daily! I have just been informed that I couldn't have any sugar because I hadn't the money to pay for it!"

"Who said that? By God, I'll kill any——"

"Tut. It wasn't his fault. It was yours. And mine. I now humbly beg you to give me the money with which to buy meals for you to eat. And hereafter to remember it. The next time, I sha'n't beg. I shall simply starve. Do you understand? I can't go on being a slave——"

Her defiance, her enjoyment of the rôle, ran out. She was sobbing against his overcoat, "How can you shame me so?" and he was blubbering, "Dog-gone it, I meant to give you some, and I forgot it. I swear I won't again. By golly I won't!"

He pressed fifty dollars upon her, and after that he remembered to give her money regularly. . . . sometimes.

Daily she determined, "But I must have a stated amount—be business-like. System. I must do something about it." And daily she didn't do anything about it.

Que sacrificio!

Francisco Goya. *What a sacrifice!*

Tell Martha Not to Moan

SHIRLEY WILLIAMS

My mamma is a big woman, tall and stout, and men like her cause she soft and fluffy-looking. When she round them it all smiles and dimples and her mouth be looking like it couldn't never be fixed to say nothing but darling and honey.

They see her now, they sho see something different. I should not even come today. Since I had Larry things ain't been too good between us. But—that's my mamma and I know she gon be there when I need her. And sometime when I come, it okay. But this ain't gon be one a them times. Her eyes looking all ove me and I know it coming. She snort cause she want to say god damn but she don't cuss. "When it due, Martha?"

First I start to say, what. But I know it ain't no use. You can't fool old folks bout something like that, so I tell her.

"Last part of November."

"Who the daddy?"

"Time."

"That man what play piano at the Legion?"

"Yeah."

"What he gon do bout it?"

"Mamma, it ain't too much he can do, now is it? The baby on its way."

She don't say nothing for a long time. She sit looking at her hands. They all wet from where she been washing dishes and they all wrinkled like yo hand be when they been in water too long. She get up and get a dish cloth and dry em, then sit down at the table. "Where he at now?"

"Gone."

"Gone? Gone where?" I don't say nothing and she start cussing then. I get kinda scared cause mamma got to be real mad foe she cuss and I don't know who she cussing—me or

Time. Then she start talking to me. "Martha, you just a fool. I told you that man wan't no good first time I seed him. A musician the worst kind of men you can get mixed up with. Look at you. You ain't even eighteen years old yet, Larry just barely two, and here you is pregnant again." She go on like that for a while and I don't say nothing. Couldn't no way. By the time I get my mouth fixed to say something, she done raced on so far ahead that what I got to say don't have nothing to do with what she saying right then. Finally she stop and ask, "What you gon do now? You want to come back here?" She ain't never liked me living with Orine and when I say no, she ask, "Why not? It be easier for you."

I shake my head again. "If I here, Time won't know where to find me, and Time coming; he be back. He gon to make a place for us, you a see."

"Hump, you just played the fool again, Martha."

"No Mamma, that not it at all; Time want me."

"Is that what he say when he left?"

"No, but . . ."

Well, like the first night we met, he come over to me like he knowed me for a long time and like I been his for awmost that long. Yeah, I think that how it was. Cause I didn' even see him when we come in the Legion that first night.

Me and Orine, we just got our checks that day. We went downtown and Orine bought her some new dresses. But the dress she want to wear that night don't look right so we go racing back to town and change it. Then we had to hurry home and get dressed. It Friday night and the Legion crowded. You got to get there early on the weekend if you want a seat. And Orine don't want just any seat; she want one right up front. "Who gon see you way back there? Nobody. You don't dance, how you gon meet people? You don't meet people, what you doing out?" So we sit up front. Whole lots a people there that night. You can't even see the bandstand cross the dance floor. We sharing the table with some more people and Orine keep jabbing me, telling me to sit cool. And I try cause Orine say it a good thing to be cool.

The set end and people start leaving the dance floor. That when I see Time. He just getting up from the piano. I like him right off cause I like men what look like him. He kind of tall and slim. First time I ever seed a man wear his hair so long and nappy—he tell me once it an African Bush—but he look good anyway and he know it. He look around all cool. He step down from the bandstand and start walking toward me. He come over to the table and just look. "You," he say, "you my Black queen." And he bow down most to the floor.

Ah shit! I mad cause I think he just trying to run a game. "What you trying to prove, fool?" I ask him.

"Ah man," he say and it like I cut him. That the way he say it. "Ah man. I call this woman my Black queen—tell her she can rule my life and she call me a fool."

"And sides what, nigga," I tell him then, "I ain't black." And I ain't, I don't care what Time say. I just a dark woman.

"What's the matter, you shamed of being Black? Ain't nobody told you Black is pretty?" He talk all loud and people start gathering round. Somebody say, "Yeah, you tell her bout it, soul." I embarrassed and I look over at Orine. But she just grinning, not saying nothing. I guess she waiting to see what I gon do so I stand up.

"Well if I is black, I is a fine black." And I walk over to the bar. I walk just like I don't know they watching my ass, and I hold my head up. Time follow me right on over to the bar and put his arm round my shoulder.

"You want a drink?" I start to say no cause I scared. Man not supposed to make you feel like he make me feel. Not just like doing it—but, oh, like it right for him to be there with me, touching me. So I say yes. "What's your name?" he ask then.

I smile and say, "They call me the player." Orine told a man that once in Berkeley and he didn't know what to say. Orine a smart woman.

"Well they call me Time and I know yo mamma done told you Time ain't nothing to play with." His smile cooler than mine. We don't say nothing for a long while. He just stand there with his arm round my shoulder looking at us in the mirror behind the bar. Finally he say, "Yeah, you gon be my Black queen." And he look down at me and laugh. I don't know what to do, don't know what to say neither, so I just smile.

"You gon tell me your name or not?"

"Martha."

He laugh. "That a good name for you."

"My mamma name me that so I be good. She name all us kids from the Bible," I tell him laughing.

"And is you good?"

I nod yes and no all at the same time and kind of mumble cause I don't know what to say. Mamma really did name all us kids from the Bible. She always saying, "My mamma name me Veronica after the woman in the Bible and I a better woman for it. That why I name all my kids from the Bible. They got something to look up to." But mamma don't think I'm good, specially since I got Larry. Maybe Time ain't gon think I good neither. So I don't answer, just smile and move on back to the table. I hear him singing soft-like, "Oh Mary don't you weep, tell yo sister Martha not to moan." And I kind of glad cause most people don't even think bout that when I tell em my name. That make me know he really smart.

We went out for breakfast after the Legion close. Him and me and Orine and German, the drummer. Only places open is on the other side of town and at first Time don't want to go. But we finally swade him.

Time got funny eyes, you can't hardly see into em. You look and you look and you can't tell nothing from em. It make me feel funny when he look at me. I finally get used to it, but that night he just sit there looking and don't say nothing for a long time after we order.

"So you don't like Black?" he finally say.

"Do you?" I ask. I think I just ask him questions, then I don't have to talk so much. But I don't want him to talk bout that right then, so I smile and say, "Let's talk bout you."

"I am not what I am." He smiling and I smile back, but I feel funny cause I think I supposed to know what he mean.

"What kind of game you trying to run?" Orine ask. Then she laugh. "Just cause we from the country don't mean we ain't hip to niggas trying to be big-time. Ain't that right, Martha?"

I don't know what to say, but I know Time don't like that. I think he was going to cuss Orine out, but German put his arm round Orine and he laugh. "He just mean he ain't what he want to be. Don't pay no mind to that cat. He always trying to blow some shit." And he start talking that talk, rapping to Orine.

I look at Time. "That what you mean?"

He all lounged back in the seat, his legs stretched way out under the table. He pour salt in a napkin and mix it up with his finger. "Yeah, that's what I mean. That's all about me. Black is pretty, Martha." He touch my face with one finger. "You let white people make you believe you ugly. I bet you don't even dream."

"I do too."

"What do you dream?"

"Huh?" I don't know what he talking bout. I kind of smile and look at him out the corner of my eye. "I dreams bout a man like you. Why, just last night, I dream—"

He start laughing. "That's all right. That's all right."

The food come then and we all start eating. Time act like he forgot all bout dreams. I never figure out how he think I can just sit there and tell him the dreams I have at night, just like that. It don't seem like what I dream bout at night mean as much as what I think bout during the day.

We leaving when Time trip over this white man's feet. That man's feet all out in the aisle but Time don't never be watching where he going no way. "Excuse me," he say kind of mean.

"Say, watch it buddy." That white man talk most as nasty as Time. He kind of old and maybe he drunk or an Okie.

"Man, I said excuse me. You the one got your feet in the aisle."

"You," that man say, starting to get up, "you better watch yourself boy."

And what he want to say that for? Time step back and say real quiet, "No, motherfucker. You the one. You better watch yourself and your daughter too. See how many babies she gon have by boys like me." That man get all red in the face, but the woman in the booth with him finally start pulling at him, telling him to sit down, shut up. Cause Time set to kill that man.

I touch Time's arm first, then put my arm round his waist. "Ain't no use getting messed behind somebody like that."

Time and that man just looking at each other, not wanting to back down. People was gon start wondering what going on in a few minutes. I tell him, "Got something for you, baby," and he look down at me and grin. Orine pick it up. We go out that place singing, "Good loving, good, good loving, make you feel so clean."

"You like to hear me play?" he ask when we in the car.

"This the first time they ever have anybody here that sound that good."

"Yeah," Orine say. "How come you all staying round a little jive-ass town like Ashley?"

"We going to New York pretty soon," Time say kind of snappy.

"Well, shit, baby, you—"

"When you going to New York?"

I ask real quick. When Orine in a bad mood, can't nobody say nothing right.

"Couple of months." He lean back and put his arm round me. "They doing so many things with music back there. Up in the City, they doing one maybe two things. In L.A. they doing another one, two things. But man, in New York, they doing everything. Person couldn't never get stuck in one groove there. So many things going on, you got to be hip, real hip to keep up. You always growing there. Shit, if you 'live and playing, you can't help but grow. Say, man," he reach and tap German on the shoulder, "let's leave right now."

We all crack up. Then I say, "I sorry but I can't go, got to take care of my baby."

He laugh, "Sugar, you got yo baby right here."

"Well, I must got two babies then."

We pull in front of the partment house then but don't no one move. Finally Time reach over and touch my hair. "You gon be my Black queen?"

I look straight ahead at the night. "Yeah," I say. "Yeah."

We go in and I check first on Larry cause sometimes that girl don't watch him good. When I come in some nights, he be all out the cover and shivering but too sleepy to get back under em. Time come in when I'm pulling the cover up on Orine two kids.

"Which one yours," he ask.

I go over to Larry bed. "This my baby," I tell him.

"What's his name?"

"Larry."

"Oh, I suppose you name him after his daddy?"

I don't like the way he say that, like I was wrong to name him after his daddy. "Who else I gon name him after?" He don't say nothing and I leave him standing there. I mad now and I go in the bedroom and start pulling off my clothes. I think, That nigga can stand up in the living room all night, for all I care; let Orine talk to German and him, too. But Time come in the bedroom and put his arms round me. He touch my hair and my face and my tittie, and it scare me. I try to pull away but he hold me too close. "Martha," he say, "Black Martha." Then he just stand there holding me, not saying nothing, with his hand covering one side on my face. I stand there trembling but he don't notice. I know a woman not supposed to feel the way I feel bout Time, not right away. But I do.

He tell me things nobody ever say to me before. And I want to tell him that I ain't never liked no man much as I like him. But sometime you tell a man that and he go cause he think you liking him a whole lot gon hang him up.

"You and me," he say after we in bed, "we can make it together real good." He laugh. "I used to think all I needed was that music, but it take a woman to make that music sing, I think. So now stead of the music and me, it be the music and me and you."

"You left out Larry," I tell him. I don't think he want to hear that. But Larry my baby.

"How come you couldn't be free," he say real low. Then, "How you going when I go if you got a baby?"

"When you going?"

He turns his back to me. "Oh, I don't know. You know what the song say, 'When a woman take the blues,

She tuck her head and cry. But when a man catch the blues, he grab his shoes and slide.' Next time I get the blues," he laugh a little, "next time the man get too much for me, I leave here and go someplace else. He always chasing me. The god damn white man." He turn over and reach for me. "You feel good. He chasing me and I chasing dreams. You think I'm crazy, huh? But I'm not. I just got so many, many things going on inside me I don't know which one to let out first. They all want out so bad. When I play—I got to be better, Martha. You gon help me?"

"Yes, Time, I help you."

"You see," and he reach over and turn on the light and look down at me, "I'm not what I am. I up tight on the inside but I can't get it to show on the outside. I don't know how to make it come out. You ever hear Coltrane blow? That man is together. He showing on the outside what he got on the inside. When I can do that, then I be somewhere. But I can't go by myself. I need a woman. A Black woman. Them other women steal your soul and don't leave nothing. But a Black woman—" He laugh and pull me close. He want me and that all I care bout.

Mamma come over that next morning and come right on in the bedroom, just like she always do. I kind of shamed for her to see me like that, with a man and all, but she don't say nothing cept scuse me, then turn away. "I come to get Larry."

"He in the other bedroom," I say, starting to get up.

"That's okay; I get him." And she go out and close the door.

I start to get out the bed anyway. Time reach for his cigarettes and light one. "Your mamma don't believe in knocking, do she?"

I start to tell him not to talk so loud cause Mamma a hear him, but that might make him mad. "Well, it ain't usually nobody in here with me for her to walk in on." I standing by the bed buttoning my house coat and Time reach out and pull my arm, smiling.

"I know you ain't no tramp, Martha. Come on, get back in bed."

I pull my arm way and start out the door. "I got to get Larry's clothes together," I tell him. I do got to get them clothes together cause when Mamma come for Larry like that on Sadday morning, she want to keep him for the rest of the weekend. But —I don't know. It just don't seem right for me to be in the bed with a man and my mamma in the next room.

I think Orine and German still in the other bedroom. But I don't know; Orine don't too much like for her mens to stay all night. She say it make a bad impression on her kids. I glad the door close anyway. If Mamma gon start talking that "why don't you come home" talk the way she usually do, it best for Orine not to hear it.

Orine's two kids still sleep but Mamma got Larry on his bed tickling him and playing with him. He like that. "Boy, you sho happy for it to be so early in the morning," I tell him.

Mamma stop tickling him and he lay there breathing hard for a minute. "Big Mamma," he say laughing and pointing at her. I just laugh at him and go get his clothes.

"You gon marry this one?" Every man I been with since I had Larry, she ask that about.

"You think marrying gon save my soul, Mamma?" I sorry right away cause Mamma don't like me to make fun of God. But I swear I gets tired of all that. What I want to marry for anyway? Get somebody like Daddy always coming and going and every time he go leave a baby behind. Or get a man what stay round and beat me all the time and have kids thinking they big shit just cause they got a daddy what stay with them, like them saddity kids at school. Shit, married or single they still doing the same thing when they goes to bed.

Mamma don't say nothing else bout it. She ask where he work. I tell her and then take Larry in the bathroom and wash him up.

"The older you get, the more foolish you get, Martha. Them musicians ain't got nothing for a woman. Lots sweet talk and babies, that's all. Welfare don't even want to give you nothing for the one you got now, how you gon—" I sorry but I just stop listening. Mamma run her mouth like a clatterbone on a goose ass sometime. I just go on and give her the baby and get the rest of his things ready.

"So your mamma don't like musicians, huh?" Time say when I get back in the bedroom. "Square-ass people. Everything they don't know about, they hate. Lord deliver me from a square-ass town with square-ass people." He turn over.

"You wasn't calling me square last night."

"I'm not calling you square now, Martha."

I get back in the bed then and he put his arm round me. "But they say what they want to say. Long as they don't mess with me things be okay. But that's impossible. Somebody always got to have their little say about your life. They want to tell you where to go, how to play, what to play, where to play it—shit, even who to fuck and how to fuck em. But when I get to New York—"

"Time, let's don't talk now."

He laugh then. "Martha, you so Black." I don't know what I should say so I don't say nothing, just get closer and we don't talk.

That how it is lots a time with me and him. It seem like all I got is lots little pitchers in my mind and can't tell nobody what they look like. Once I try to tell him bout that, bout the pitchers, and he just laugh. "Least your head ain't empty. Maybe now you got some pictures, you get some thoughts." That make me mad and I start cussing, but he laugh and kiss me and hold me. And that time, when we doing it, it all—all angry and like he want to hurt me. And I think bout that song he sing that first night bout having the blues. But that the only time he mean like that.

Time and German brung the piano a couple days after that. The piano small and all shiny black wood. Time cussed German when German knocked it against the front door getting it in the house. Time went to put it in the bedroom but I want him to be thinking bout me, not some damn piano when he in there. I tell him he put it in the living room or it don't come in the house. Orine don't want it in the house period, say it too damn noisy

—that's what she tell me. She don't say nothing to Time. I think she half-way scared of him. He pretty good bout playing it though. He don't never play it when the babies is sleep or at least he don't play loud as he can. But all he thinking bout when he playing is that piano. You talk to him, he don't answer; you touch him, he don't look up. One time I say to him, "Pay me some tention," but he don't even hear. I hit his hand, not hard, just playing. He look at me but he don't stop playing. "Get out of here, Martha." First I start to tell him he can't tell me what to do in my own self's house, but he just looking at me. Looking at me and playing and not saying nothing. I leave.

His friends come over most evenings when he home, not playing. It like Time is the leader. Whatever he say go. They always telling him how good he is. "Out of sight, man, the way you play." "You ought to get out of this little town so somebody can hear you play." Most times, he just smile and don't say nothing, or he just say thanks. But I wonder if he really believe em. I tell him, sometime, that he sound better than lots a them men on records. He give me his little cool smile. But I feel he glad I tell him that.

When his friends come over, we sit around laughing and talking and drinking. Orine like that cause she be playing up to em all and they be telling her what a fine ass she got. They don't tell me nothing like that cause Time be sitting right there, but long as Time telling me, I don't care. It like when we go to the Legion, after Time and German started being with us. We all the

time get in free and then get to sit at one a the big front tables. And Orine like that cause it make her think she big-time. But she still her same old picky self; all the time telling me to "sit cool, Martha," and "be cool, girl." Acting like cool the most important thing in the world. I finally just tell her, "Time like me just the way I am, cool or not." And it true; Time always saying that I be myself and I be fine.

Time and his friends, they talk mostly bout music, music and New York City and white people. Sometime I get so sick a listening to em. Always talking bout how they gon put something over on the white man, gon take something away from him, gon do this, gon do that. Ah shit! I tell em. But they don't pay me no mind.

German say, one night, "Man, this white man come asking if I want to play at his house for—"

"What you tell him, man, 'Put money in my purse'?" Time ask. They all crack up. Me and Orine sit there quiet. Orine all swole up cause Time and them running some kind of game and she don't know what going down.

"Hey, man, yo all member that time up in Frisco when we got fired from that gig and wan't none of our old ladies working?" That Brown, he play bass with em.

"Man," Time say, "all I remember is that I stayed high most of the time. But how'd I stay high if ain't nobody had no bread? Somebody was putting something in somebody's purse." He lean back laughing a little. "Verna's mamma must have been sending her some money till she got a job. Yeah, yeah man, that was it. You remember the first time her mamma sent that money and she gave it all to me to hold?"

"And what she wanna do that for? You went out and gambled half a it away and bought pot with most of the rest." German not laughing much as Time and Brown.

"Man, I was scared to tell her cause you remember how easy it was for her to get her jaws tight. But she was cool, didn't say nothing. I told her I was going to get food with the rest of the money and asked her what she wanted, and—"

"And she say cigarettes," Brown break in laughing, "and this cat, man, this cat tell her, 'Woman, we ain't wasting this bread on no non-essentials!'" He doubled over laughing. They all laughing. But I don't think it that funny. Any woman can give a man money.

"I thought the babe was gon kill me, her jaws was so tight. But even with her jaws tight, Verna was still cool. She just say, 'Baby, you done fucked up fifty dollars on nonessentials; let me try thirty cents.'"

That really funny to em. They all cracking up but me. Time sit there smiling just a little and shaking his head. Then, he reach out and squeeze my knee and smile at me. And I know it like I say; any woman can give a man money.

German been twitching round in his chair and finally he say, "Yeah, man, this fay dude want me to play at his house for fifty cent." That German always got to hear hisself talk. "I tell him take his fifty cent and shove it up his ass—oh scuse me. I forgot that baby was here—but I told him what to do with it. When I play for honkies, I tell him, I don't play for less than two hundred dollars and he so foolish he gon pay it." They all laugh, but I know German lying. Anybody offer him ten cents let lone fifty, he gon play.

"It ain't the money, man," Time say. "They just don't know what the fuck going on." I tell him Larry sitting right there. I know he ain't gon pay me no mind, but I feel if German can respect my baby, Time can too. "Man they go out to some little school, learn a few chords, and they think they know it all. Then, if you working for a white man, he fire you and hire him. No, man, I can't tie shit from no white man."

"That where you wrong," I tell him. "Somebody you don't like, you supposed to take em for everything they got. Take em and tell em to kiss yo butt."

"That another one of your pictures, I guess," Time say. And they all laugh cause he told em bout that, too, one time when he was mad with me.

"No, no," I say. "Listen, one day I walking downtown and this white man offer me a ride. I say okay and get in the car. He start talking and hinting round and finally he come on out and say it. I give you twenty dollars, he say. I say okay. We in Chinatown by then and at the next stop light he get out his wallet and give me a twenty-dollar bill. 'That what I like bout you colored women,' he say easing all back in his seat just like he already done got some and waiting to get some more. 'Yeah,' he say, 'you all so easy to get.' I put that money in my purse, open the door and tell him, 'Motherfucker, you ain't got shit here,' and slam the door."

"Watch your mouth," Time say, "Larry sitting here." We all crack up.

"What he do then?" Orine ask.

"What could he do? We in Chinatown and all them colored folks walking round. You know they ain't gon let no white man do nothing to me."

Time tell me after we go to bed that night that he kill me if he ever see me with a white man.

I laugh and kiss him. "What I want with a white man when I got you?" We both laugh and get in the bed. I lay stretched out waiting for him to reach for me. It funny, I think, how colored men don't never want no colored women messing with no white mens but the first chance he get, that colored man gon be right there in that white woman's bed. Yeah, colored men sho give colored womens a hard way to go. But I know if Time got to give a hard way to go, it ain't gon be for scaggy fay babe, and I kinda smile to myself.

"Martha—"

"Yeah, Time," I say turning to him.

"How old you—eighteen? What you want to do in life? What you want to be?"

What he mean? "I want to be with you," I tell him.

"No, I mean really. What you want?" Why he want to know, I wonder. Everytime he start talking serious-like, I think he must be hearing his sliding song.

"I don't want to have to ask nobody for nothing. I want to be able to take care of my own self." I won't be no weight on you, Time, I want to tell him. I won't be no trouble to you.

"Then what are you doing on the Welfare?"

"What else I gon do? Go out and scrub somebody else's toilets like my mamma did so Larry can run wild like I did? No. I stay on Welfare awhile, thank you."

"You see what the white man have done to us, is doing to us?"

"White man my ass," I tell him. "That was my no good daddy. If he'd gone out and worked, we woulda been better off."

"How he gon work if the man won't let him?"

"You just let the man turn you out. Yeah, that man got yo mind."

"What you mean?" he ask real quiet. But I don't pay no tention to him.

"You always talking bout music and New York City, New York City and the white man. Why don't you forget all that shit and get a job like other men? I hate that damn piano."

He grab my shoulder real tight. "What you mean, 'got my mind?' What you mean?" And he start shaking me. But I crying and thinking bout he gon leave.

"You laugh cause I say all I got in my mind is pitchers but least they better some old music. That all you ever think about, Time."

"What you mean? What you mean?"

Finally I scream. "You ain't gon no damn New York City and it ain't the white man what gon keep you. You just using him for a scuse cause you scared. Maybe you can't play." That the only time he ever hit me. And I cry cause I know he gon leave for sho. He hold me and say don't cry, say he sorry, but I can't stop. Orine bamming on the door and

Time yelling at her to leave us lone and the babies crying and finally he start to pull away. I say, "Time . . ." He still for a long time, then he say, "Okay, Okay, Martha."

No, it not like he don't want me no more, he—

"Martha. Martha. You ain't been listening to a word I say."

"Mamma." I say it soft cause I don't want to hurt her. "Please leave me lone. You and Orine—and Time too, sometime—you all treat me like I don't know nothing. But just cause it don't seem like to you that I know what I'm doing, that don't mean nothing. You can't see into my life."

"I see enough to know you just get into one mess after another." She shake her head and her voice come kinda slow. "Martha, I named you after that woman in the Bible cause I want you to be like her. Be good in the same way she is. Martha, that woman ain't never stopped believing. She humble and patient and the Lord make a place for her." She lean her hands on the table. Been in them dishes again, hands all wrinkled and shiny wet. "But that was the Bible. You ain't got the time to be patient, to be waiting for Time or no one else to make no place for you. That man ain't no good. I told you—"

Words coming faster and faster. She got the cow by the tail and gon on down shit creek. It don't matter though. She talk and I sit here thinking bout Time. "You feel good . . . You gon be my Black queen? . . . We can make it together . . . You feel good . . ." He be back.

Down in the Reeds
by the River

Victoria Lincoln

Why are we never prepared, why do all the books and all the wisdom of our friends avail us nothing in the final event? How many deathbed scenes we have read, how many stories of young love, of marital infidelity, of cherished ambition fulfilled or defeated. There is nothing that can happen to us that has not happened again and again, that we have not read over a thousand times, closely, carefully, accurately recorded; before we are fully launched on life, the story of the human heart has been opened for us again and again with all the patience and skill of the human mind. But the event, when it comes, is never anything like the description; it is strange, infinitely strange and new, and we stand helpless before it and realize that the words of another convey nothing, nothing.

And still we cannot believe that personal life is, in its essence, incommunicable. We, too, having lived the moment, are impelled to convey it, to speak the words so honest in intent, so false in the final effect. Now, after so many years, I want to tell you about Mr. deRocca, although it is a queer story—not a story at all, really, only an incident in the life of a young girl—simply to show that it was not what you would have expected. It was not like the books or the whispered, ugly confidence that you remember from your school days; it was quite, quite different. I want to tell you, although I know from the outset that I shall fail, as we all fail.

But now that I come to it, I hesitate. It should have been evil, frightening, all wrong; of course it should. It should have been the repellent accident that can queer an emotional development for years to come. And still, when it was happening, it was not like that at all.

I was fourteen, a wiry, red-headed, unimaginative little tomboy, fond of sand-lot baseball. My parents were dead, killed in an accident a year before, and I lived with an aunt and uncle in Braeburn Heights, a suburb of a small city in Kansas. Bereft, rudely transplanted from the life I had known—a happy-go-lucky life in the brown hills of California—I was lonely beyond words. I had grown up in the careless warmth of love, and for my Aunt Elsa's genuine, if worried, kindness I could feel nothing but ingratitude. The house was strange, the neighboring children were strange, with their neat, pretty bedrooms, their queer talk of dates, and formals, and going steady. I felt dry and hard and empty inside myself, day after day. I used to take my bicycle and ride out into the country, but the country was strange, too, and ugly to my eyes, all flat and dull.

And then, one day, I found White Creek Row. It was the town's Hooverville, a row of shanties between the creek and the railroad, little huts like the playhouse that I had built back in the hills with the children of our Mexican gardener—a tragic, shocking, sordid shantytown, as I see it now. But to my enchanted eyes it was romantic and delightful and, more than that, comprehensible, as my aunt's house in Braeburn Heights was not.

It was in White Creek Row that, unknown to Aunt Elsa, I made my first real friends in Kansas. The squatters in the row were shy of me at first, as I was shy of the people in Braeburn Heights. My decent clothes, my bicycle, made me alien, an object for suspicion and resent-ment. And still, somehow or other, I managed to scrape an acquaintance with Posy Moreno, an acquaintance that grew into love.

She was a gentle creature with a mop of soft black curls piled high on her head and a womanliness, at sixteen, that made me feel, for the first time, glad that I, too, was growing near to womanhood. She lived in the last shanty in the row with her little brother Manuel, and next door was Mrs. Grimes, her self-appointed duenna. She was very proud of Mrs. Grimes' watchfulness.

"Me, I'm never chasing with the feller," she used to say, "but if I was to chase with the feller, Mrs. Grimes she's knock me down, you bet. She's not let anybody get fresh with Posy Moreno."

"I wouldn't want anyone bossing me like that," I said once. And Posy, lifting her head in the pride of her womanhood, replied, "You not need. You just a kid." But as we became better acquainted she treated me less and less like a kid.

Through our long afternoons on the creek bank, listening to her conversation, I would sit spellbound, infinitely flattered that she considered me a girl and not a child, feeling within myself a new softening, a shy preening, a tremulousness delicious and unfamiliar.

Besides Posy and Manuel, the only other child on the row was Chuck Hansen, who was twelve. I liked him, too, and I used to let him ride my bicycle while Posy and I talked. I could never hear enough about life in the row, and the people who lived in it. They had everything, I used to tell myself, everything that anybody could want, for I was too young to understand the need for security, for dignity. They had everything, and they had got it all free—even a church.

Mrs. Grimes had wanted the church, and Mr. deRocca, who had been a carpenter in Italy, had built it for her, although he was a free-thinker and had accompanied every hammer blow, so Posy told me, with a lot of bad talk about religion being made up by rich people to keep poor people quiet.

How I wished I might have been there to see him, sitting on the roof, pounding down the shingles that were made from flattened tin cans, with his delicate, hard little old hands, and shouting all the time, "Opium of the people. You getta pie in a sky when you die!" The church even had a piano, with a good many keys that still sounded, nice and loud, if not true, and Mrs. Grimes played gospel hymns on it by ear.

Mr. deRocca would not go to the prayer meetings. He lived in the best shanty in the row, and in his front yard was a beautiful American flag laid out in bits of broken brick and slate and white stones. I admired it intensely and used to stop before his house, the better to enjoy it, but Posy would shy off and draw me away, throwing up her head with a sort of wild-pony elegance. "Better we're not hanging around here," she would say. "Mr. deRocca, he's liking the girl."

I did not understand. Would anyone so old want a wife as young as Posy, I wondered. It must be that, I decided, when Posy told me that Mrs. Grimes had not let Mr. de-Rocca help with the building of Posy's shack. I supposed they thought it would not be fair to encourage him. But I saw no reason why the caution should also apply

to me. I was charmed by the little I had managed to see of Mr. deRocca. He seemed to be a very clever, very nice old man.

And now I come to my story, and it is hard to tell. It is hard to tell because I should have been so different. Perhaps there were undertones that I have forgotten. That is likely, for the memory has a curiously clear and classic air, quite unlike life as I have since found it—the nymph and the old satyr frozen in attitudes of timeless innocence under the box elders by the creek bank, the sacred grove where liquid Peneus was flowing and all dark Tempe lay. And still, still, I remember it like that. If there was fear, if there was guilt, they came later.

One afternoon, Chuck Hansen met me on the cinder track, looking wistful. "I don't guess you'll want to stay today, Connie," he said. "Mrs. Grimes and Posy, they went uptown." He rubbed the handle bars of my bicycle with his hands, hard, as if he were fondling a horse. "Guess you won't have much to stick around for," he said humbly.

How nice he was, I thought, never teasing.

"Well, listen, Chuck," I said. "I'm tired, a little. I'll go down and walk around a while and sit on the creek bank."

His grin made me feel warm and pleasant. I began to saunter along the front of the row. Mr. deRocca was sitting on a packing case by his door, eating an onion. His face, lifted to the sky, wore the blank, peaceful expression of one enjoying the quiet of a village street after a procession has passed, the look of remembering in quietness.

I came along very slowly, watching

Mr. deRocca from the corners of my eyes. He wore a plaid flannel shirt, ragged and, of course, unironed, but fairly clean, and the neck was unbuttoned. I noticed how the flesh under his chin was firm and didn't hang down in wattles, and the cords in his neck didn't stick out. He looked harder and nicer than other old men.

How old was he, really? About fifty, I should guess now, looking back; maybe a little less. But if I had known it then, it would not have changed my picture of him at all. Fifty to eighty in those days were all of a piece in my mind. Mr. deRocca was an old man. And he was nice. As I came very close, I realized with a sudden throb of excitement that he had been watching me all along, just as I had been watching him. Watching me and waiting for the moment to speak, just as I had been, with him. I turned, pretending to have seen him for the first time. I smiled at him. The white teeth gleamed in the thin, brown face; the elegant, small, brown paw that held the onion described a vast semicircle of greeting. "Hi, kid," he said. "Looka for da Posy? She's a not home."

I did not answer. I realized, quite abruptly, that it was the sight of him sitting down there below me, fully as much as Chuck's longing hands rubbing the handle bars, that made up my mind for me up there on the embankment, and I turned shy, hoping that he would not guess it.

"I always like to look at that flag, Mr. deRocca," I said.

"Come on in a yard," he said. "Looka good. It's a pretty, hey?"

We stood together, eyeing the charming sight in a sort of shared pride. He pulled out another packing case from the corner of the house and waved me to it with the flattering charm of a courtier.

"Please to sit," he said. "Scusa." He went in the house for a second and returned, extending his hand with the same grave courtesy. "You like-a onion?"

I looked at it dubiously. Father had disliked salads, saying firmly that hay was for God-damned Frenchmen, and Aunt Elsa's were of the pineapple, cream cheese, and mayonnaise school. Raw onions were new to me, and alarming. But it was so lovely, being treated like a lady, that I could not disappoint him.

I took it and bit into it gingerly. The sharp, pungent, biting juice ran over my tongue, the firm, fleshy layers crunched between my teeth in a stinging, breathtaking ecstasy of delicious pain.

"Oh!" I cried in sincere delight. "It's good!" Then, with the snobbery of the young guest who does not wish his host to think him ignorant of the wines he is offered, I added, "It's one of the best onions I ever ate."

"Sure," he said proudly. "Sure, you bet it's a good, it's a fine. I grow."

I regarded him happily, rejoicing in his kingly acceptance of the compliment, so unlike the mincing, genteel self-depreciation which, of all the mannered compulsions of the Heights, I found most unfamiliar and most dismal.

I went on with my compliments, sincerely, but also eager for the continuing pleasure of his openness. "You have a wonderful house," I said. "The church is wonderful, too. You're a fine carpenter."

His eyes glowed and he swayed his head from side to side, like someone keeping time to music. "You bet I'm a good," he replied. "I'm a learn in a Old Country, worka slow, take-a pain, think for the job, for looka pretty, not think for hurry up, getta money. I'm a good like nobody's business."

"I should think you'd get lots of jobs," I said, "and be rich."

He shrugged. "Bad a time," he said. "Everywhere bad a time. Smart a man everywhere hungry, no work. Someday come a good time." He finished the onion and wiped his thin lips on the backs of his neat little fingers. "Someday, different time, all be good, not graba, graba, be man and man together, not dog and dog. First a big fight, maybe, then all be good."

I remembered something we had studied in social science. I leaned forward, trying to look intelligent and grownup. "You mean a revolution?" I said. "Are you a Communist, Mr. deRocca?"

"Pah!" he replied. "Not!" He spat to one side, to emphasize his attitude. Then, with a flashing, all-embracing smile: "Lots good in de Communista, lots smart. I read, I like, good. Only alla time boss, boss. Boss so bad like we got here, now. I'm a no like all a time boss. I am Anarchista, me."

"What's that?" I asked.

"Everyone's treat everyone else right. No push around, no boss. People no gotta lot of stuff, graba, graba. No law, no boss, everyone a same. Treata them right, they treata you right. All good."

It sounded lovely.

"What do you call that? Anarchista? I guess I'm Anarchista, too," I said.

He threw both arms wide, embracing me in the universal fellowship. "That's a fine. You smart a kid."

Master and disciple, we sat happily together in the blissful country of utopian anarchy, regarding the flag of America spread out at our feet with absent, gently admiring eyes. Gradually, the conversation took a personal turn.

"You name a Constansia?"

"Constance."

"Pretty name," he said. "Pretty name for pretty girl! Nice when a pretty girl have a pretty name."

No one had told me I was pretty since my mother died. I was grateful to him, but unbelieving. "I have awful red hair," I said.

"Pretty," he said. "Pretty hair, pretty eye, pretty shape. How old?"

"Going on fifteen."

He smiled, as if I could not possibly have been a nicer age, as if it were a peculiar grace and wisdom in me to be going on fifteen.

"Last year, da little kid," he said. "Next year, da woman, look at da fella, think for da fella. Now she not know what she think—that right?"

I was deeply struck with the truth of his words. It was what I had been feeling in my inarticulate way all the time I was sitting with Posy on the creek bank, admiring her womanly young beauty, listening to her sternly virtuous, so very sex-conscious conversation, hoping that she did not still think of me as just a little kid.

I looked earnestly at Mr. deRocca sitting on his packing case, as if I could discover in the glowing, friendly eyes the source of his remarkable understanding. He was old, but I thought suddenly that he was handsome, as handsome as my father had been. His features were so sharp and delicate, his body so fine-boned, the shoulders so narrow, compared with the Mexicans with whom I unconsciously classed him. A fleeting wonder passed through my mind if all Italians were like him, so little and handsome and wise.

He held out his hand toward me, palm up and slightly cupped, almost as if he were coaxing a tame bird with seed. "That right?" he said again, quite soft.

I was surprised at my voice when I answered. It was unfamiliar—low and a little unsteady. "That's right," I said.

He stood up, smiling more than ever. "Come on down a creek bank," he said. "I show you where I gotta good catfish net. Other guy wait to fish, watch, work. Me, I sit and they come."

Thinking back, remembering, I wonder for the first time if he spoke in any conscious analogy. I do not believe that he did.

I followed Mr. deRocca trustfully down the creek bank, under the box-elder trees. At the water's edge, he turned and looked at me, and I saw the changed look in his eyes. It was as if the door had opened and I were looking upon a landscape that was both strange and familiar. I glanced around me, and I saw that the box elders grew thick where we stood, that we were in a place that was private, sheltered from the eyes of the world. Suddenly, I understood everything that Posy had said. I knew what she meant when she said, "He's liking the girl."

"Show me the net," I said nervously.

His eyes smiled at me, reassuring, his voice quieted me. "Pretty soon," he said. "Right down here." But he made no move toward going on. Instead, he put out a lean, brown paw and touched my head. "Pretty," he said. "Pretty hair."

His hand slipped down my back and around my waist, the fingers firm and hard against me, warm through my cotton dress. And again he paused, his eyes still smiling with that same gentle reassurance.

He was old at the game, I see now, and grown wise in method, wise and patient. If he had hurried, if he had let me see his eagerness, I should have been terribly frightened, I should have run away crying. I should have run away full of fear and hate, and the fear and hate would have lived in me a long time.

But he stood, smiling at me, until I was used to his arm, his hand, feeling it not as a sexual advance but as warm, human affection in my body that was aching for human affection, for the demonstrative love on which I had thrived through a warm, loving childhood. He was quiet until I felt my fear dissolve in gratitude for the kindness of his arm, his firm, affectionate hand.

It was easy, then, for him to turn me against him, to hold me firm and close, stroking my hair, firm and close against him, waiting till his accustomed, patient hands should tell him that I was ready for more.

I knew that I must be doing something bad, and still I could not feel that it was bad yet, not yet. And his slowness made me confident that I was free to decide if it was really bad, that he would let me go quickly the minute I thought it had begun to be bad. It still did not seem bad when he kissed me, or when his kissing changed and made me feel all soft and strange inside, or when his hands began to describe all the differences that the year had made in my body, and to tell me silently that they were beauties, richness, a bounty of which to be proud.

Once he made a little motion to draw me down in the thick grass, and I had the sense to be frightened, but he felt it at once and waited, and I waited, too, sure that I would know when I should run away, growing softer and stranger by the moment, forgetting everything outside me. I was wholly lost when I heard Posy's shrill voice calling my name, and heard her pushing through the branches down the creek bank.

Mr. deRocca let me go and dropped to his knees at the water's edge. "Like a this," he said. "I'm a tie right here, da fish swim right in. Some net, hey?"

He looked over his shoulder and saw Posy. She was white and out of breath. "Connie!" she cried. "I don't know where you are. I'm scaring." She snatched at my hand, too relieved, too wrought up, to look at my revealing face. "Come along outa here," she said. Then, remembering her manners, "Hello, Mr. deRocca."

She yanked me back to the row. "You crazy," she scolded me. "What you think, you go down there with deRocca? I'm telling you he's liking the girl."

"You said I was just a kid. That's what you said," I repeated.

"I know," she said. "Well, I'm crazy. Just as soon Chuck he tell me you down here, I'm knowing I'm crazy. You no kid, not for looks. No more. Was a little while ago, now no more. Mother of God, I'm scar-

ing." She paused, momentarily suspicious. "What you going down in there with deRocca for?"

"He said he was going to show me his catfish net."

"Ha, I bet! You poor kid, you got no sense. What he say? He talk dirty?"

"No," I replied with perfect truth. "He talked just as nice as you and Mrs. Grimes."

"Thanks God," said Posy, over and over again. "Thanks God."

In the unpleasant shock of nearly being caught out, all the new feeling that I had learned—the lovely, soft, flowing, flowering openness—was driven back in me, and the present moment closed above it so completely that the afternoon might have been lived years before, or not at all, by anything I felt in myself. Instead, I was troubled by an unwilling anger against Posy, as if she were making a disproportionate fuss.

Something of this she must have felt, or perhaps she now decided that my unwary innocence had been scolded long enough, for she took my hand, smiling again, as if, for her, too, the incident had suddenly dropped away out of sight.

"Come now," she said. "Is early yet, you don't got to going home, come now down to the house. We don't say nothing from this to Mrs. Grimes."

"No, Posy, no, I've got to get home," I said.

All the way home, I pedalled hard, as if I were very late—so hard that there was no room in me for anything else. Even before I saw the letters lying on the hall rug, where they had fallen from the mail slit in the door, I could tell from the silence that the house was empty. I stood in the sun that poured in at the open doorway, absorbing gratefully the quality of an empty house. I had not realized at all, as I forced myself home, faster and faster, how I would need, once I had got there, to be alone. I shut my eyes and sighed heavily, feeling the silence, the aloneness all through me like a merciful, unexpected blessing.

What had happened that afternoon, what had really happened? It wasn't only that I had let Mr. deRocca kiss me and touch me like that. It was something that had happened in me. There was something in me—and in the world, too—that I had never known was there before, something powerful and lovely, something powerful and new.

I stood there alone in the quiet house, in the sunshine, with my eyes closed. "I wish," I thought slowly, "that Posy hadn't come. I wish . . ."

Suddenly, I knew that I had begun to be bad right there in Mr. deRocca's front yard, before we had ever gone down to the creek. I knew that I had been bad all along, terribly bad. Fear and guilt rose in me like a storm, shaking my body until my teeth chattered and I had to sit on the bottom step of the stairs and lean against the wall to hold myself still.

"If Posy knew," I thought, "if she knew about me, if she knew what I did, I'd die. I should die, I'd die."

Aunt Elsa found me like that when she came in a few minutes later. "Why, Connie!" she cried. "What is it, dear? You're sick."

"I got a chill," I said. "Just right now."

"Let me hang up my coat, dear," she said, "and I'll get you right into bed. Why, you poor baby!"

I let her help me up the stairs. I clung to her motherly warmth all the way, hungry for it, like a child that has been lost and found again. "Oh, Aunt Elsa," I cried. "I'm so glad you're home." And her gentle voice soothed me again and again. "There, dear, there. You're going to be all right. There, poor little girl. Aunt Elsa'll put you to bed. Yes, she will. Of course she will."

In the complex agony of the moment, I was broken wide open. She's real, too, I thought in slow wonder; Aunt Elsa is real, too. She was my mother's sister.

I caught at her light, smooth dress, hiding my face in it. She smelled nice, clean and fresh with a light perfume. I let my head fall against her shoulder, and it was soft and firm, comforting, comforting.

"Oh, Aunt Elsa," I cried, wondering because it was true, because it had not been true before, at all, and now it was wholly true. "Aunt Elsa, I love you."

That is the story, and that is all. When I woke in the morning, the ecstasy and the shame alike were gone. I had shut my mind upon them, as I had learned earlier to shut it upon grief and loss.

Oddly enough—for the defense mechanism seldom works that way —I still liked Mr. deRocca. Apparently, his attempted seduction had been quite impersonal, for, as I used to pass his yard, walking up the row to Posy's house in the warm, dusty August afternoons, he would always wave his little paw at me and say, "Hi ya, kid," warmly, but with no attempt to detain me.

For my own part, I always felt a tingling as I passed him; not enough to be unpleasant—just a sort of shy, quickening self-consciousness. It made me avoid his face as I replied,

"Hello, Mr. deRocca." My voice, as I spoke, was always a trifle breathless. I told myself that it was funny how I hardly remembered that afternoon by the creek at all. But as I passed his house, I always stood up straight and moved slowly, and tried to look grownup.

Saul

MICHAEL MASON

"You see that guy," Saul said to me; "I'm going to get him." I looked across the coffee house to where Susi Wood was getting a doughnut from the self-service box.

"You mean the guy in the brown shirt?" I said.

"Yeah. I'm going to get that boy. I have my suspicions about him."

"Who is he?" I said. I really knew him. Susi was a sensitive boy with homosexual problems he couldn't solve in any acceptable way. He didn't play it defiantly, but, although committed, kept an icy control of himself in social situations, which was broken only by guarded and hesitant moves for contact, or in situations where he could be sure, of tolerance, of sympathy. That reason was why he talked to me. His trust in me was pretty great. Not many people, for instance, knew that he had changed his name to Susi in high school to please his chemistry teacher.

"His name's Wood." said Saul. "I have sneaky irritated feelings about him. He irks me." Saul slapped the table and slouched down in his chair with his legs stuck out. His eyes followed Susi as he moved about the cafeteria and sat down with a group of people across the room. Most of the students at Potter College met at a coffee house in the evenings, which became pretty crowded and noisy with talk, chess and checkers, music and dancing.

"I don't like the way he walks," said Saul angrily. "He walks funny." He looked at me suddenly and said, "Don't you think?"

"In a way," I said.

"The screw in a way," said Saul. "You don't think so, you're just being agreeable. I *know* he walks funny."

41

"Maybe you're right," I said. He pouted and looked away. Saul was a big, light-skinned guy, with very light blue eyes and black hair. He wore very clean T-shirts with the sleeves rolled up, and levis, no matter how cold it got. "When you're getting your coffee, you get me an orange," he said.

I got up and wound between the tables and brought a coffee for myself and an orange for Saul. When I came back, he pulled a knife out of his pocket, clicked it open, and then said, "I got a letter from my mother and sister today."

"How are they?" I said.

He closed the knife and pressed the button that opened it, nervously. "Okay. Mama's sick a little in her neck; something funny that she doesn't want to see the doctor. She's gotta go see the doctor or something'll really happen. Sis is okay. She's studying hard. She's extremely bright in high school."

"How old is she?" I said.

"Eighteen." He clicked the knife again. "She's very pure. She's good-looking, has a very pale forehead; very real slender arms."

"She very popular?"

He looked at me and compressed his lips. "Sure; hell, yes."

"What do you mean, hell, yes?"

"What I mean, there's a lot of little peckers that'd like to get their schmucky hands on her," he said. "But they won't, buddy. Not when I know it."

"What do you do," I said, "chaperone her?"

He grinned and stuck the knife in the table. "I wield an invisible influence." I frowned. "You don't like that either, do you—" I shook my head—"and my answer to that is screw you. Look at that prick." I turned my head. "I don't even like the way he moves his face." His voice shook. Susi Wood was leaning across his table and talking very fast and earnestly to the rest of his company, gesturing with his hands.

"What have you got against him?" I said.

"Well—you understand these are small signs," he said, sitting up straight, "and by themselves wouldn't mean very much." He paused.

"Yeah," I said.

"But I've got a lot of them collected. The first time was—" A girl came over to our table and said, "Hello Saul," and sat down.

"Hello, chick," he said, leaning back again. "This is Jack Valente, a buddy of mine."

"I'm glad to meet you," she said, smiling.

"Charmed," I said.

"How have you been getting along lately, Saul?" she said.

"Okay," said Saul, glancing briefly at her from under his eyebrows and tilting back in his chair. He began to gouge a hole in the end of his orange with his knife, his head bent and his eyes lowered.

"That's nice," she said.

"Yeah," he said.

"Have you been pretty busy?" she said.

"Yeah, boy," he said. He pulled the plug out of the end of his orange and began to suck hungrily at the hole.

"We've got an open hall Sunday afternoon, why don't you come over and see me?" she said.

"Un huh, maybe I can make it; look chick, Jack and I were talking about something pretty important, why don't I see you some other time?"

She looked at me. "All right." She

stood up quickly and left.

"What's with her?" I asked.

He twisted in his chair and hooked his feet behind the legs. "Oh, I banged her once, and now she has this idea we're married and I'm deserting, or something." He laughed, and leaned forward. "Lemme tell you about this guy now."

"Yeah?"

"Well, you know how some guys stare at you until you begin to get some idea of something—you know? Well, he did that for one; I was eating lunch in here, it was a couple of weeks ago—whenever I looked at him, he'd look away. The second time was Wednesday last week, at the coffee shop, he was sitting with some guys I know, and gave me this very intense look when he was passing cigarettes." He paused for a long time, thinking, "and the last time was last night here, when I was getting something from the counter, and he came up behind me and took me by the arm, and reached out and got something with his other hand. Did you notice he wears a ring? That's another thing. You know, I knew he was right there behind me, and I just had this idea he would try something that way, and then expecting it that way—it was very annoying." He slapped the table a couple of times, and looked away, frowning. "There he is now, sitting there." I looked over; the people who had been sitting with Susi had left him, and he sat with his fists supporting his head and his feet tucked under his chair.

Saul stood up. "Come on," he said.

"What?"

"Come on, let's go over there." He continued to watch Susi, who glanced at him once and away again.

"What's the idea?" I said.

He pulled my arm strongly. "Come ahn, will ya?" I stood up and followed him across the room. The tables were very crowded, with eight or nine chairs around some, and their tops covered with ashes, napkins, and pools of coffee and tea. "I'll show you something," Saul said over his shoulder. We sat down at Susi's table.

"Hello," said Susi. He looked at Saul steadily.

Saul stared back, expressionless, with his arms folded on the table. "Hello," he said.

"My name is Susi Wood," Susi said, extending his hand.

"I know it," said Saul.

"I know your name too," said Susi hesitantly. He laid his hand on the table.

"Yeah, well, listen," said Saul.

"What?"

"This is a very important matter. I have this idea that you feel a certain way about me." Saul leaned on the table and rested two fingers on the back of Susi's wrist.

"That's right," said Susi.

"Yeah, well, what way is that?"

Susi paused, glancing at me. "I don't know if I ought to tell you."

"Yeah, but I want you to."

"Well—" Susi wrote on a napkin and passed it to Saul.

Saul looked at it. "You're not kidding; you don't mean platonically or fraternally or anything?"

"No to all of that." Susi's voice was shaking.

"Well, get this straight, buddy." Saul's face began to flush and his voice rose. "I don't want any part of it. Nothing. You and your swish habits. You can keep it."

"I beg your pardon—" said Susi.

"Screw it. You've got the wrong

guy completely. Do you know what I mean? Do you understand what I'm trying to tell you? Is it clear?"

"Yes—" People were turning to look.

"What do you want me to do?" said Saul. "Act like a woman? Is that it? Is that what you want?"

"No," said Susi. "Like a man."

"I don't want to be bothered any more," Saul continued more loudly. "You get me? No more. Jam it."

Saul slammed the table. "Don't forget what I told you" —and stood up and left. I sat and watched Susi, sitting with his head in his hands, hiding his crimson face. Finally he pulled a handkerchief from his pocket and blew his nose and wiped his eyes. "Do you see that?" he said.

"What," I said, "his shitfulness?"

"No, no, that's not it. He didn't act like that out of indifference. Do you understand? I didn't mind it in a way—it was very pleasant for him, you see? He gets to hurt me because he thinks I'm wrong, and it proves what a man he is."

"Oh."

"You see what I mean? *I* didn't mind," said Susi, smiling. "Anytime, anytime he wants, it's all right with me. Do you see what it means, knowing the kind of man he is?"

"Yeah, I guess so," I said.

"I've almost got him," said Susi, staring toward the door, his hands tightly gripped.

Paul Veronese. *The Wedding at Cana.*

The Barretts of Wimpole Street

RUDOLF BESIER

ELIZABETH: *(quickly and nervously)* Oh, It's hard to explain to someone who doesn't know all the circumstances. You see, Papa is very devoted to me, and—

BROWNING: Devoted?

ELIZABETH: He's very devoted to me and depends a great deal on my companionship. He hasn't many points of contact with my brothers and sisters. If I were away for six months, he—

BROWNING: *(visibly restraining himself, rising and going toward her)* Miss Barrett, may I speak plainly?

ELIZABETH: Oh, do you think you'd better? I know—more or less—how you feel about this. But you see, you don't quite understand all the situation. How should you?

BROWNING: *(walking up to center window)* Oh, very well—then I'll say nothing. *(His control suddenly gives way, and his words pour out. Returns down to upper right corner of sofa.)* You tell me I don't understand. You are quite right I don't. You tell me he is devoted to you. I don't understand a devotion that demands favors as if they were rights, demands duty and respect, and obedience and love, demands all and takes all, and gives nothing in return. I don't understand a devotion that spends itself in petty tyrannies and gross bullying. I don't understand a devotion that grudges you any ray of light and glimpse of happiness, and doesn't even stop at risking your life to gratify its colossal selfishness. Devotion! Give me good sound, honest hatred, rather than devotion like that! At our first meeting you forbade me to speak of love—there was nothing more than friendship between us. I obeyed you, but I knew very well—we both knew

—that I was to be much more than just your friend. Even before I passed that door, and our eyes first met across the room, I loved you, and I've gone on loving you—and I love you more now than words can tell—and I shall love you to the end and beyond. You know that? You've always known?

ELIZABETH: *(brokenly)* Yes—I've always known. And now, for pity's sake—for pity's sake—leave me. *(Rising.)*

BROWNING: *(with a firm grasp of both her hands, rises, comes around right end of sofa)* No!

ELIZABETH: Oh, please—please—let me go! Leave me. We must never see each other again.

BROWNING: *(maintaining his grasp)* I shall never let you go—I shall never leave you! *(Draws her into his arms.)* Elizabeth—Elizabeth!

ELIZABETH: *(struggling feebly in his embrace)* No—no—Oh, Robert, have mercy on me—

BROWNING: Elizabeth, my darling— *(He kisses her, and at the touch of his lips her arms go round his neck.)*

ELIZABETH: Oh, Robert—I love you —I love you—I love you.

They kiss again, then she sinks onto sofa and he sits right of her, holding her hands.

BROWNING: And yet you ask me to take my marching orders, and go out of your life?

ELIZABETH: Yes, Robert, for what have I to give you? I have so little of all that love asks for. I have no beauty and no health—and I'm no longer young—

BROWNING: I love you.

ELIZABETH: *(speaking with restrained spiritual ecstasy)* I should have refused to see you after our first meeting. For I loved you then, though I denied it even to myself. Oh, Robert, I think Eve must have felt as I did when her first dawn broke over Paradise—the terror—the wonder—the glory of it. I had no strength to put up any kind of resistance, except the pitiful pretense of mere friendship. I was paralyzed with happiness that I had never dreamt it was possible to feel. That's my only excuse—and God knows I need one—for not having sent you away from me at once.

BROWNING: I love you.

ELIZABETH: *(continuing as before)* My life had reached its lowest ebb. I was worn out, and hope was dead. Then you came. Robert, do you know what you have done for me? I could have laughed when Dr. Chambers said that I had cured myself by wanting to live. He was right— oh, he was right. I wanted to live— eagerly, desperately, passionately— and all because life meant you—you —*(He leans down to kiss her hands.)* —and the sight of your face, and the sound of your voice, and the touch of your hand. Oh, and so much more than that! Because of you the air once more was sweet to breathe, and all the world was good and green again.

BROWNING: *(rising from kissing her hands)* And with those words singing in my ears, I'm to turn my back on you and go?

ELIZABETH: But, Robert, can't you see how impossible—?

BROWNING: I've never yet turned my back on a friend or an enemy. Am I likely to turn it on you?

ELIZABETH: But how is it all to end? What have we to look forward to? And how—?

BROWNING: I love you, and I want you for my wife.

ELIZABETH: Robert, I can't marry you. How can I, when—?

BROWNING: Not today or tomorrow. Not this year, perhaps, or next. Perhaps not for years to come—

ELIZABETH: I may never be able to marry you.

BROWNING: What then? If you remain to the last beyond my reach I shall die proud and happy in having spent a lifetime fighting to gain the richest prize a man was ever offered.

ELIZABETH: Oh, Robert, put aside your dream of me and look on me as I am. I love you too well to let you waste your manhood pursuing the pale ghost of a woman.

BROWNING: Do you think I'm a boy to be swept off my feet by an impulse, or a sentimental dreamer blind to reality? There's no man alive who sees things clearer than I do, or has his feet more firmly planted on the earth. And I tell you in all soberness that my need of you is as urgent as your need of me. If your weakness asks my strength for support, my abundant strength cries out for your weakness to complete my life and myself.

ELIZABETH: *(after a pause, turning to lie down)* Robert, have you thought what your position here would be like if you went on seeing me after today?

BROWNING: Yes.

ELIZABETH: *(continuing)* We should have to keep our love secret from everyone lest a whisper of it get to my father's ears.

BROWNING: I know.

ELIZABETH: If he had the least suspicion that you were more than a friend, the door would be slammed in your face, my letters supervised, and my life made unbearable.

BROWNING: I know.

ELIZABETH: And you, my dear, you're as frank and open as the day. How would you enjoy coming here under false pretenses, and all the subterfuges and intrigues we'd be forced to use?

BROWNING: *(smiling)* I shall detest it—I shall hate it with all my heart and soul—and I thank God for that.

ELIZABETH: But, Robert—

BBOWNING: For it's splendid and right that I should suffer some discomfort at least for such a reward as you. The immortal garland was never won without dust and heat.

Normal Love

JOYCE CAROL OATES

Downtown

I park my car in a high-rise garage, three floors up. Everything is silent. The garage is gray, the color of concrete blocks and metal. Many cars are parked here, in silence, but no one is around. A small tension rises in me, an alarm. Is there anyone around? Anyone? Our city is not a large city, there is no danger. There might be danger late at night for a woman alone. Now it is a winter afternoon, a weekday, overcast, too cold for anyone to make trouble. . . . I lock the car door, I put the keys in my purse, I walk quickly to the elevator and press the button for down.

The elevator is slow. Is it out of order? Why is there no one around? A sudden noise behind me . . . behind me a man is walking this way, putting something in his pocket. Car keys, probably. His footsteps make brisk noises on the concrete. The air is cold. My heart begins to pound absurdly; I know there is no danger, and yet my muscles stiffen as if in expectation of danger. The man waits with me for the elevator. I don't look at him. He doesn't look at me. He is wearing a fairly good overcoat, he is no danger to me. There is no danger in this city; the very coldness of the air on this December day makes everything abrupt and undramatic; there is no tension, nothing. The elevator comes. The door opens. I step inside, the man steps quickly inside; for a moment I feel a sense of panic, as if inside me a door were opening suddenly upon nothing, upon blackness. The elevator takes us down. The man says nothing, makes no movement, does not take his hands out of his pockets as if . . . as if to take them out would be a sign and he dare not make a sign. I wait with my heart hammering. I wait. The elevator stops, the door opens, a woman and some children are waiting to get inside. I step out quickly and escape. . . .

I spend the afternoon shopping. I am not followed.

My husband

sits alone downstairs after we have all gone to bed. Secrets rise in him at this time. If I come to the landing to say, "It's after two o'clock," he will stare up at me, startled. He sits on the sofa with an ashtray beside him. He is smoking and thinking. He is sitting there in a kind of troubled peace, a man of forty, six feet two and lean, unmuscular, a city man with dark hair thinning on top, the tension and bewilderment of the city in the lines of his face. He is thinking, dreaming, a terrible sadness fills him, he is sitting there alone and will glance up at me, startled, if I come to the landing. The newspaper is out back, folded to be thrown out tomorrow, the newspaper with the story about the missing girl.

My children

are eager to get out of the doors of the house and eager to get back through them again. They jerk one way, scrambling for freedom. Then they are hurrying back. They are hungry. The boys have long unpredictable legs. They are always knocking against tables. They spill milk, drop plates; their nostrils have been raw with colds for weeks. They like hot dogs and hamburgers. The meat comes processed into strange shapes they never notice—tied neatly with a tiny knot of intestine, pink, or

ground to an intricate maze of wormlike red tissue. It is all tissue. The boys like this meat very much. One of my daughters is melancholy and selfish, remote and spiteful, thirteen years old . . . the corners of her eyes narrow at things I can't imagine. She is always thinking. She looks like her father, with the same pinched, calculating face; a smile can transform such a face. The other girl is only nine, a good-natured child; she loves us all and can't understand why her father is drifting from us, at the age of forty, a mysterious, stubborn drifting we can't understand.

My neighborhood supermarket

has a tinsel Christmas tree inside. A fluffy angel blows a horn at its very top. The tree is pretty but difficult to look at; the metallic branches catch light and reflect it painfully. A radio is playing "Jingle Bells." Which day of the week is this? I do most of my shopping on Tuesday. But sometimes on Wednesday we have already run out of something, or I forgot to buy something, and I go back to the store. By the time Thursday comes I need something more, usually milk, and on Friday I have my hair done, and the hairdresser's is just down the block so it's no trouble to drop in again. Christmas carols are being played now. But the sounds are scrapy, as if the angels singing such songs were distracted, glancing over their shoulders at something. During the week the supermarket is not very crowded. Sometimes I come back again on Saturday and it's crowded then, but more girls are stationed at the checkout counters, so I suppose everything works out.

Most things are familiar—these

cans and packages—but still I look carefully, to see if there is any change, to see if there are special things that I want. I have to buy a lot of groceries. Two shopping carts are necessary sometimes, so one of the boys comes along. As long as he doesn't knock anything over he's helpful. I should be proud of him, a son of my own, but I don't have time to be proud of him. His nose is running, he wipes it on the side of his hand . . . on his sleeve. . . . "Take this," I tell him and give him a Kleenex. He accepts it. He wipes his nose as if his nose had become suddenly delicate.

But I prefer to shop alone. I take my time with everything. I take my time buying meat, inspecting the shapes of meat. The radio is now playing "Let It Snow." Far beyond my hearing are the cries of amazed animals stunned by hammer blows, their hooves skidding in the dirt, their shoulders and heads wrenching to get free of the horns that imprison them—men have hold of these horns! They are herded into trucks. Their flanks and sides are carried frozen out of trucks, big refrigerated trucks so long they can hardly make the turns of our old-fashioned little intersections.

Our house
    is on a street that is partly good, partly bad. There are boardinghouses at the corner. This is a college town, quiet and unexciting, a nice place to live, only in the last several years has the crime rate begun to rise, but nothing has happened to us. Students living in those boardinghouses sometimes make trouble, but inside the houses only; they have never bothered us. The

street has potholes. I drive automatically around them now, not even seeing them.

The university's president lives a few blocks away, in a large, old home.

Mornings in our house are quiet. All my children go to school now. The telephone rings suddenly, and my mouth goes dry; I hurry to answer the phone, I am anxious, wondering . . . could it be a wrong number? Or is it a friend? Is it someone inviting us out? There is so much cruel power in that person calling me, in his anonymity! But the telephone must ring also when I am not home, and then, then I am the one who is in power, then the caller (probably another wife, like myself) must stand listening sadly to the ringing of a telephone in an empty house, denied a few minutes' conversation. I work around my house thinking to myself about the mystery of a house, the lives dreamt out in it. My children are eager to get away in the morning and eager to come back in the afternoon. I catch myself up quick, dreaming of them, their bouncing impatient limbs somehow inside me, damaging me. . . . Do they want to damage me, my flesh? No. Does my husband want to damage me? No. The house, which I wanted so badly eight years ago, is very silent in the morning. I walk through the rooms, buttoning my car coat, getting ready to go out and shop. . . . I like to shop, I go shopping every day. I cannot locate myself precisely in this house, so I go out. I have bought everything for the rooms myself, choosing the pieces of furniture carefully, worrying over them, studying magazines like *House Beautiful*. Our sofa is dark brown, our rug is light beige. The coffee table is a long modern oval, of dark walnut. The room used to seem striking to me, even beautiful, but now it looks a little worn and cheap, I don't know why. My husband shows no sign of himself here. He puts everything of his away, as his own father did. Nothing remains of him downstairs; he is a professor at the university, and most of his books are at school; his real life is somewhere else, he is invisible here.

Across the street is a house like this one, of dark red brick, two stories high with a big attic, where, as in our house, two boys have their room. These houses were built in the forties. I look out the front window. It seems to me that something moves against the windows of that house—another woman, looking out? Is she looking across the street at me?

The purse

came into our lives by accident. My husband took the car to have the brakes fixed, and on his way home he cut across a vacant lot. I have seen him out walking, alone, and the strength of his walk has always impressed me—a man with somewhere to get to, a stubbornness that women need in men. But he is not really like that. He was wearing his trench coat, a soiled tan coat; his hands were stuck in his pockets, he was walking fast and with his head bowed as usual (thinking of what? of his students? of the bill for the car?), his eyes drifting along the ground . . . and something caught his eye, the corner of his eye.

He saw a woman's purse. It had been thrown into the frozen grass, a few yards from the path, a black patent leather purse. He paused. He leaned over and picked it up. . . .

Around him in the field was frozen milkweed.

"I found this," he said to me, coming in the back door. He looked worried, slightly embarrassed, as if I would blame him for something. Inside the purse was a wallet of some brown cheap plastic material, and inside the wallet some snapshots, a few dollar bills, some change, an identification card.

Linda Slater, 1463 St. Clair. In case of emergency notify Mr. and Mrs. Frank Slater, 1463 St. Clair.

At the dentist's   at the hair-dresser's   at the supermarket

Betty has three cavities this time, She won't brush her teeth. I check her toothbrush at night: sometimes it is wet, but what does that mean? All my children tell lies.

The dental assistant is about nineteen years old, with her hair in a big frothy mess, bleached. I glance through my checkbook ahead of time and see that I have forgotten already today to record one check . . . what was that check? A small storm rises in me, irritation and alarm. My husband never makes mistakes with the checkbook or with money or figures of any kind; he does them in his head. He doesn't make much money as a professor, but he never makes mistakes adding up that money or subtracting it.

Outside in the waiting room while I wait for Betty I notice last night's newspaper still here. From across the room I can see the headline on the left-hand side of the page, GIRL MISSING. Last night we read that story. Linda Slater, twenty, was reported missing and her whereabouts not known. Her purse was found in a vacant lot late Monday afternoon by Dr. Norman York, Professor of History at the University. He telephoned her parents. He brought the purse over to them. My eye darted at his name again and again—his name is my name, that is my name in the newspaper about a man who found a girl's purse. The girl is now missing. She is five feet three inches tall, weighs one hundred ten pounds, dark brown hair, blue eyes, I can't remember. . . .

I leaf through a magazine and look at the photographs of food. Christmas is coming. We will all make Christmas cookies, the children and I. I will plan meals, a week of meals for Christmas week, I will make up things ahead of time and freeze them . . . and on Christmas Day we will have a ham, I think, instead of a turkey. . . . The table settings in the magazine are very beautiful, decorated with holly and pine boughs.

Finally I pick up the newspaper and look at her picture again—Linda Slater, twenty, dark hair and firm, staring, curious eyes, a very short upper lip. A posed photograph. It probably exaggerates her beauty.

Friday, the hairdresser's. Glenda pins my hair up in big rollers. There are four chairs before the big mirror, two others occupied, women having their hair pinned up. This place is not very clean, but it is reasonable. The air is chatty and warm. All the girls are friendly. Glenda has a big, robust air about her; I can smell the gum she is chewing. Her diamond ring looks much too big to be real; I know her husband is a factory worker.

"Hey, you ever seen a thalidomide baby?" Glenda asks the girl who is working next to her.

"What kind?"

"Thalidomide, you know—that sleeping pill they had."

"Oh, yeah. No. Where is there one?"

"He's not a baby now, he's pretty grown up."

"Where?"

"My mother-in-law's street, across the street from her."

The woman whose hair is being done next to me twists her head around. I see her here often; she has a bleached-out, staring face, a redhead with pale freckles. She is about my age, forty. "There's a thalidomide girl lives down the street from us," she said.

"How old?"

"Twelve."

"This boy, this one I was telling you about, is real nasty. He's maybe six foot tall, he goes to high school already and is a real brat. His mother spoiled him."

"The girl has little arms, real short arms, little flippers."

"*He's* got flippers. He wears some wool things, like mittens, up around them in the winter. He can use them flippers like they were arms."

"Is he smart?"

"He's a smart aleck."

"The girl is pretty smart, I guess. But she's a show-off too. I seen her once in a store downtown acting up. Her and two other girls. They're about in seventh grade, these girls, and flirting with some guy. This girl's got real short little flippers, just like baby arms, and she was touching some salesguy's chin with one of them. It was a shoe store."

"She was what?"

"She was kidding around, flirting

—that guy's face was so red, it wasn't even funny."

"Flirting, she was flirting?"

"Yeah, with those little arms of hers. They say she's real smart, but she's nasty."

"*He's* nasty. Down at school the girls are just crazy for him, and they call him up all the time, because he's cute, and they don't seem to care about the flippers or anything. But around the house his mother has to do everything for him . . . he can't even go to the bathroom by himself, he thinks he's so smart, but he has to have help. They asked him if he would like some artificial arms but he wouldn't. He said no. They tried to get him to take them but he wouldn't. He gets all this attention because of them short arms. . . ."

At the supermarket I go back to the dairy products and get a carton of milk, lifting it up to see if it's leaking underneath. It seems all right. I carry it pressed up against my chest. At the check-out counter I notice some watery milk on my coat.

Texture of wet snow

The snow is stubborn, won't melt. It has turned gray and wet. I was not always forty years old. I remember looking out the front window of my parents' house in Indiana, watching the rain. I had long blond hair, I always dressed well; I was waiting for something to happen to me, and it happened. Once I went for a walk, alone, when I was visiting my grandmother in West Bend. I went for a walk into a little park. I looked at the roses because they had gone to all that trouble to plant roses, I stopped and looked at a sundial . . . but the sun was not shining, I couldn't see what time it was.

When I get home Susan says, "Betty is sort of sick."

Betty is throwing up in the bathroom upstairs. While I am with her the telephone rings, one of the children answers it. I wait but no one calls me; it must be for one of the children.

Dinner

always takes place at six. Everyone is ready to eat. My husband has been home since five, has had a drink, has looked through the paper. His name no longer appears in those stories about Linda Slater. I want to ask him about it, how does he feel, what is he thinking . . . ? but his silence baffles me. I resent this silence in him though it has always been in him, since we met twenty years ago.

He comes into the kitchen, when Susan gets him. Susan hangs onto him, teasing. She loves him, and her love is a torment to all of us and to herself, making her forehead rise in childish angry frowns. "Daddy, you're not listening!" she often says, throwing herself around as if trying to damage herself; or she gets up from the table with dignity and walks away, and Norman calls her back, and she says *no,* and he makes a sudden movement to push his chair back and get her, and pretending to be frightened, perhaps a little frightened, she does come back. . . .

We sit. Tonight we are having creamed chicken with carrots and peas. It is a familiar dish, they like it, they are hungry. Now we are beginning dinner at last, sitting around the big kitchen table, all of us eased into our places as if at the start of a boat race . . . floating with difficulty on the element of our lives, which is love. Is it love? We are here, around this particular table, because two people loved each other and got married. On that day I stared at the sundial I was positive no one would marry me . . . but I needn't have worried, like most women. It happened.

Sometimes I lean over the bathroom sink, alone, feeling nauseated, clutching my head and thinking *What is going to happen?* Am I going to throw up? But I never throw up. I can't bring it up out of me, whatever is inside. Even when I was pregnant I had a strong stomach. The nausea passes, and my head is filled suddenly with activity, the pictures of things I must use—the colored sponges I use in the kitchen, one for dishes exclusively and the other for wiping counters and the table; the sheets and pillowcases I must fold, fold, again and again; the beds I must make, the small rugs I must straighten, the cans I must open with the can opener. My head is filled suddenly with a love for these things. Plates, forks, spoons, knives, paper napkins, glasses . . . these achieve a secret meaning, placed on the table. Everyone sits. A Friday night dinner. Do they understand this bouquet of love I have set out for them? What, precisely, do they see? If the carrots were missing from the main dish, and there were only peas in it, along with pepper and salt and a few other spices, would this make any difference? In the salad I have put two kinds of lettuce, tomatoes, radishes, cucumber slices, bits of celery and green pepper. If I had left out the green pepper, would they notice? Do they notice that it is there? I have made biscuits from a mix. They all put the biscuits on their plates and the chicken and sauce over it, except for Susan, who can't stand soggy things; if the biscuits were missing . . . ? Bobby drops his fork onto the floor. No matter, wipe it off. The dinner has begun. The race is on, no going back. My husband is saying something. Stern, or smiling? He smiles. Good. My husband is saying something to me about this evening. "What time are they coming?" He always asks that question when we have someone over for the evening; they are always invited for eight thirty. "Eight thirty," I tell him. My husband is an intelligent man, and his intelligence is kindly, gentle; he has perfected small attentions over the years while his imagination drifts from me.

What can I say to him?

Friends

come in, smile at us, take off their coats. Norm makes drinks, scotch and water. We sit. Arnold and Brenda look a little tired? No, Brenda looks good. She is wearing those new stockings that have a wet look to them, and her shoes look new too. People sometimes mistake Brenda for me, and me for Brenda, though my hair is lighter than hers and I am taller than she is. I don't mind being mistaken for someone else. We are both faculty wives. Our husbands talk together, we talk together. I look over at Norman to see if he is still so distracted, but I can't tell. He avoids people's eyes, an old habit of his. Around eleven I go out into the kitchen, and Brenda comes with me, carrying her drink. I take slices of cheese and meat out of the refrigerator, which I have prepared earlier and wrapped in cellophane paper. I put some bread in a little wicker basket with a cloth napkin in it. I set out some pickles and olives. Brenda leans her stomach against the counter, looking into her drink. "I'm sorry Arnold is acting so funny tonight," she says. She has creases in her neck that show when she looks down, her chin creasing into her throat. "He started drinking when he got home from school. He thinks they're easing him out. . . ."

Norm and I sleep heavily at night, in the two halves of our bed. Vividly I can remember the past years, those months when I was pregnant. But I can't remember how it came to be that I was pregnant. I can remember being in my earlier twenties, a new wife, sitting in an erotic daze somewhere . . . on a train? . . . my loins dazzled with the memory of our love, the unbearable dazzling of what my young husband had done to me, again and again, but the girl on the train seems to be in a movie, being taken away from me and not me at all . . . she stares dreamily out the window. I am not that girl.

I could never remember why I was pregnant, precisely. It had seemed important, it had seemed sacred, that I remember the precise day, the precise night . . . but I never could, I didn't have time. I had a small baby, I was going to have another; I had two babies, and I was going to have another; we had to pack, I was pregnant and afraid of a miscarriage, we couldn't afford to pay movers, we spent all day packing dishes and books . . . we drove across the country to a school where Norm taught for one year; then we drove back across the country to another school, where he taught for five years, thinking he had found his place; then he decided that that school wasn't good enough for him, and so we came to this school, nine years ago, or per-

haps ten years ago, a school that seems to me precisely like the other two we were at.

"We heard on the radio, coming over, that part of a woman's torso was found on the shore, downriver," Brenda says, making a face. "Some kids found it. Wouldn't that be awful to have one of our own kids find something like that? They were playing down by the river. . . ."

So she is dead.

Saturday

I take Betty to the doctor. Asian flu. That means they will all get it, all the children. Susan goes for her piano lessons. The car heater is broken, I must tell Norman, I am afraid to tell Norman . . . we can't afford to pay for it. I have started to buy Christmas presents. My head is dazzled suddenly with the thought of presents, Christmas presents . . . days of buying presents stretch out before me. And then Christmas week. Some presents I will mail all the way back to Indiana; it is important to keep up these traditions. Even my old uncles, they appreciate being remembered. Everyone appreciates being remembered. It is terrible not to be remembered.

The supermarket is crowded. Everyone is in a hurry, but it is a pleasant hurry. The cans on the shelves stretch up over my head, so many different sizes and colors . . . I put cans in the cart, and then on to the dairy counter—four cartons of milk. On another counter are bathroom things. Pills, hair shampoo, soap, deodorants . . . the pills remind me of an advertisement on television for cold tablets. The capsules detonate gradually over a period of twelve hours, I think. They release themselves in tiny fragments into the bloodstream. I think of the cells of my body with the seeds of my future inside them, unreadable. They have the seeds of cancer inside them, death itself, the particular way in which I will wear out and die, everything contained secretly in them and ready to go off at a certain time. But that time is a secret.

At the drugstore there is a pile of newspapers, and one of the headlines is GIRL'S BODY FOUND IN RIVER. The picture of that girl appears again, a beautiful girl, staring out at the camera with her perky upper lip, lipsticked and pretty, very sure of herself. Parts of a body were found along the river and in the river, a woman's torso, a head. The face mutilated. And so her face is no longer that face, the one in the paper . . . ? I stare down at her and I feel panic inside me, in the back of my head, behind my knees. What is this threat to me? Am I going to break down? Am I going to scream? A yellowish cell threatens to burst inside me; like sperm it is yellow and living.

A man
         in the corner of a woman's eye paralyzes the entire eye. A woman wants to rake her body with her nails, streaming blood, she wants to gash her face so that no man need look at it, she wants to be finished and safe. But why does my heart pound so? We are not at war. Yes, we are at war somewhere, soldiers somewhere "at war," but we ourselves are not at war and should therefore be at peace. Why am I not at peace, being forty years old?

This man has a weak face, he looks very young. The photograph is blurred. Why, he is only a boy, his eyes are a boy's eyes . . . but there is no youth in him, only finality. He has come to the end of something. Identified by a motel proprietor, last seen with Linda Slater on Saturday evening, the two of them came to a certain motel out on the highway and there, in a room, they argued, and then they left. . . . What did they argue about? Why will it never be known precisely what they argued about on that night?

Imagine the strength behind a knife that could sever a head from a body, so beautiful a head! Imagine the torrents of blood that would gush from the throat! There must have been confusion at the end, madness, not love or hate. Things are speeded up as they approach the end of something. The boy must have been hurried, making mistakes, whimpering to himself, everything speeded up and dazzling and crazy beyond his imagination. . . . At one o'clock neither of them, the girl or the boy, knew what would happen at two o'clock. Perhaps they knew that something would happen, some strange thing, but perhaps they had sensed such events earlier in their lives, falsely, when nothing did happen. This time it happened.

He lets the newspaper fall from him. A stunned, vacuous sorrow shows in his face; I watch him from the darkened dining room. I see the dreaminess in him, the stunned clarity of some final perception—I would like to shout in his face, "Why are you surprised? She had to die like everyone else!" But I say nothing. He came home late from school today, looking a little sick. If he gets the flu along with Betty I will have to take care of him as if he were a child, worse than a child, and a stab of pleasure comes to me. . . . I am in a hurry, I can't feel anything. Time is snatched from me in handfuls. The people who laugh over the television set, in the other room, are laughing in a terrible unison like tiny people with tiny lungs, laughing at me. I stand here watching a man I have been married to for many years, and I can never possess him, my husband. I can never be that girl's age. My head and torso are connected. He will never look at me as he has looked at her, at her photograph.

I have to run out to the store again before six. The metallic Christmas tree looks the same, the angel looks the same; I don't bother with a cart but hurry back to the dairy counter. I pick up a carton of eggs, Grade A eggs I forgot to buy earlier, and a few other things, I hurry back to the check-out counter.

The parking lot is nearly empty at this time of day. Rough ice on the pavement, a white and blue container marked "Salvation Army Pickup," a kid's jalopy idling noisily at the curb, with some high school kids in it, frozen weeds and trash between the parking lot and the sidewalk. . . . I hurry to my car. I am not followed.

from

The Street

ANNE PETRY

It was three rooms. The wind held it still for an instant in front of her and then swooped it away until it was standing at an impossible angle on the rod that suspended it from the building. She read it rapidly. Three rooms, steam heat, parquet floors, respectable tenants. Reasonable.

She looked at the outside of the building. Parquet floors here meant that the wood was so old and so discolored no amount of varnish or shellac would conceal the scars and the old scraped places, the years of dragging furniture across the floors, the hammer blows of time and children and drunks and dirty, slovenly women. Steam heat meant a rattling, clanging noise in radiators early in the morning and then a hissing that went on all day.

Respectable tenants in these houses where colored people were allowed to live included anyone who could pay the rent, so some of them would be drunk and loud-mouthed and quarrelsome; given to fits of depression when they would curse and cry violently, given to fits of equally violent elation. And, she thought, because the walls would be flimsy, why, the good people, the bad people, the children, the dogs, and the godawful smells would all be wrapped up together in one big package—the package that was called respectable tenants.

The wind pried at the red skullcap on her head, and as though angered because it couldn't tear it loose from its firm anchorage of bobby pins, the wind blew a great cloud of dust and ashes and bits of paper into her face, her eyes, her nose. It smacked against her ears as though it were giving her a final, exasperated blow as proof of its displeasure in not being able to make her move on.

Lutie braced her body against the wind's attack determined to finish thinking about the apartment before she went in to look at it. Reasonable —now that could mean almost anything. On Eighth Avenue it meant tenements—ghastly places not fit for humans. On St. Nicholas Avenue it meant high rents for small apartments; and on Seventh Avenue it meant great big apartments where you had to take in roomers in order to pay the rent. On this street it could mean almost anything.

She turned and faced the wind in order to estimate the street. The buildings were old with small slit-like windows, which meant the rooms were small and dark. In a street running in this direction there wouldn't be any sunlight in the apartments. Not ever. It would be hot as hell in summer and cold in winter. 'Reasonable' here in this dark, crowded street ought to be about twenty-eight dollars, provided it was on a top floor.

The hallways here would be dark and narrow. Then she shrugged her shoulders, for getting an apartment where she and Bub would be alone was more important than dark hallways. The thing that really mattered was getting away from Pop and his raddled women, and anything was better than that. Dark hallways, dirty stairs, even roaches on the walls. Anything. Anything. Anything.

* * *

No. She didn't want to see the apartment—the dark, dirty three rooms called an apartment. Then she thought of where she lived now. Those seven rooms where Pop lived with Lil, his girl friend. A place filled with roomers. A place spilling over with Lil.

There seemed to be no part of it that wasn't full of Lil. She was always swallowing coffee in the kitchen; trailing through all seven rooms in housecoats that didn't quite meet across her lush, loose bosom; drinking beer in tall glasses and leaving the glasses in the kitchen sink so the foam dried in a crust around the rim—the dark red of her lipstick like an accent mark on the crust; lounging on the wide bed she shared with Pop and only God knows who else; drinking gin with the roomers until late at night.

And what was far more terrifying giving Bub a drink on the sly; getting Bub to light her cigarettes for her. Bub at eight with smoke curling out of his mouth.

Only last night Lutie slapped him so hard that Lil cringed away from her dismayed; her housecoat slipping even farther away from the fat curve of her breasts. 'Jesus!' she said. 'That's enough to make him deaf. What's the matter with you?'

But did she want to look at the apartment? Night after night she'd come home from work and gone out right after supper to peer up at the signs in front of the apartment houses in the neighborhood, looking for a place just big enough for her and Bub. A place where the rent was low enough so that she wouldn't come home from work some night to find a long sheet of white paper stuck under the door: 'These premises must be vacated by——' better known as an eviction notice. Get out in five days or be tossed out. Stand by and watch your furniture pile up on the sidewalk. If you could call those broken beds, worn-out springs, old chairs with the stuffing crawling out from under, chipped porcelain-topped kitchen table, flimsy kitchen chairs with broken rungs—if you could call those things furniture. That was an important point—now could you call fire-cracked china from the five-and-dime, and red-handled knives and forks and spoons that were bent and coming apart, could you really call those things furniture?

'Yes,' she said firmly. 'I want to look at the apartment.'

'I'll get a flashlight,' he said and went back into his apartment, closing the door behind him so that it made a soft sucking sound. He said something, but she couldn't hear what it was. The whispering voice inside the apartment stopped and the dog was suddenly quiet.

Then he was back at the door, closing it behind him so it made the same soft, sucking sound. He had a long black flashlight in his hand. And she went up the stairs ahead of him thinking that the rod of its length was almost as black as his hands. The flashlight was a shiny black—smooth and gleaming faintly as the light lay along its length. Whereas the hand that held it was flesh—dull, scarred, worn flesh—no smoothness there. The knuckles were knobs that stood out under the skin, pulled out from hauling ashes, shoveling coal.

But not apparently from using a mop or a broom, for, as she went up and up the steep flight of stairs, she saw that they were filthy, with wastepaper, cigarette butts, the discarded wrappings from packages of snuff, pink ticket stubs from the movie houses. On the landings there were empty gin and whiskey bottles.

She stopped looking at the stairs, stopped peering into the corners of the long hallways, for it was cold, and she began walking faster trying to keep warm. As they completed a flight of stairs and turned to walk up another hall, and then started climbing another flight of stairs, she was aware that the cold increased. The farther up they went, the colder it got. And in summer she supposed it would get hotter and hotter as you went up until when you reached the top floor your breath would be cut off completely.

* * *

A crowd of people surged in to the Eighth Avenue express at 96th Street. By elbowing other passengers in the back, by pushing and heaving, they forced their bodies into the coaches, making room for themselves where no room had existed before. As the train gathered speed for the long run to 125th Street, the passengers settled down into small private worlds, thus creating the illusion of space between them and their fellow passengers. The worlds were built up behind newspapers and magazines, behind closed eyes or while staring at the varicolored show cards that bordered the coaches.

Lutie Johnson tightened her clutch on an overhead strap, her tall long-legged body swaying back and forth as the train rocked forward towards its destination. Like some of the other passengers, she was staring at the advertisement directly in front of her and as she stared at it she became absorbed in her own thoughts. So that she, too, entered a small private world which shut out the people tightly packed around her.

For the advertisement she was looking at pictured a girl with incredible blond hair. The girl leaned close to a dark-haired, smiling man in a navy uniform. They were standing in front of a kitchen sink—a sink whose white porcelain surface gleamed under the train lights. The faucets looked like silver. The linoleum floor of the kitchen was a crisp black-and-white pattern that pointed up the sparkle of the room. Casement windows. Red geraniums in yellow pots.

It was, she thought, a miracle of a kitchen. Completely different from the kitchen of the 116th Street apartment she had moved into just two weeks ago. But almost exactly like the one she had worked in in Connecticut.

So like it that it might have been the same kitchen where she had washed dishes, scrubbed the linoleum floor and waxed it afterward. Then gone to sit on the small porch outside the kitchen, waiting for the floor to dry and wondering how much longer she would have to stay there. At the time it was the only job she could get. She had thought of it as a purely temporary one, but she had ended up by staying two years—thus earning the money for Jim and Bub to live on.

Every month when she got paid she walked to the postoffice and mailed the money to Jim. Seventy dollars. Jim and Bub could eat on that and pay the interest on the mortgage. On her first trip to the postoffice, she realized she had never seen a street like that main street in

Lyme. A wide street lined with old elm trees whose branches met high overhead in the center of the street. In summer the sun could just filter through the leaves, so that by the time its rays reached the street, it made a pattern like the lace on expensive nightgowns. It was the most beautiful street she had ever seen, and finally she got so she would walk to the little postoffice hating the street, wishing that she could get back to Jamaica, back to Jim and Bub and the small frame house.

In winter the bare branches of the trees made a pattern against the sky that was equally beautiful in snow or rain or cold, clear sunlight. Sometimes she took Little Henry Chandler to the postoffice with her and she couldn't help thinking that it wasn't right. He didn't need her and Bub did. But Bub had to do without her.

And because Little Henry Chandler's father manufactured paper towels and paper napkins and paper handkerchiefs, why, even when times were hard, he could afford to hire a Lutie Johnson so his wife could play bridge in the afternoon while Lutie Johnson looked after Little Henry. Because as Little Henry's father used to say, 'Even when times are hard, thank God, people have got to blow their noses and wipe their hands and faces and wipe their mouths. Not quite so many as before, but enough so that I don't have to worry.'

Her grip on the subway strap tightened until the hard enameled surface cut into her hand and she relaxed her hand and then tightened it. Because that kitchen sink in the advertisement or one just like it was what had wrecked her and Jim. The sink had belonged to someone else

—she'd been washing someone else's dishes when she should have been home with Jim and Bub. Instead she'd cleaned another woman's house and looked after another woman's child while her own marriage went to pot; breaking up into so many little pieces it couldn't be put back together again, couldn't even be patched into a vague resemblance of its former self.

Yet what else could she have done? It was her fault, really, that they lost their one source of income. And Jim couldn't get a job, though he hunted for one—desperately, eagerly, anxiously. Walking from one employment agency to another, spending long hours in the musty agency waiting-rooms, reading old newspapers. Waiting, waiting, waiting to be called up for a job. He would come home shivering from the cold, saying 'God damn white people anyway. I don't want favors. All I want is a job. Just a job. Don't they know if I knew how I'd change the color of my skin?'

There was the interest to be paid on the mortgage. It didn't amount to much, but they didn't have anything to pay it with. So she answered an advertisement she saw in the paper. The ad said it was a job for an unusual young woman because it was in the country and most help wouldn't stay. 'Seventy-five dollars a month. Modern house. Own room and bath. Small child.'

She sat down and wrote a letter the instant she saw it; not telling Jim, hoping against hope that she would get it. It didn't say 'white only,' so she started off by saying that she was colored. And an excellent cook, because it was true—anyone who could fix good meals on practically no money at all was an excellent cook. An efficient housekeeper—because it was easy to keep their house shining, so she shouldn't have any trouble with a 'modern' one. It was a good letter, she thought, holding it in her hand a little way off from her as she studied it—nice neat writing, no misspelled words, careful margins, pretty good English. She was suddenly grateful to Pop. He'd known what he was doing when he insisted on her finishing high school. She addressed the envelope, folded the letter, and put it inside the envelope.

She was about to seal it when she remembered that she didn't have any references. She couldn't get a job without them, and as she'd never really had a job, why, she didn't have any way of getting a reference. Somehow she had been so sure she could have got the job in the ad. Seventy-five dollars a month would have meant they could have saved the house; Jim would have got over that awful desperate feeling; that bitterness that was eating him up; and there wouldn't have been any need to apply for relief.

Mrs. Pizzini. That was it. She'd go to Mrs. Pizzini where they bought their vegetables. They owed her a bill, and when she explained that this job would mean the bill would be paid, why, Mrs. Pizzini would write her out a reference.

Business was slow and Mrs. Pizzini had plenty of time to listen to Lutie's story, to study the advertisement in the paper, to follow the writing on Lutie's letter to Mrs. Henry Chandler, line by line, almost tracing the words on the page with her stubby fingers.

'Very good,' she said when she

finished reading it. 'Nice job.' She handed the letter and the newspaper to Lutie. 'Me and Joe don't write so good. But my daughter that teaches school, she'll write for me. You can have tomorrow.'

And the next day Mrs. Pizzini stopped weighing potatoes for a customer long enough to go in the back of the vegetable store and bring the letter out carefully wrapped up in brown paper to keep it clean. Lutie peeled off the brown paper and read the letter through quickly. It was a fine letter, praising her for being hard-working and honest and intelligent; it said that the writer hated to lose Lutie, for she'd worked for her for two years. It was signed 'Isabel Pizzini.'

The handwriting was positively elegant, she thought, written with a fine pen and black ink on nice thick white paper. She looked at the address printed on the top and then turned to stare at Mrs. Pizzini in astonishment, because that part of Jamaica was the section where the houses were big and there was lawn around them and evergreen trees grew in thick clusters around the houses.

Mrs. Pizzini nodded her head. 'My daughter is a very smart woman.'

And then Lutie remembered the letter in her hand. 'I can't ever thank you,' she said.

Mrs. Pizzini's lean face relaxed in a smile, 'It's all right. You're a nice girl. Always known it.' She walked toward her waiting customer and then, hesitating for the barest fraction of a second, turned back to Lutie. 'Listen,' she said. 'It's best that the man do the work when the babies are young. And when the man is young. Not good for the woman to work when she's young. Not good for the man.'

Curiously enough, though she only half-heard what Mrs. Pizzini was saying, she remembered it. Off and on for the past six years she had remembered it. At the time, she hurried home from the vegetable store to put the precious reference in the letter to Mrs. Henry Chandler and mail it.

After she had dropped it in the mail box on the corner, she got to thinking about the Pizzinis. Who would have thought that the old Italian couple who ran the vegetable store would be living in a fine house in a fine neighborhood? How had they managed to do that on the nickels and dimes they took in selling lettuce and grapefruit? She wanted to tell Jim about it, but she couldn't without revealing how she knew where they lived. They had a fine house and they had sent their daughter to college, and yet Mrs. Pizzini had admitted she herself 'couldn't write so good.' She couldn't read so good either, Lutie thought. If she could find out how the Pizzinis had managed, it might help her and Jim.

Then she forgot about them, for Mrs. Chandler wrote to her sending the train fare to Lyme, telling her what train to take. When she showed Jim the letter, she was bursting with pride, filled with a jubilance she hadn't felt in months because now they could keep the house. And she need no longer feel guilty about having been responsible for losing the State children that had been their only source of income.

'How'm I going to look after Bub

and him only two?' he asked, frowning, handing the letter back to her, not looking at her.

Even on the day she was to leave he was sullen. Not talking. Frowning. Staring off into space. He came into the bedroom where she was putting carefully ironed clothes into her suitcase. He stood in front of the window and looked out at the street, his back turned to her, his hands in his pockets as he told her he wouldn't be going to the station with her.

'We can't afford that extra dime for carfare,' he explained briefly.

So she went by herself. And feeling the suitcase bump against her legs when she walked down the long ramp at Grand Central to get on the train, she wished that Jim had been along to carry it. So that she could have kissed him goodbye there in the train shed and thus carried the memory of his lips right onto the train with her—so that it could have stayed with her those first few days in Lyme, helping her to remember why she had taken the job. If he'd come to the train with her, he would have lost that pretended indifference; the sight of her actually getting on the train would have broken down the wall of reserve he had built around himself. Instead of that quick hard peck at her forehead, he would have put his arms around her and really kissed her. Instead of holding his body rigid, keeping his arms hanging limp and relaxed at his sides, he would have squeezed her close to him.

As the train left the city, she stopped thinking about him, not forgetting him, but thrusting him far back in her mind because she was going to a new strange place and she didn't want to get off the train wrapped in gloom, and that's exactly what would happen if she kept on thinking about Jim. It was important that Mrs. Henry Chandler should like her on sight, so Lutie carefully examined the countryside as the train went along, concentrating on it to shut out the picture of Jim's tall figure.

There was low, marshy land on each side of the train tracks. Where the land was like that, there were very few houses. She noticed that near the cities the houses were small and meanlooking, for they were built close to the railroad tracks. In Bridgeport the houses were blackened with soot and smoke from the factories. Then the train stopped in New Haven and stayed there for all of ten minutes. She looked at the timetable and saw that it was a scheduled stop for that length of time. Saybrook was the next stop. That's where she was to get off. And she began to worry. How would Mrs. Chandler recognize her? How would she recognize Mrs. Chandler? Suppose they missed each other. What would she do stranded in some little jerk-water town? Mrs. Chandler had said in her letter that she lived in Lyme, and Lutie began to wonder how she could get to Lyme if Mrs. Chandler didn't meet her or missed her at the station.

But almost the instant she stepped on the platform at Saybrook, a young blond woman came toward her smiling and saying, 'Hello, there. I'm Mrs. Chandler. You must be Lutie Johnson.'

Lutie looked around the platform. Very few people had got off the train, and then she wanted to laugh. She needn't have worried about Mrs. Chandler recognizing her; there wasn't another colored person in sight.

'The car's over there.' Mrs. Chandler waved in the direction of a station wagon parked in the dirt road near the platform.

Walking toward the car, Lutie studied Mrs. Chandler covertly and thought, What she's got on makes everything I'm wearing look cheap. This black coat fits too tightly and the velvet collar is all wrong, just like these high-heeled shoes and thin stockings and this wide-brimmed hat. For Mrs. Chandler wore ribbed stockings made of very fine cotton and flat-heeled moccasins of a red-brown leather that caught the light. She had on a loose-fitting tweed coat and no hat. Lutie, looking at the earrings in her ears, decided that they were real pearls and thought, Everything she has on cost a lot of money, yet she isn't very much older than I am—not more than a year or so.

Lutie didn't say anything on the ride to Lyme, for she was thinking too hard. Mrs. Chandler pointed out places as they rode along. 'The Connecticut River,' she said with a wave of her hand toward the water under the bridge they crossed. They turned off the road shortly after they crossed the river, to go for almost a mile on a country road where the trees grew so thickly Lutie began to wonder if the Chandlers lived in a forest.

Then they entered a smaller road where there were big gates and a sign that said 'private road.' The road turned and twisted through thick woods until finally they reached a large open space where

there was a house. Lutie stared at it, catching her lip between her teeth; it wasn't that it was so big; there were houses in certain parts of Jamaica that were just as big as this one, but there weren't any so beautiful. She never quite got over that first glimpse of the outside of the house—so gracious with such long low lines, its white paint almost sparkling in the sun and the river very blue behind the house.

'Would you like to sort of go through the inside of the house before I show you your room?' Mrs. Chandler asked.

'Yes, ma'am,' Lutie said quietly. And wondered how she had been able to say 'yes, ma'am' so neatly and so patly. Some part of her mind must have had it already, must have already mapped out the way she was to go about keeping this job for as long as was necessary by being the perfect maid. Patient and good-tempered and hard-working and more than usually bright.

Later she was to learn that Mrs. Chandler's mother and father regarded this house as being very small. 'The children's house.' The very way they said it told her they were used to enormous places ten times the size of this and that they thought this doll-house affair cute and just right for children for a few years. Mr. Chandler's father never commented on it one way or the other. So it was impossible for Lutie to tell what he thought about it when he came to stay for an occasional week-end.

But to Lutie the house was a miracle, what with the four big bedrooms, each one with its own bath; the nursery that was as big as the bedrooms, and under the nursery a room and bath that belonged to her. On top of that there was a living room, a dining room, a library, a laundry. Taken all together it was like something in the movies, what with the size of the rooms and the big windows that brought the river and the surrounding woods almost into the house. She had never seen anything like it before.

That first day when she walked into Mrs. Chandler's bedroom her breath had come out in an involuntary 'Oh!'

'You like it?' Mrs. Chandler asked, smiling.

Lutie nodded and then remembered and said, 'Yes, ma'am.' She looked at the room, thinking there wasn't any way she could say what this bedroom looked like to her when all her life she had slept on couches in living rooms, in cubicles that were little more than entrances to and exits from other rooms that were rented out to roomers; when the first real bedroom she ever had was that small one in Jamaica, where if you weren't careful you would bump your head on the lowhanging ceiling, for the dormer window only raised the ceiling right where the window was.

No, she decided, there wasn't any way to explain what this room looked like to her. It ran across one whole end of the house so that windows looked out on the river, out on the gardens in front, out on the woods at the side. It was covered from wall to wall with thick red carpet and right near the fourposter bed was a round white rug—a rug with pile so deep it looked like fur. Muted chintz draperies gleamed softly at the windows, formed the petticoat on the bed, covered the chaise longue in front of the win-dows that faced the river and the pair of chairs drawn up near the fireplace.

The rest of the house was just as perfect as Mrs. Chandler's bedroom. Even her room—the maid's room with its maple furniture and vivid draperies—that, too, was perfect. Little Henry Chandler, who was two years older than Bub, was also perfect—that is, he wasn't spoiled or anything. Just a nice, happy kid, liking her at once, always wanting to be with her. The Chandlers called him Little Henry because his father's name was Henry. She thought it funny at first, because colored people always called their children 'Junior' or 'Sonnie' when the kid's name was the same as his father's. But she had to admit that calling a kid Little Henry gave him a certain dignity and a status all his own, while it prevented confusion, for there was no mistaking whom you were talking about.

Yes. The whole thing was perfect. Mr. Chandler was young and attractive and obviously made plenty of money. Yet after six months of living there she was uneasily conscious that there was something wrong. She wasn't too sure that Mrs. Chandler was overfond of Little Henry; she never held him on her lap or picked him up and cuddled him the way mothers do their children. She was always pushing him away from her.

Mr. Chandler drank too much. Most people wouldn't have noticed it, but having lived with Pop who had an unquenchable thirst it was easy for her to recognize all the signs of a hard drinker. Mr. Chandler's hands were shaking when he came down for breakfast and he had to have a pick-up before he could even face a cup of coffee. When he came in the house at night, the first thing he did was to get himself a good-sized drink. It was almost impossible for her to keep full bottles in the liquor cabinet, their contents disappeared in no time at all.

'Guess Lutie forgot to put a new supply in the bar,' Mr. Chandler would say when she came in in answer to his ring.

'Yes, sir,' she would say quietly and go to get more bottles.

The funny thing about it was that Mrs. Chandler never noticed. After a while Lutie discovered that Mrs. Chandler never noticed anything about Mr. Chandler anyway. Yet she was awfully nice; she was always laughing; she had a great many young friends who dressed just like she did—some of them even had small children about the age of Little Henry.

But she didn't like Mrs. Chandler's friends much. They came to the house to luncheon parties or to bridge parties in the afternoon. Either they ate like horses or they didn't eat at all, because they were afraid they would get too fat. And she never could decide which irritated her the most, so see them gulp down the beautiful food she had fixed, eating it so fast they really didn't taste it, or to see them toy with it, pushing it around on their plates.

Whenever she entered a room where they were, they stared at her with a queer, speculative look. Sometimes she caught snatches of their conversation about her. 'Sure, she's a wonderful cook. But I wouldn't have any good-looking colored wench in my house. Not with John.

You know they're always making passes at men. Especially white men.' And then, 'Now I wonder—'

After that she continued to wait on them quietly, efficiently, but she wouldn't look at them—she looked all around them. It didn't make her angry at first. Just contemptuous. They didn't know she had a big handsome husband of her own; that she didn't want any of their thin unhappy husbands. But she wondered why they all had the idea that colored girls were whores.

It was, she discovered slowly, a very strange world that she had entered. With an entirely different set of values. It made her feel that she was looking through a hole in a wall at some enchanted garden. She could see, she could hear, she spoke the language of the people in the garden, but she couldn't get past the wall. The figures on the other side of it loomed up life-size and they could see her, but there was this wall in between which prevented them from mingling on an equal footing. The people on the other side of the wall knew less about her than she knew about them.

She decided it wasn't just because she was a maid; it was because she was colored. No one assumed that the young girl from the village who came in to help when they had big dinner parties would eagerly welcome any advances made toward her by the male guests. Even the man who mowed the lawn and washed the windows and weeded the garden didn't move behind a wall that effectively and automatically placed him in some previously prepared classification. One day when he was going to New Haven, Mrs. Chandler drove him to the railroad station in

Saybrook, and when he got out of the car Lutie saw her shake hands with him just as though he had been an old friend or one of her departing week-end guests.

When she was in high school she had believed that white people wanted their children to be president of the United States; that most of them worked hard with that goal in mind. And if not president—well, perhaps a cabinet member. Even the Pizzinis' daughter had got to be a school-teacher, showing that they, too, had wanted more learning and knowledge in the family.

But these people were different. Apparently a college education was all right, and seemed to have become a necessity even in the business world they talked about all the time. But not important. Mr. Chandler and his friends had gone through ̗Yale and Harvard and Princeton, casually, matter-of-factly, and because they had to. But once these men went into business they didn't read anything but trade magazines and newspapers.

She had watched Mr. Chandler reading the morning newspaper while he ate his breakfast. He riffled through the front pages where the news was, and then almost immediately turned to the financial section. He spent quite a while reading that, and then, if he had time, he would look at the sports pages. And he was through. She could tell by looking at him that the effort of reading had left him a little tired just like Pop or Mrs. Pizzini. Mr. Chandler's father did the same thing. So did the young men who came up from New York to spend the weekend.

No. They didn't want their children to be president or diplomats or anything like that. What they wanted was to be rich—'filthy' rich, as Mr. Chandler called it.

When she brought the coffee into the living room after dinner, the conversation was always the same. 'Richest damn country in the world——'

'Always be new markets. If not here in South America, Africa, India ——Everywhere and anywhere——'

'Hell! Make it while you're young. Anyone can do it——'

'Outsmart the next guy. Think up something before anyone else does. Retire at forty——'

It was a world of strange values where the price of something called Tell and Tell and American Nickel and United States Steel had a direct effect on emotions. When the price went up everybody's spirits soared; if it went down they were plunged in gloom.

After a year of listening to their talk, she absorbed some of the same spirit. The belief that anybody could be rich if he wanted to and worked hard enough and figured it out carefully enough. Apparently that's what the Pizzinis had done. She and Jim could do the same thing, and she thought she saw what had been wrong with them before—they hadn't tried hard enough, worked long enough, saved enough. There hadn't been any one thing they wanted above and beyond everything else. These people had wanted only one thing—more and more money—and so they got it. Some of this new philosophy crept into her letters to Jim.

When she first went to work for the Chandlers, Mrs. Chandler had suggested that, instead of her taking one day off a week, it would be a good idea if she took four days off right together all at once at the end of a month; pointing out that that way Lutie could go home to Jamaica and not have to turn right around and come back. As Lutie listened to the conversations in the Chandlers' house, she came more and more under the influence of their philosophy. As a result she began going home only once in two months, pointing out to Jim how she could save the money she would have spent for train fare.

She soon discovered that the Chandlers didn't spend very much time at home in spite of their big perfect house. They always went out in the evening unless they had guests of their own. After she had been there a year and a half, she discovered, too, that Mrs. Chandler paid a lot more attention to other women's husbands than she did to her own. After a dinner party, Mrs. Chandler would walk through the garden with someone else's husband, showing him the river view, talking to him with an animation she never showed when talking to Mr. Chandler. And, Lutie observed from the kitchen window, leaning much too close to him.

Once, when Lutie went into the living room, Mrs. Chandler was sitting on the window-seat with one of the dinner guests and his arms were tight around her and he was kissing her. Mr. Chandler came right in behind Lutie, so that he saw the same thing. The expression on his face didn't change—only his lips went into a straight thin line.

Two weeks before Christmas, Mrs. Chandler's mother came for a visit. A tall, thin woman with cold gray eyes and hair almost exactly the same color as her eyes. She took one look at Lutie and hardly let her get out of the door before she was leaning across the dining-room table to say in a clipped voice that carried right out into the kitchen: 'Now I wonder if you're being wise, dear. That girl is unusually attractive and men are weak. Besides, she's colored and you know how they are——'

Lutie moved away from the swinging door to stand way over by the stove so she couldn't hear the rest of it. Queer how that was always cropping up. Here she was highly respectable, married, mother of a small boy, and, in spite of all that, knowing all that, these people took one look at her and immediately got that now-I-wonder look. Apparently it was an automatic reaction of white people—if a girl was colored and fairly young, why, it stood to reason she had to be a prostitute. If not that—at least sleeping with her would be just a simple matter, for all one had to do was make the request. In fact, white men wouldn't even have to do the asking because the girl would ask them on sight.

She grew angrier as she thought about it. Of course, none of them could know about your grandmother who had brought you up, she said to herself. And ever since you were big enough to remember the things that people said to you, had said over and over, just like a clock ticking, 'Lutie, baby, don't you never let no white man put his hands on you. They ain't never willin' to let a black woman alone. Seems like they all got a itch and a urge to sleep with 'em. Don't you never let any of 'em touch you.'

Something that was said so often and with such gravity it had become a part of you, just like breathing, and you would have preferred crawling in bed with a rattlesnake to getting in bed with a white man. Mrs. Chandler's friends and her mother couldn't possibly know that, couldn't possibly imagine that you might have a distrust and a dislike of white men far deeper than the distrust white women had of you. Or know that, after hearing their estimation of you, nothing in the world could ever force you to be even friendly with a white man.

And again she thought of the barrier between her and these people. The funny part of it was she was willing to trust them and their motives without questioning, but the instant they saw the color of her skin they knew what she must be like; they were so confident about what she must be like they didn't need to know her personally in order to verify their estimate.

* * *

She had been at the Chandlers' exactly two years on the day she got the letter from Pop. She held it in her hand before she opened it. There was something terribly wrong if Pop had gone to all the trouble of writing a letter. If the baby was sick, he would have phoned. Jim couldn't be sick, because Pop would have phoned about that, too. Because he had the number of the Chandlers' telephone. She had given it to him when she first came here to work. Reluctantly she opened the envelope. It was a very short note: *'Dear Lutie: You better come home. Jim's carrying on with another woman. Pop.'*

It was like having the earth suddenly open up so that it turned everything familiar into a crazy upside down position, so that she could no longer find any of the things that had once been hers. And she was filled with fear because she might not ever be able to find them again. She looked at the letter for a third, a fourth, a fifth time, and it still said the same thing. That Jim had fallen for some other woman. And it must be something pretty serious if it so alarmed Pop that he actually wrote her a letter about it. She thought Pop can't suddenly have turned moral—Pop who had lived with so many Mamies and Lauras and Mollies that he must have long since forgotten some of them himself. So it must be that Jim had admitted some kind of permanent attachment for this woman whoever she was.

She thrust the thought away from her and went to tell Mrs. Chandler that she had to go home that very day because the baby was seriously ill. She couldn't bring herself to tell her what the real trouble was because, if Mrs. Chandler was anything like her mother, she took it for granted that all colored people were immoral and Lutie saw no reason for providing further evidence.

On the train she kept remembering Mrs. Pizzini's words: 'Not good for the woman to work when she's young. Not good for the man,' Queer. Though she hadn't paid too much attention at the time, just remembering the words made her see the whole inside of the vegetable store again. The pale yellow color of the grapefruit, dark green of mustard greens and spinach. The patient brown color of the potatoes. The delicate green of the heads of lettuce. She could see Mrs. Pizzini's dark weather-beaten skin and remembered how Mrs. Pizzini had hesitated and then turned back to say: 'It's best that the man do the work when the babies are young.'

She forgot that Jim wasn't expecting her as she hurried to the little frame house in Jamaica, not thinking about anything except the need to get there quickly, quickly, before every familiar thing she knew had been destroyed.

Still hurrying, she opened the front door and walked in. Walked into her own house to find there was another woman living there with Jim. A slender, dark brown girl whose eyes shifted crazily when she saw her. The girl was cooking supper and Jim was sitting at the kitchen table watching her.

If he hadn't held her arms, she would have killed the other girl. Even now she could feel rage rise inside her at the very thought. There she had been sending practically all her wages, month after month, keeping only a little for herself; skimping on her visits because of the carfare and because she was trying to save enough money to form a backlog for them when she quit her job. Month after month and that black bitch had been eating the food she bought, sleeping in her bed, making love to Jim.

He forced her into a chair and held her there while the girl packed and got out. When Lutie finally cooled off enough to be able to talk coherently, he only laughed at her. Even when he saw that she was getting into a red rage at the sight of his laughter.

'What did you expect?' he asked. 'Maybe you can go on day after day with nothing to do but just cook meals for yourself and a kid. With just enough money to be able to eat and have a roof over your head. But I can't. And I don't intend to.'

'Why didn't you say so?' she asked fiercely. 'Why did you let me go on working for those white people and not tell me——'

He only shrugged and laughed. That was all she could get out of him—laughter. What's the use—what's the point—who cares? If even once he had put his arms around her and said he was sorry and asked her to forgive him, she would have stayed. But he didn't. So she called a moving man and had him take all the furniture that was hers. Everything that belonged to her: the scarred bedroom set, the radio, the congoleum rug, a battered studio couch, an easy-chair—and Bub. She wasn't going to leave him behind for Jim to abuse or ignore as he saw fit.

She and Bub went to live with Pop in that crowded, musty flat on Seventh Avenue. She hunted for a job with a grim persistence that was finally rewarded, for two weeks later she went to work as a hand presser in a steam laundry. It was hot. The steam was unbearable. But she forced herself to go to night school—studying shorthand and typing and filing. Every time it seemed as though she couldn't possibly summon the energy to go on with the course, she would remind herself of all the people who had got somewhere in spite of the odds against them. She would think of the Chandlers and their young friends—'It's the richest damn country in the world.'

Mrs. Chandler wrote her a long letter and Jim forwarded it to her from Jamaica. *'Lutie dear: We haven't had a decent thing to eat*

*since you left. And Little Henry
misses you so much he's almost sick—'*
She didn't answer it. She had more
problems than Mrs. Chandler and
Little Henry had and they could always find somebody to solve theirs
if they paid enough.

It took a year and a half before
she mastered the typing, because at
night she was so tired when she went
to the business school on 125th Street
she couldn't seem to concentrate on
what she was doing. Her back ached
and her arms felt as though they had
been pulled out of their sockets. But
she finally acquired enough speed so
that she could take a civil service examination. For she had made up her
mind that she wasn't going to wash
dishes or work in a laundry in order
to earn a living for herself and Bub.

Another year dragged by. A year
in which she passed four or five
exams each time way down on the
list. A year that she spent waiting
and waiting for an appointment and
taking other exams. Four years of
the steam laundry and then she got
an appointment as a file clerk.

That kitchen in Connecticut had
changed her whole life—that kitchen
all tricks and white enamel like this
one in the advertisement. The train
roared into 125th Street and she
began pushing her way toward the
doors, turning to take one last look
at the advertisement as she left the
car.

On the platform she hurried
toward the downtown side and el-
bowed her way toward the waiting
local. Only a few minutes and she
would be at 116th Street. She didn't
have any illusions about 116th Street
as a place to live, but at the moment
it represented a small victory—one
of a series which were the result of

her careful planning. First the white-
collar job, then an apartment of her
own where she and Bub would be
by themselves away from Pop's bois-
terous friends, away from Lil with
her dyed hair and strident voice,
away from the riff-raff roomers who
made it possible for Pop to pay his
rent. Even after living on 116th
Street for two weeks, the very fact of
being there was still a victory.

As for the street, she thought, getting up at the approaching station
signs, she wasn't afraid of its influence, for she would fight against it.
Streets like 116th Street or being
colored, or a combination of both
with all it implied, had turned Pop
into a sly old man who drank too
much; had killed Mom off when she
was in her prime.

In that very apartment house in
which she was now living, the same
combination of circumstances had
evidently made the Mrs. Hedges who
sat in the street-floor window turn to
running a fairly well-kept whore-
house—but unmistakably a whore-
house; and the superintendent of the
building—well, the street had pushed
him into basements away from light
and air until he was being eaten up
by some horrible obsession; and still
other streets had turned Min, the
woman who lived with him, into a
drab drudge so spineless and so limp
she was like a soggy dishrag. None
of those things would happen to her,
Lutie decided, because she would
fight back and never stop fighting
back.

She got off the train, thinking that
she never felt really human until she
reached Harlem and thus got away
from the hostility in the eyes of the
white women who stared at her on
the downtown streets and in the sub-

way. Escaped from the openly appraising looks of the white men whose eyes seemed to go through her clothing to her long brown legs. On the trains their eyes came at her furtively from behind newspapers, or half-concealed under hatbrims or partly shielded by their hands. And there was a warm, moist look about their eyes that made her want to run.

These other folks feel the same way, she thought—that once they are freed from the contempt in the eyes of the downtown world, they instantly become individuals. Up here they are no longer creatures labeled simply 'colored' and therefore all alike.

*   *   *

So day by day, month by month, big broad-shouldered Jim Johnson went to pieces because there wasn't any work for him and he couldn't earn anything at all. He got used to facing the fact that he couldn't support his wife and child. It ate into him. Slowly, bit by bit, it undermined his belief in himself until he could no longer bear it. And he got himself a woman so that in those moments when he clutched her close to him in bed he could prove that he was still needed, wanted. His self-respect was momentarily restored through the woman's desire for him. Thus, too, he escaped from the dreary monotony of his existence.

She examined this train of thought with care, a little surprised to realize that somehow during the last few years she had stopped hating him, and finally reached the point where she could think about him objectively. What had happened to them was, she supposed, partly her fault. And yet was it? They had managed

to live on the income from having State children there in the house in Jamaica and it had been her fault that they lost them.

She began to go over the whole thing step by step. Jim's mother died when Bub was not quite two. There was a mortgage on the house and the mortgage money had to be paid.

'We don't have to worry about a thing, Lutie,' Jim had said. 'Mom left a thousand dollars' insurance money.'

So he didn't put too much effort into looking for a job. Somehow the thousand dollars melted away—interest on the mortgage, and taxes and gas and light bills nibbled at it. Mom's funeral took three hundred and fifty dollars of it. They had to have clothes and food.

Six months after the funeral there wasn't any money left in the bank. She found the bank book on the kitchen table. Its pages were neatly perforated with the words 'Account Closed.' The last entry left a nice row of zeroes where the balance would normally have been. Jim started hunting for a job in dead earnest and couldn't find one.

Finally they went into Harlem to consult Pop. It was on a Sunday—a warm spring day. Irene, Pop's girl friend at the moment, gave them beer and they sat around the kitchen table drinking it.

'You got the house,' Pop offered. He spoke slowly as though he were thinking hard. 'Tell you what you do. You get some of these State chillern. They pay about five dollars a week apiece for 'em. You get four or five of 'em and you can all live on the money.'

Lutie sipped the beer and thought about the house. There was an unfinished room in the attic and three small bedrooms on the second floor. Put two kids in each room and they could take six of them—six times five would mean thirty dollars a week.

'He's right, Jim,' she said. 'It would be about thirty bucks a week.' She took a big swallow of beer. 'We could manage on that.'

There were papers to be filled out and investigators to be satisfied, but finally the children arrived. Lutie was surprised at how easy it was. Surprised and a little chagrined because theirs wasn't a completely honest setup. For they had said Jim worked in Harlem and a friend had verified the fact when inquiries were made. So the State people didn't know that the children were their only source of income. It made her uneasy, for it didn't seem quite right that two grown people and another child should be living on the money that was supposed to be used exclusively for the State children.

She had to work very hard to make ends meet. She tried to make all the meals good, appetizing ones and that meant spending most of her time hunting bargains in the markets and preparing dishes that required long, careful cooking. It was during that period that she learned about soups and stews and baked beans and casserole dishes. She invented new recipes for macaroni and spaghetti and noodles.

It had been nothing but work, work, work—morning, noon, and night—making bread, washing clothes and ironing them, looking after the children, and cleaning the house. The investigator used to compliment her, 'Mrs. Johnson, you do a wonderful job. This house and the children fairly shine.'

She had to bite her lips to keep from saying that that wasn't half the story. She knew she was doing a fine job. She was feeding eight people on the money for five and squeezing out what amounted to rent money in the bargain. It got so at night she couldn't go to sleep without seeing figures dancing before her eyes, and mornings when she got up, she was so tired she would have given anything just to lie still in bed instead of getting up to cook quanties of oatmeal because it was cheap and filling, to walk twelve blocks to get cooperative milk because it cost less.

She could hear the word 'cheap,' 'cheap,' 'cheap,' whether she was asleep or awake. It dominated all her thinking. Cheap cuts of meat, cheap yellow laundry soap, yeast in bulk because it was cheap, white potatoes because they were cheap and filling, tomato juice instead of orange juice because it was cheaper; even unironed sheets because they saved electricity. They went to bed early because it kept the light bill down. Jim smoked a pipe because cigarettes were a luxury they couldn't afford. It seemed to her their whole lives revolved around the price of things and as each week crawled by she grew a little more nervous, a little more impatient and irritable.

Jim finally stopped looking for work entirely. Though to be fair about it he did help around the house—washing clothes, going to the market, cleaning. But when there wasn't anything for him to do, he would read day-old newspapers and play the radio or sit by the kitchen stove smoking his pipe until she felt, if she had to walk around his long legs just one more time, get just one more whiff of the rank, strong smell of his pipe, she would go mad.

Then Pop almost got caught selling the liquor he concocted in his apartment. So he stopped making it. He couldn't get a job either, so he couldn't pay the rent on his apartment and he came home one night to find one of those long white eviction notices under his door.

He came all the way to Jamaica to tell her about it.

'You can come stay with us till things look up if you don't mind sleeping in the living room,' she offered.

'You'll never regret it, Lutie darlin',' he said fondly. 'I'll make it up to you.' His lips brushed against her cheek and she caught the strong smell of the raw whiskey he had been drinking.

She stood on the little glass-enclosed front porch and watched him walk down the path. He didn't seem to grow any older, no stoop in his shoulders; his step was firm. As a matter of fact he held himself more stiffly erect with each passing year. She sighed as she watched him cross the street heading toward the bus stop. He might hold himself up straighter and straighter as the years slipped by, but he drank more and more as he grew older.

That night after dinner she told Jim. 'Pop's been put out of his apartment. He's coming here to stay with us.'

'He can't stay here,' Jim protested. 'He drinks and carries on. He can't stay here with these kids.'

She remembered that she had been washing dishes at the kitchen sink and the dishwater slopped over on her legs from the sudden abrupt movement she made. She never wore stockings in the house because it was cheaper not to and the dishwater was lukewarm and slimy on her bare legs, so that she made a face and thought

again of the word 'cheap.' She was tired and irritable and the least little thing upset her.

She couldn't stop herself from answering him and she was too exhausted to be persuasive about it, too incensed by the criticism of Pop to let the whole thing drop and bring it up again later, leading around to it, not arguing but gently showing how she really couldn't do anything else.

'He's my father and he hasn't any other place to go. He's going to stay here with us.' Her voice was insistent and she threw the words at him bluntly, using no tact.

Jim got up from the chair where he'd been sitting and stood over her, newspaper in hand. 'You're crazy!' he shouted.

Then they were both shouting. The small room vibrated with the sound of their anger. They had lived on the edge of nothing for so long that they had finally reached the point where neither of them could brook opposition in the other, could not or would not tolerate even the suggestion of being in the wrong.

It ended almost as swiftly as it began. Because she said, 'All right, I'm crazy.' Her voice was tight with rage. 'But either he comes or I go.'

So Pop came. At first he was apologetic about being there and so self-effacing that she was only aware of him as a quiet, gray-haired figure doing the marketing, wiping the dishes, playing gently with the children. And she thought it was working out beautifully. Jim was wrong as usual.

After the first couple of weeks, Pop started drinking openly. She would meet him at the door with a brown-paper package in his hand. He would come downstairs from the bathroom holding himself very straight and eat supper in a genial expansive mood smelling to high heaven of raw whiskey.

He urged her and Jim to go out at night. 'You're nothin' but kids,' he said, loftily waving his hands to give his words emphasis and managing to get into the gesture a sense of the freedom and joyousness that belonged to the young. 'Shouldn't be shut up in the house all day. Go out and have yourselves a time. I'll look after the kids.'

Somehow he always managed to have a little money and he would take two or three limp dollar bills from his pocket and shove them into her reluctant hands. When she protested, his invariable answer was, 'Oh, call it room rent if you gotta be formal about it.'

They would head straight for Harlem. The trip rarely included anything more than an evening spent drinking beer in someone's living room and dancing to a radio. But it was like being let out of jail to be able to forget about the houseful of kids, forget about not having any money. Sometimes they would stop at Junto's Bar and Grill, not so much to drink the beer as to listen to the juke-box and the warm, rich flow of talk and laughter that rippled through the place. The gay, swirling sounds inside the Junto made both of them believe that one of these days they would be inside a world like that to stay for keeps.

Going home on the subway, Jim would put his arm around her and say, 'I'll make it all up to you some day, Lutie. You just wait and see. I'm going to give you everything you ever wanted.'

Just being close to him like that, knowing that they were both thinking much the same thing, shut out the roar and rush of the train, blotted out the other passengers. She would ride home dreaming of the time when she and Jim and Bub would be together—safe and secure and alone.

They were always very late coming home. And walking through the quiet little street they lived on, past the small houses that seemed to nudge against each other in the darkness, she used to imagine that the world at that hour belonged to her and Jim. Just the two of them alone traveling through a world that slept. It was easy to believe it, for there was no sound except that of their own footsteps on the sidewalk.

They would tiptoe into the house so as not to wake Pop and the kids. The living room always smelt strongly of whiskey.

'The place smells like a gin mill,' she would say, and giggle as they went up the stairs toward their bedroom. Because somehow the fact of having been away for a while, the lateness of the hour, the stealth with which they had come into the house, made her feel young and carefree.

As they went up the stairs, Jim would put his arm around her waist. His silence, the bulky feel of his shoulders in the darkness, turned their relationship into something mysterious and exciting, and she wanted to put off the moment when she would undress and get in bed beside him, wanted to defer it at the same time that she wanted to hurry it.

It got so they went to Harlem two or three times a week. They wanted to go, anyway, and Pop made it very easy, for he insisted that they go and he invariably proffered a crumpled bill or two with which to finance the trip.

And then all the fun went out of it. Mrs. Griffin, who lived next door, banged on the kitchen door early one morning. She was filled with an indignation that thrust her mouth forward in such an angry pout that Lutie was prepared for something unpleasant.

'There's so much noise over here nights 'at my husband and me can't sleep,' she said bluntly.

'Noise?' Lutie stared at her, not certain that she had heard correctly. 'What kind of noise?'

'I dunno,' she said. 'But it's got to stop. Sounds like wild parties to me. Las' night it kept up till all hours. An' my husband say if somp'n ain't done about it he's going to complain.'

'I'm sorry. I'll see that it doesn't happen any more.' She said it quickly because she knew what it was. Pop had been having parties on the nights they went to Harlem.

As soon as Mrs. Griffin left the kitchen, slamming the door behind her, Lutie asked Pop about it.

'Parties?' he said innocently. His forehead wrinkled as though he were trying to figure out what she was talking about. 'I ain't had no parties. Some of my friends come out a coupla times. But I ain't had no parties.' His voice sounded hurt.

'You must have made a lot of noise,' she said, ignoring his denial of having had any parties. 'We gotta be careful, Pop. The neighbors might complain to the State people.'

She tried to eliminate the trips to Harlem after that. But Jim was unaccountably and violently suspicious

of the headaches that came on suddenly just before they were to leave, of the other thin excuses that she found for not going with him. She couldn't bring herself to tell him that she was scared to go off and leave Pop in the house alone.

'You think I'm too shabby to go out with,' Jim had said. And then later, 'Or have you got yourself another boy friend?'

She wouldn't swallow her pride and tell him about Pop, so they continued going to Harlem two or three times a week. Besides, every time Jim said something about her having a boy friend his face turned resentful, sullen, and she couldn't bear to see him like that, so she stopped making excuses and pretended an anticipation and an enthusiasm for the trips that she didn't feel.

Now when she came home she was filled with a fear that made her walk faster and faster, hastening toward their street in her eagerness to ascertain that the house was dark and quiet. Once in bed, she twisted and turned the rest of the night, impatiently waiting for morning to come when the neighbors would soon inform her if anything had happened while they were gone.

She could remember so vividly the night they returned to find the house blazing with light. She got a sick feeling deep in her stomach, for there was an uproar coming from it that could be heard way up the street.

* * *

All Jim said to her that night was, 'You wanted that whiskey-soaked bum here. Now you've got him, I hope you're satisfied.'

The next afternoon a disapproving

white woman arrived and took the children away with her. 'They can't stay in a place where there's any such goings-on as there is here,' she said.

Lutie pleaded with her, promised her that everything would be different; it couldn't possibly happen again if she would just let the children stay.

The woman was unmoved. 'These children belong to your own race, and if you had any feelings at all you wouldn't want them to stay here,' she said, going out the door. In less than half an hour she had the youngsters packed and was putting them into a station wagon. She moved competently with no waste motion.

Lutie watched her from the front porch. Damn white people, she thought. Damn them. And then—but it isn't that woman's fault. It's your fault. That's right, but the reason Pop came here to live was because he couldn't get a job and we had to have the State children because Jim couldn't get a job. Damn white people, she repeated.

* * *

'Where are the kids?' he demanded.

'Did you have any supper?' she asked.

'What the hell—you heard me. Where are the kids?'

'They're gone. The State woman came and got them this morning.'

'I suppose you figured if those little bastards were taken away, I'd have to find a job. That I'd go out and make one. Buy one, mebbe.'

'Oh, Jim, don't——' she protested.

'You knew what would happen when you brought that old booze hound here to live.'

Perhaps if she kept quiet, and let him go on raving without answering him, he would get tired and stop. She bit her lip, looked away from him, and the words came out in spite of her, 'Don't you talk about my father like that.'

'A saint, ain't he?' he sneered. 'He and those old bitches he sleeps with. I suppose I'm not good enough to talk about him.'

'Oh, shut up,' she said wearily.

'Mebbe it runs in the family. Mebbe that's why you had him come here. Because you figured with him here you'd be able to get rid of the kids. And that would give you more time to sleep with some Harlem nigger you've got your eye on. That's it, ain't it?'

'Shut up!' This time she shouted. And she saw Pop go quietly down the hall, his worn old traveling bag in his hand. The sight shocked her. He didn't have any place to go, yet he was leaving because of the way she and Jim were carrying on. 'Oh, Jim,' she said, 'Don't let's fight. It's all over and done with. There isn't any point in quarreling like this.'

'Oh, yeah? That's what you think.' He leaned over the bed. His eyes were bloodshot, angry. 'I oughtta beat you up and down the block.' He slapped her across the face.

She was out of bed in a flash. She picked up a chair, the one chair in the bedroom—a straight-backed wooden one. She had painted it with bright yellow enamel shortly after they were married. And she had said, 'Jim, look. It makes sunlight walk right into the room.'

He had looked at her squatting on the floor, paintbrush in hand, her face glowing as she smiled up at him. He had leaned over and kissed her forehead, saying, 'Honey, you're all the sunlight I'll ever need.'

It was the same chair. And she aimed it at his head as she shouted. 'You come near me and so help me I'll kill you.'

It had been a loud, bitter, common fight. It woke Bub up and set him to crying. And it was more than a week afterward before they were able to patch it up. In the meantime the mortgage money was due and, though Jim didn't say so, she felt that if they lost the house by not being able to pay the interest, it would be her fault.

The Fifth Avenue bus lurched to a stop at 116th Street. She climbed down the steep stairs from the top deck, thinking that if they hadn't been so damn poor she and Jim might have stayed married. It was like a circle. No matter at what point she started, she always ended up at the same place. She had taken the job in Connecticut so they could keep the house. While she was gone, Jim got himself a slim dark girl whose thighs made him believe in himself again and momentarily released him from his humdrum life.

She had never seen him since the day she had gone to the house in Jamaica and found that other woman there. The only time she had heard from him was when he had forwarded the letter from Mrs. Chandler—and then all he had done was put Pop's address on the envelope. There had been no messages, no letters—nothing for all these years.

Once Pop had said to her, 'Hear Jim's left town. Nobody knows where he went.'

And she was so completely indifferent to anything concerning Jim that she had made no comment. She watched the bus until it disappeared out of sight where Seventh Avenue joins 110th Street. This clear understanding she had of what caused Jim to acquire that other woman was because the same thing was happening to her. She was incapable of enduring a bleak and lonely life encompassed by those three dark rooms.

Love Is Not All

(Sonnet XXX)

Edna St. Vincent Millay

Love is not all: it is not meat nor drink
Nor slumber nor a roof against the rain;
Nor yet a floating spar to men that sink
And rise and sink and rise and sink again;
Love can not fill the thickened lung with breath,
Nor clean the blood, nor set the fractured bone;
Yet many a man is making friends with death
Even as I speak, for lack of love alone.
It well may be that in a difficult hour,
Pinned down by pain and moaning for release,
Or nagged by want past resolution's power,
I might be driven to sell your love for peace,
Or trade the memory of this night for food.
It well may be. I do not think I would.

part **2**

Parenting...

Amedeo Modigliani. *Gypsy Woman with Baby.*

A Daughter of My Own

Merrill Joan Gerber

Every girl I know who has ever had a baby has had her mother come and stay with her for two weeks—or her mother and a nurse—and each one tells me how wonderful it was, and how she never could have survived without her mother's advice and help and soothing presence.

I just don't know. I think I love my mother as much as any girl loves hers, and I think I have always gotten along better with my mother, on the average, than most girls do with theirs. But if I never have another baby it won't be because I didn't like being pregnant or being in labor or losing six-months' worth of sleep—it will be because I won't know what to do about my mother.

When I found out I was pregnant in September, and told my mother I was due to have the baby in April, she immediately began making plans to fly up "in February or March, whenever you want me." It had never occurred to me that I would want her at all—I mean, it was my baby, and we were very far away from my parents and they didn't have the money to toss away on airplane trips, and my husband was a student at the time and at home nearly all day to give me any help I might need, and there just wasn't room for anyone. There was hardly room for the baby. Bill and I lived in a three-room apartment near the University, and had figured that by gouging out the shelves from a built-in-the-wall livingroom bookcase, we could just manage to squeeze in a tiny crib for the baby. We hadn't exactly planned on enlarging our family just yet, and there was quite a bit of arranging and arithmetic to do in those early months. Not that we were unhappy about it—far from it—it just took some adjusting.

My husband was happy to adjust to a baby, but not to a mother-in-law. What bothered him most, to begin with, was that if she came up *before* I had the baby, he would be left alone with her for five days in the house while I was in the hospital. I could see what he meant: my mother wasn't one to discuss Aristotle, and Bill wasn't one to discuss the upbringing of children, at least not with my mother and not the upbringing of *his* child, and that seemed to be all she thought about these days, as far as we could judge from her letters.

It was hard to tell her not to come before the event, because I knew what was on her mind. To put it rather simply, she wanted to hold my hand. My mother had always held my hand through crises—I had been a very sick child: four pneumonias, one broken arm, a heart murmur, an infected appendix, crooked teeth, crossed eyes, and nosebleeds. I'm quite sure I couldn't have survived any of those blights without my mother's five fingers around mine— while they were giving me oxygen, while setting my arm, while taking an electrocardiogram (which she ruined the results of because she interfered in some way with the electrical current by touching me), while I was being put under ether, while the braces were being tightened, and so on. I think that in those frightening years when doctors loomed everywhere, the grip of my mother's fingers when needles were piercing my very soul was all that sustained me.

Now I was having a baby. People died in childbirth, I knew, and even if you made it through it it was bound to be pretty rough. My mother nearly died having me, and I knew how worried she was, and why she wanted to be around to whisper encouragement at the hardest moments. The only difference, I felt, was that this was not a *crisis*. I was not sick, I was not frightened, I was not even worried. I was in perfect health, I liked and trusted my doctor, and I loved the little thumps I was beginning to feel low down where my appendix had once been. Most important of all, though, was that I *had* someone to hold my hand if necessary, someone who would fill the bill very well in the absence of my mother: my husband.

I don't suppose my mother gave him much credit for being useful in any situation. After all, here we had been married for three years and I had been supporting *him,* while all he did, as far as my mother could tell, was lounge around in unpressed pants all day reading big musty books that left yellow crumbs of paper everywhere on the floor. Which her daughter then had to vacuum!

I tried to show my mother how calm and unworried I was. I sent her clippings of the fœtus in different stages of development with long technical explanations about its growth. I recommended books on childbirth for her to read, which told how simple and safe the procedure was these days. I wrote her with the gain of every new ounce, knowing that she believed gaining weight meant being healthy, since I had always been a very skinny, unhealthy child. I had Bill take pictures of me in different smiling poses, my immense belly foremost, my jolly wave saying "See how fine I am? Nothing is wrong with *me.*"

Finally I was able to persuade my

mother to come right *after* the baby was born. When she agreed, though reluctant and somewhat hurt, I was finally able to relax and enjoy the remainder of my pregnancy. I saw every old Tarzan movie on TV that had ever been made. I sat at the window on gray afternoons with all the lights off and watched the snow come down, eating from a plate on my lap that usually held green olives and grapefruit halves. I spent long hours staring up at the ceiling from my bed, imagining my baby, and imagining *me* with my baby.

Bill was home nearly all that winter, studying for his comprehensives, and we talked to each other mostly at dinnertime, usually about what names we would like the baby to have, and how we wouldn't do to him the bad things that we thought had been done to us. It was a fine snowy happy winter—wet and icy outside, warm and steamy inside. I bought eight-dozen diapers in my ninth month, and washed them all three times, by hand, so they would be soft for the baby. Bill bought a used crib and we painted it and got a new mattress. We read Dr. Spock, we read *Childbirth Without Fear,* we read Gesell. We were ready.

The baby started to come on the very day he was scheduled to come, which made us think that he had a very reliable character. We saw the doctor in the afternoon and he told us to go home and call him when there was "some real action." So we went home and Bill fed me strained chicken soup, and we timed all the warming-up pains, and soon it was dark out and I was warming up a little faster. We played all six Brandenburg Concertos while I lay on the living-room couch, and Bill sat on the floor next to me with a pencil and paper and his wristwatch in his lap. At eleven we called the doctor who said "This sounds like it, come on down."

The snow had changed to rain, and it was a very appropriate, dramatic night for racing to the hospital. I knew though, from the hospital brochure I had, that if we arrived before midnight we would be charged for the entire day, so I cautioned Bill to drive as slowly as he could, and we meandered to the hospital—even circling a few extra blocks to kill time—until the baby made it known we had better meander no more, and we covered the last five miles in about thirty seconds.

We parked in the hospital lot at four minutes to midnight. I was game to sit it out till twelve, but Bill was beginning to get glassy-eyed and I was not as confident as I sounded, so we went in and signed the admittance form and they wrapped adhesive tape around my wedding ring.

Bill was permitted to stay with me as long as I wanted to be awake, and we stayed together till nearly dawn, having a very sweet dreamy time, holding hands in fact, and making faces at each other every time a nurse came in to listen to the baby's heartbeat, or time a contraction.

When he went away, I was floated into an elevator on a very soft high bed, and in the morning I had a little girls, six pounds, two ounces.

I had truly never felt better. It was a bright sunny morning, and I was cranked up on my neat white bed and the light was coming in right on my knees, making them warm and comfortable. I had just seen my baby in a tiny plexiglass cart, pink and perfect, asleep, and more beautiful than any beautiful thing I had ever seen in the world. They wouldn't let me hold her because I was still rather groggy. In fact, no one would believe me later when I told them I had seen a nurse come into my room a few minutes before, and take a swig of whisky straight from a bottle she kept in her deep white pocket.

I drifted about in the bright sunlight for a while and then Bill peeked his head into the room, grinning like mad, and we had a big kiss and a tremendous long smile together and then I sent him off to see his daughter.

He was with me all afternoon, but occasionally went out into the hall when a nurse came in to poke my stomach and take my pulse and temperature.

He came back one time with roses for me, and I felt like a queen there in the sunshine, all loved and loving, and I thought we had just begun to get happier than we had ever been.

When they came in with my dinner, Bill got up to leave, and I remembered that he hadn't called my mother. I told him to do that right away, and he said he would, but it seems he drove home first and had dinner and fell asleep, and it wasn't till about ten at night that he remembered to call, and the baby had been alive nearly twelve hours by then.

Everything started falling apart. My mother called me the next day at the hospital when I could walk to the phone, and the first thing she said was "Why didn't Bill call me right away?" So I made up something about his wanting to make sure I was all right before he called, but it wasn't very pleasant to argue and to come out of that gentle haze I'd been in, and I resisted it. I said I had to hang up and go back to bed, I was getting dizzy, but my mother managed to mention that she was coming in two days, when I would be ready to go home.

I spent the rest of the afternoon worrying about her coming and feeling very helpless and unhappy. They brought the baby to me only twice for the first two days, to get her used to me, and me to her. There was no sense in her coming oftener, since she wanted mostly to sleep, and my milk hadn't come in yet. It was very fine to hold her, and each time the nurse left I would unbutton her little kimono and examine her tiny body and count all her toes and fingers.

After they took her back to the nursery I'd get worried again about my mother, and feel bad till the next time they brought the baby to me. When I told Bill my mother was really coming, that she had bought her plane ticket, tears came to my eyes, and he tried to cheer me up by saying that as long as she *had* to come, she would be a great help and not to worry a bit, but he didn't sound very convinced himself.

My mother arrived the evening of my fourth day at the hospital. I was to leave the next morning. Bill had a night class, so no one met her at the airport, and she took a taxi to the hospital, and dragged her suitcase by herself, and when she came up to the desk they told her it was final feeding time and no visitors allowed now till morning. She told them she hadn't seen me for a year, and some kind lady let her up, and she came into my room just as the baby finished nursing, and without a word, I held out the baby to her and she took her, and we both were crying, because it had been so long and we loved each other so much, and now I had a daughter of my own.

But even though I was so happy to see her, that perfect moment couldn't last very long, and it didn't, because immediately she was asking me if we could pay the hospital bill, since Bill was obviously not earning any money, and then asking again why Bill had waited so long to call her, and asking if I had had "too terrible a time." Nothing had seemed wrong till then, and suddenly I was worrying about the hospital bill, and feeling very sorry for myself because it was hard to sit down and my breasts ached while they were getting used to the baby's nursing schedule. I didn't want to feel bad, I wanted to stay feeling like that queen in the sunshine for a while, but it was too complicated, we were talking all about practical things and old rifts were coming up, and I wished I had someplace to go back into like where my baby had been all those months.

When my mother left to meet Bill downstairs and go home with him, she said "Do you want me to come with Bill in the morning when he picks you and the baby up?" and I said "Whatever you like," hoping that she would understand I really meant This Is A Private Time And It Would Be Nicer For The Three Of Us To Be Alone.

But the next morning she was there, very proud and pleased, giving directions to everyone. "Bill, you go down to the cashier, and I'll stay here and help Molly pack up all her things and then we'll have the nurse dress the baby," and then she asked me if I had a nursing brassiere and I said no, so she went down to the gift shop and bought me two for five dollars each, and I knew she couldn't afford it and neither could we, and everyone was bustling around so that I could hardly think of what time in my life it was—the time that I was taking my little baby to her home where I would be her mother for the rest of my life.

We drove home, my mother sitting in the back seat, leaning over my shoulder all the way, looking at the sleeping baby in my lap and touching her little curled fingers, and saying how beautiful she was, which was true, but somehow seemed false with my mother saying it aloud like that. Bill didn't speak all the way home, and when we got to the house, he took my suitcase and went inside and my mother and I sat in the car waiting for him to come out and open the door for me, which he didn't do. After five foolish minutes, I had to open the car door myself, nearly dropping the baby and nearly crying, and my mother gave me a look which said all she had always thought of Bill, and it nearly broke my heart, to have everything ruined when it could have been so nice at a time like this.

I went inside and put the baby in her crib, and didn't know what to do then. Bill had gone into the kitchen with a book, and was sitting at the table reading. I wanted to see him so badly, but he hardly looked at me. I wanted him to admire the baby and tell me what a fine child I had made and what a good brave girl I was, but he never raised his eyes. It seemed as though I had not seen him in years, and I was missing him because I had been away from him for five days in the hospital and, in a way, for all the months before that when, if he so much as gave me a warm kiss, my already overburdened heart would begin to palpitate and shudder to remind me that warm kisses would have to wait till the baby did not demand so much blood and energy of it.

And now we were further apart than ever.

My mother took over. "You get into bed," she said, "I'll take care of the baby. You need to rest."

"But I want to look at her," I protested.

"You'll look later—you just got out of the hospital"—and she took off my coat and led me into the bedroom she'd never ever seen and tucked me into bed and closed the door to leave me aching and open-eyed and missing my baby and my husband.

The baby, because she was small, had to nurse every two and a half hours, and each feeding lasted nearly an hour, so I was never able to sleep for much more than an hour at a time. The nursing which I had loved so much in the hospital became a terrible ordeal at home, because neither Bill nor my mother could be in the room together with me and the baby at that time without becoming very embarrassed. If one was in the room and the other inadvertently came in, they would both avert their eyes from me, as though neither would acknowledge to the other his intimate relation with me. The baby and I had done beautifully in the hospital, but now with everyone avoiding everyone in our three rooms, and doors being closed as they went in or out, and me being so exhausted and tense, the baby sucked less and cried more, and made me desperate for relief of some kind, sleep at least, or a little privacy and

quiet. Privacy was what we lacked most—I wanted to be alone with my baby, I wanted to be alone with Bill, and I wanted the three of us, so newly a family, to have some time alone. My mother, though, was everywhere. If Bill came over to the crib to look at the infant, my mother would appear and look too—and look at Bill to see his reaction, and he would mumble and walk away.

On the third day home my mother said, "I've never seen Bill kiss the baby . . . doesn't he like her?" and what could I answer? "You haven't seen him kiss me either"—or—"He'd kiss her if you weren't watching all the time"? So I just sighed and asked my mother to bring me a drink of water. "Nursing makes you very thirsty," I told her.

That week, my mother prepared all the meals and called us in to eat when they were ready. She washed the baby's diapers every day and hung them up outside. She rocked the baby so I could sleep, she bought me a rubber ring to sit on so I would be more comfortable, she cleaned the apartment from ceiling to floor, she baked my favorite kind of chocolate cake, she sewed hems on my skirts that needed them, she let out the waist of everything that no longer fit me, she ironed, she mended, she labored like ten mules.

And it was horrible. One night, in bed (my mother was sleeping on the couch in the living room), Bill whispered to me, "I'm sorry, Molly, if I seem so awful to her and to you, but I can't stand this. I don't feel as though this is my home anymore. I feel as though I were courting you again and calling for you at your mother's house—the way she calls us in to meals and is so polite, and the way she just goes into our closets and drawers as if it were her own house. I feel like I don't belong here."

I took his head in my arms and held him, but he said, "I really can't stand it, Molly," and then he asked me to please do some of the cooking and dishwashing so my mother would remember it was she who was visiting us and not us visiting her. "I know you're tired and still a little sick, but she won't even let me into the kitchen, and you have to show her what she's doing to us."

So the next day I started to wash the dishes after breakfast though I hadn't slept three hours in twenty-four, and my mother asked me if I were crazy, to get back into bed this minute, and I said no, I felt fine, I was getting stronger every day, and then I fainted.

Which made Bill even sadder than he had been, and soon he just left the house in the morning and went to the library and didn't come back till suppertime. He had yet to hold the baby.

So I stayed in bed, and my mother brought my meals in to me, and brought the baby in for me to nurse, and did the changing and dressing, and soon she would not even wake me if it were feeding time, but would make up a formula bottle and give it to the baby, so as not to disturb me, and soon I wasn't having enough milk because she had destroyed the pattern and the breasts didn't think they needed to make any more.

One night when the baby started crying, I leaped out of bed and lifted her from her crib, and carried her back to my bed with me, where I was going to wake Bill and tell him to look at his daughter finally. But suddenly my mother was right in our bedroom, white and disheveled in her nightgown, her eyes not yet focused, her gray hair disordered from sleep, her arms out for the baby—"Give her to me, Molly, I'll get her quiet, you go back to sleep."

I couldn't help what I said, and it was wrong of me, but I said "Why on earth do you have to come poking around every minute? Why can't you leave us alone?" and my mother, horrified, went right out of our bedroom, and I heard her walking around in the living room in the dark all the rest of the night, while I sat in the bedroom with the baby in my lap till the sun rose.

In the morning my mother's eyes were red, and she said it was because she had a cold, and she was going to fly home because she didn't want the baby to catch anything from her. It was only the end of the first week, and she had planned to stay two. I knew all I had to do was ask her to please stay, and she needed me to, her eyes were on my face so pitifully, but all I did was say that I was surely much stronger now and could manage alone easily, and Bill spoke for the first time in days, to volunteer to take her to the airport any time she wanted to be there.

My mother and I couldn't look at each other, both of us had tears on our cheeks all morning, and she packed and I pretended to be busy in my bedroom.

On the way to the airport she said from the back seat, "I had to borrow on Daddy's life insurance policy to get enough money for plane fare. I suppose I should have stayed home. You didn't need me."

"I did, oh I *did*, mother," I cried, and then recognizing the lie exposed by this solemn, unhappy trip to the airport, I fell silent, while my mother fumbled in her purse for a handkerchief.

"I thought you would need me," she said, and the grief in her voice was so deep, I reached back for her hand, to hold it, but she pulled it away and looked out the window. "A good day for a flight," she said, making her voice steady.

My hand dropped to my lap, where my daughter lay asleep wrapped in a blanket, and I touched her little cheek, thinking, Will You And I Ever Come To This?

At the airport, Bill wanted to go up on the Observation Deck because it was early, and he paid three dimes to get us through the turnstile. As soon as we got there, my mother said "It's too windy here for the baby. Let's go down," and Bill said, "A little wind can't hurt her," and my mother said, "Wind is the worst thing for an infant, not even two weeks old yet," and Bill said "It's a warm wind," and my mother said to me "Are you going to let her catch pneumonia because of him?" and I said "Oh please, let's not fight, it *is* a warm wind, Mother," and she said "Have it your way. I have to get on the plane."

She went down, and in about two minutes we went down, too, and she had checked her luggage and was ready to leave.

"Goodbye, Mother," I said. "Thank you for coming."

She stood stiffly, looking at my face as though she didn't know me, and then she began to walk toward the gate. I ran after her and threw my arms around her and hugged her, crying "I love you, Mother, I love you," and we embraced desperately, as though this were the last time we could ever express our love, and then she went through the gate, her head down, her hand to her eyes.

I went back to Bill who stood to the side with the baby, and shouted at him, *"Why couldn't you have let her kiss the baby goodbye? Why couldn't you? Why did you have to carry her away like that?"* and he didn't answer me, just put the baby in my arms and then put his arm around my shoulder, and began to lead me to the car. When we got outside, I did not even look at the big plane my mother was flying away in. We walked to the car, and the wind blew the baby's cap off, and I yelled at Bill, "It *is* too windy for the baby. My mother was right! Don't you see? She was *right!*"

The cap, filling with wind, flew and bumped across the parking lot, and Bill and I watched it until it got tangled under the wheel of a car, and then we just left it there and drove home.

Hilaire-Germain Edgard Degas. *The Bellelilli Family.*

from

The Father

AUGUST STRINDBERG

CHARACTERS

THE CAPTAIN
LAURA, *his wife*
BERTHA, *their daughter*
DR. ÖSTERMARK
THE PASTOR
THE NURSE
NÖJD, *a servant*
THE CAPTAIN'S BATMAN

ACT ONE

A room in the CAPTAIN'S *house.*

Scene 1

The CAPTAIN *rings. The* BATMAN *enters.*

BATMAN: Sir?
CAPTAIN: Is Nöjd out there?

BATMAN: He's waiting for orders in the kitchen, sir.
CAPTAIN: In the kitchen again! Send him here at once!
BATMAN: Sir! *(Goes.)*
PASTOR: What's the matter now?
CAPTAIN: Oh, the blackguard's been mucking about with one of the girls again. Damned nuisance, that fellow!
PASTOR: Nöjd? Why, you had trouble with him last year too!
CAPTAIN: You remember? Perhaps you'd give him a friendly talking-to —that might have some effect. I've sworn at him, and given him a tanning, but it doesn't do any good.
PASTOR: So you want me to read him a sermon! Do you think the Word of God will have any effect on a cavalryman?
CAPTAIN: Well, brother-in-law, it doesn't have much effect on me, as you know——

PASTOR: Yes, I know!

CAPTAIN: But on him——? Try, anyway.

Scene 2

The CAPTAIN. *The* PASTOR. NÖJD.

CAPTAIN: Well, Nöjd, what have you been up to now?

NÖJD: God bless you, Captain, I couldn't tell you before his Reverence.

PASTOR: Come, come, don't be bashful, my lad!

CAPTAIN: Own up, or you know what'll happen!

NÖJD: Well, sir, it was like this, you see. We was dancing up at Gabriel's, and then, yes, well, Louis said——

CAPTAIN: What's Louis got to do with it? Stick to the facts!

NÖJD: Well, Emma suggested we should go to the barn.

CAPTAIN: I see! So it was Emma who seduced you!

NÖJD: Not far off. And I'll say this—if a girl ain't willing, she don't run no danger.

CAPTAIN: Out with it! Are you the child's father or not?

NÖJD: How should I know?

CAPTAIN: What! You don't know!

NÖJD: Well, you can never be sure.

CAPTAIN: Weren't you the only one, then?

NÖJD: I was that time, but that don't mean to say I was the only one.

CAPTAIN: Are you saying Louis's to blame? Is that it?

NÖJD: It ain't easy to say who's to blame.

CAPTAIN: But you've told Emma you want to marry her.

NÖJD: Yes, well, you have to tell them that.

CAPTAIN (*to* PASTOR): This is monstrous!

PASTOR: It's the old story. Now, look here, Nöjd, surely you're man enough to know whether you're the father!

NÖJD: Well, I did go with her, but, as your Reverence knows, that don't necessarily mean anything need happen.

PASTOR: Come, come, my lad, don't start trying to evade the issue! You surely don't want to leave the girl alone with the child! Of course we can't force you to marry her, but you must accept responsibility for the child. You must!

NÖJD: All right, but Louis must pay his share.

CAPTAIN: Oh, very well, it'll have to go to court. I can't unravel the rights and wrongs of this, and I don't feel inclined to try. Right, get out!

PASTOR: Nöjd! One moment. Hm! Don't you regard it as dishonourable to leave a girl high and dry like that with a child? Eh? Well? Don't you feel such behaviour would be —hm——?

NÖJD: Oh, yes, if I knew I was the child's father. But that's something a man can never be sure of, your Reverence. And it's no joke spending your whole life sweating for other men's children. I'm sure you and the Captain'll both appreciate that.

CAPTAIN: Get out!

NÖJD: Sir! (*Goes.*)

CAPTAIN: And keep out of the kitchen, damn you!

Scene 3

CAPTAIN: Well, why didn't you lay into him?

PASTOR: What? I thought I spoke very strictly.

CAPTAIN: Oh, you just sat there mumbling to yourself.

PASTOR: To be honest, I don't know what one ought to say. It's bad luck on the girl, yes. But it's bad luck on the boy, too. Suppose he isn't the father? The girl can suckle the child for four months at the orphanage, and then she'll be shot of him, but the boy can't dodge his responsibility like that. She'll get a good job afterwards in some decent home, but if he gets thrown out of his regiment, he's finished.

CAPTAIN: Yes, I wouldn't like to be the magistrate who has to judge this case. I don't suppose the lad's completely innocent—one can't be sure. But one thing you can be sure of. The girl's guilty—if you can say anyone's guilty.

PASTOR: Yes, yes! I'm not condemning anyone! But what were we speaking about when this blessed business intervened? Bertha's confirmation, wasn't it?

CAPTAIN: It's not just her confirmation. It's the whole question of her upbringing. This house is stuffed with women every one of whom wants to bring up my child. My mother-in-law wants to make her a spiritualist, Laura wants her to be a painter, her governess wants her to be a Methodist, old Margaret wants her to be a Baptist, and the maids are trying to get her into the Salvation Army. Well, you can't patch a soul together like a damned quilt.

I have the chief right to decide her future, and I'm obstructed whichever way I turn. I've got to get her out of this house.

PASTOR: You've too many women running your home.

CAPTAIN: You needn't tell me that. It's like a cage full of tigers—if I didn't keep a red-hot iron in front of their noses, they'd claw me to the ground the first chance they got. Yes, you can laugh, you old fox! It wasn't enough that I married your sister, you had to palm your old stepmother off on me too!

PASTOR: Well, good heavens, one can't have one's stepmother living under one's roof.

CAPTAIN: But one's mother-in-law is all right! Yes, under someone else's roof.

PASTOR: Well, well. We all have our cross to bear.

CAPTAIN: Yes, but I've a damned sight too many. I've got my old nurse too, and she treats me as though I was still in a bib! Oh, she's a dear old soul, heaven knows, but she doesn't belong here!

PASTOR: You should keep your women in their place, Adolf. You let them rule you.

CAPTAIN: My dear brother-in-law, will you kindly tell me how one keeps women in their place?

PASTOR: To speak frankly, Laura —I know she's my sister, but—well, she was always a little difficult.

CAPTAIN: Oh, Laura has her moods, but she's not too bad.

PASTOR: Ah, come on! I know her!

CAPTAIN: Well, she's had a romantic upbringing, and has a little difficulty in accepting life, but, after all, she is my wife——

PASTOR: And is therefore the best

of women. No, Adolf, she's the biggest stone round your neck.

CAPTAIN: Yes, well, anyway, now the whole house has become impossible. Laura doesn't want to let Bertha out of her sight. But I can't let her stay in this asylum.

PASTOR: So? Laura won't—? Hm, then I'm afraid things aren't going to be easy. When she was a child, she used to lie absolutely still like a corpse until she'd got what she wanted. And when she'd got it, she'd give it back, explaining that it wasn't the *thing* she wanted, simply the fact of having her will.

CAPTAIN: I see, she was like that already, was she? Hm! She gets so emotional sometimes that I become frightened, and wonder if she isn't —well—sick.

PASTOR: But what is it you want for Bertha that she finds so unacceptable? Can't you meet each other halfway?

CAPTAIN: You musn't imagine I want to build the child into a prodigy, or a copy of myself. But I don't want to play the pimp and educate her just simply for marriage—if I do that and she stays single, she'll become one of these embittered spinsters. On the other hand, I don't want to train her for some masculine vocation that'll need years of study and be completely wasted if she does get married.

PASTOR: What do you want, then?

CAPTAIN: I'd like her to become a teacher. Then, if she stays single she'll be able to look after herself, and won't be worse off than these wretched schoolmasters who have to support a family. And if she does marry, she can use the knowledge she's gained in bringing up her own children. That's logical, isn't it?

PASTOR: Perfectly. But hasn't she shown a great talent for painting? Wouldn't it be bad for her to repress that?

CAPTAIN: No, no. I've shown her efforts to a prominent artist, and he says it's only the kind of thing people learn to do at schools. But then some young ass came here last summer who knew more about such matters, and said she was a genius —and as far as Laura was concerned, that settled it.

PASTOR: Was he in love with the girl?

CAPTAIN: I presume so.

PASTOR: Then God help you, my dear fellow, for there'll be nothing you can do about that! But this is a sad business, and of course Laura has allies—in there.

CAPTAIN: Oh, yes, never you fear! The whole household is up in arms —and, between you and me, they're not fighting strictly according to the rules of chivalry.

PASTOR: *(gets up)* Do you think I haven't been through all this?

CAPTAIN: You too?

PASTOR: Are you surprised?

CAPTAIN: But the worst is, it seems to me Bertha's future is being decided in there from motives of hatred. They keep dropping hints that men will see that women can do this and do that. Man versus woman, that's their theme, all day long. Must you go now? No, stay for supper. I can't offer you much, but, did I tell you, I'm expecting the new doctor to pay a call? Have you seen him?

PASTOR: I caught a glimpse of him on my way here. He looks a pleasant, straightforward chap.

CAPTAIN: Does he? Good! Think he might be on my side?

PASTOR: Who knows? It depends how much he's had to do with women.

CAPTAIN: Oh, come on, do stay!

PASTOR: No thanks, my dear fellow. I've promised to be home for supper, and my old lady gets so worried if I'm late.

CAPTAIN: Worried? Angry, you mean! Well, as you wish. Let me give you a hand with your coat.

PASTOR: It's certainly very cold tonight. Thank you. You want to look after yourself, Adolf. You look nervy.

CAPTAIN: Do I?

PASTOR: You're not quite yourself, are you?

CAPTAIN: Has Laura given you that idea? She's been treating me like a budding corpse for twenty years.

PASTOR: Laura? No, no, I just wondered—— Take care of yourself! That's my advice. Well, goodbye, old chap. But didn't you want to talk to me about confirmation?

CAPTAIN: No, that'll have to take its course. Chalk that one up to society's conscience. I don't intend to be a martyr for the sake of truth. I'm past all that. Goodbye! Give my regards to your wife!

PASTOR: Goodbye, brother! Give mine to Laura!

Scene 4

The CAPTAIN. *Then* LAURA.

CAPTAIN: *(opens the secretaire, sits down at it and starts counting)* Thirty-four—nine—forty-three—seven, eight—fifty-six.

LAURA: *(enters from the main part of the house)* Would you mind——

CAPTAIN: In a moment. Sixty-six, seventy-one, eighty-four, eighty-nine, ninety-two, one hundred. What is it?

LAURA: Perhaps I'm disturbing you.

CAPTAIN: Not at all. The housekeeping money, I suppose?

LAURA: Yes. The housekeeping money.

CAPTAIN: Leave the bills there, and I'll go through them.

LAURA: Bills?

CAPTAIN: Yes.

LAURA: Oh, you want bills now?

CAPTAIN: Of course I want bills. We are financially embarrassed, and if things come to a head I've got to be able to produce accounts. Otherwise I can be punished as a negligent debtor.

LAURA: It isn't my fault if we're financially embarrassed.

CAPTAIN: That's just what the bills will establish.

LAURA: I'm not to blame if our tenant won't pay the lease of his farm.

CAPTAIN: Who recommended him? You. Why did you recommend such a—what shall we call him? Drone?

LAURA: If he's such a drone, why did you take him?

CAPTAIN: Because you wouldn't let me eat in peace, sleep in peace or work in peace until you'd got him here. You wanted to have him because your brother wanted to be rid of him, your mother wanted to have him because I didn't want to have him, the governess wanted to have him because he was a Methodist, and old Margaret wanted to have him because she'd known his grandmother since they were children. So we took him, and if I hadn't I should now be either sitting in an asylum or lying in the family vault. However, here is your household allowance, and some pin money. You can give me the bills later.

LAURA: *(curtseys)* Thank you, sir. Do you keep bills for your private expenses?

CAPTAIN: That's none of your business.

LAURA: True; no more than my child's upbringing. Have you gentlemen reached a decision now, after your evening session?

CAPTAIN: I had already made my decision. I merely wished to impart it to the only friend whom I and my family have in common. Bertha is to live in town. She will leave in a fortnight.

LAURA: And where is she to live, if I may be allowed to ask?

CAPTAIN: I have arranged for her to lodge with my lawyer, Mr. Saevberg.

LAURA: That freethinker!

CAPTAIN: The law states that a child is to be brought up in her father's faith.

LAURA: And the mother has no say in the matter?

CAPTAIN: None. She has sold her birthright by legal contract, and has surrendered all her claims. In return, the husband supports her and her children.

LAURA: So she has no rights over her own child?

CAPTAIN: None whatever. Once you have sold something, you can't get it back and keep the money.

LAURA: But if the father and mother should agree on a compromise——?

CAPTAIN: How is that possible? I want her to live in town, you want her to stay at home. The arithmetical mean would be that she should live at the railway station, halfway between. This is a situation which cannot be resolved by compromise.

LAURA: Then it must be resolved by force. What did Nöjd want here?

CAPTAIN: That is my professional secret.

LAURA: The whole kitchen knows.

CAPTAIN: Then you should.

LAURA: I do.

CAPTAIN: And have passed judgment?

LAURA: The law is quite explicit on the matter.

CAPTAIN: The law is not explicit as to who is the child's father.

LAURA: No. But one usually knows.

CAPTAIN: Wise men say one can never be sure about such things.

LAURA: Not be sure who is a child's father?

CAPTAIN: They say not.

LAURA: How extraordinary! Then how can the father have all these rights over her child?

CAPTAIN: He only has them if he accepts responsibility for the child— or has the responsibility forced upon him. And in marriage, of course, the question of paternity does not arise.

LAURA: Never?

CAPTAIN: I should hope not.

LAURA: But if the wife has been unfaithful?

CAPTAIN: That is not relevant to our discussion. Are there any other questions you want to ask me?

LAURA: None whatever.

CAPTAIN: Then I shall go to my room. Please be so good as to inform me when the Doctor comes. *(Shuts the secretaire and rises.)*

LAURA: Very well.

CAPTAIN: *(going through concealed door right)* The moment he arrives!

I don't wish to insult him. You understand? *(Goes.)*

LAURA: I understand.

Scene 5

LAURA *alone. She looks at the banknotes she is holding in her hand.*

GRANDMOTHER: *(offstage)* Laura!

LAURA: Yes?

GRANDMOTHER: Is my tea ready?

LAURA: *(in the doorway left)* I'll bring it in a moment.

Goes towards the door upstage. Just before she reaches it, the BATMAN *opens it.*

BATMAN: Dr. Östermark!

DOCTOR: *(enters)* Mrs. Lassen?

LAURA: *(goes to greet him, and stretches out her hand)* How do you do, Doctor! Welcome to our home. The Captain is out, but he will be back shortly.

DOCTOR: Please forgive me for coming so late. I've already had to visit some patients.

LAURA: Won't you sit down?

DOCTOR: Thank you, Mrs. Lassen, thank you.

LAURA: Yes, there's a lot of illness around here just now. However, I do hope you'll be happy here. We lead such a lonely life out here in the country, so it's important for us to have a doctor who takes an interest in his patients. And I've heard many flattering reports of you, so I hope we shall see a good deal of each other.

DOCTOR: You are too kind, Mrs. Lassen. But I trust, for your sake, that my visits will not always have to be professional! Your family enjoys good health——?

LAURA: We've never had any seri-

ous illnesses, I am glad to say. But things aren't quite as they should be——

DOCTOR: Indeed?

LAURA: I'm afraid they are not at all as we could wish.

DOCTOR: Really? You alarm me!

LAURA: There are certain domestic matters which a woman's honour and conscience require her to conceal from the world——

DOCTOR: But not from her doctor.

LAURA: Precisely. So I feel it is my painful duty to be quite open with you from the start.

DOCTOR: Could we not postpone this conversation until I have had the pleasure of making the Captain's acquaintance?

LAURA: No. You must hear what I have to say before you see him.

DOCTOR: It concerns him, then?

LAURA: Yes—my poor, beloved husband!

DOCTOR: You alarm me, Mrs. Lassen. Believe me, I am deeply touched by your distress.

LAURA: *(takes out her handkerchief)* My husband is mentally unbalanced. Now you know. You will be able to judge for yourself later.

DOCTOR: What! But I have read with admiration the Captain's excellent dissertations on mineralogy, and have always received the impression of a powerful and lucid intelligence.

LAURA: Indeed? I should be most happy if it could be proved that we have all been mistaken.

DOCTOR: It is of course possible that his mind may be unhinged where other matters are concerned. Pray proceed.

LAURA: That is what we fear. You see, sometimes he has the most extraordinary ideas, which we would gladly indulge if they didn't threaten the existence of his whole family. For example, he has a mania for buying things.

DOCTOR: That is unfortunate. But what does he buy?

LAURA: Whole crates of books, which he never reads.

DOCTOR: Well, it isn't so unusual for a scholar to buy books.

LAURA: You don't believe me?

DOCTOR: Yes, Mrs. Lassen, I am sure that what you say is true.

LAURA: But is it reasonable for a man to claim that he can see in a microscope what is happening on another planet?

DOCTOR: Does he say that?

LAURA: Yes.

DOCTOR: In a microscope?

LAURA: Yes, in a microscope.

DOCTOR: If that is so, it is serious——

LAURA: *If* it is so! You don't believe me, Doctor. And I sit here telling you all our family secrets——

DOCTOR: Now listen, Mrs. Lassen. I am honoured that you should confide in me. But as a doctor, I must investigate the matter thoroughly before I can make a diagnosis. Has the Captain shown any symptoms of capriciousness or vacillation?

LAURA: Any symptoms! We've been married for twenty years, and he has never yet taken a decision without reversing it.

DOCTOR: Is he stubborn?

LAURA: He always insists on having his own way, but once he has got it he loses interest and begs me to decide.

DOCTOR: This is serious. I must observe him closely. You see, my dear Mrs. Lassen, will is the backbone of the mind. If the will is impaired, the mind crumples.

LAURA: God knows I've done my best to meet his wishes during all these long years of trial. Oh, if you knew the things I have had to put up with! If you knew!

DOCTOR: Mrs. Lassen, your distress moves me deeply, and I promise you I will see what can done. But after what you have told me, I must ask you one thing. Avoid touching on any subject that might excite your husband. In a sick brain, fancies grow like weeds, and can easily develop into obsessions or even mono-mania. You understand?

LAURA: You mean I must take care not to make him suspicious?

DOCTOR: Exactly. A sick man is receptive to the slightest impression, and can therefore be made to imagine anything.

LAURA: Really? I understand. Yes. Yes. *(A bell rings within the house.)* Excuse me, my mother wishes to speak with me. Wait a moment—this must be Adolf!

Scene 6

The DOCTOR. *The* CAPTAIN *enters through the concealed door.*

CAPTAIN: Ah, you here already, Doctor? Delighted to meet you!

DOCTOR: Good evening, Captain. It is a great honour for me to make the acquaintance of so distinguished a scientist.

CAPTAIN: Oh, nonsense. My military duties don't allow me much time for research. All the same, I think I'm on to a new discovery.

DOCTOR: Indeed?

CAPTAIN: You see, I've been submitting meteorites to spectral analysis, and I've discovered carbon! Evidence of organic life! What do you say to that?

DOCTOR: You can see that in the microscope?

CAPTAIN: Microscope? Good God, no—spectroscope!

DOCTOR: Spectroscope? Ah—yes, of course, I mean spectroscope. Well, then, you'll soon be able to tell us what is happening on Jupiter.

CAPTAIN: Not what *is* happening, but what *has* happened. If only that damned shop in Paris would send those books! I really believe all the booksellers in the world have entered into a conspiracy against me. Would you believe it, for two months I haven't had a reply to a single order, letter or even telegram! It's driving me mad. I just don't understand it.

DOCTOR: Oh, that's just common laziness. You mustn't take it too seriously.

CAPTAIN: Yes, but, damn it, I won't be able to get my thesis ready in time—I know there's a fellow in Berlin working on the same lines. Still, we haven't met to talk about that, but about you. If you'd care to live here, we have a small apartment in the wing—or would you rather take over your predecessor's lodgings?

DOCTOR: Just as you please.

CAPTAIN: No, as *you* please. Say, now.

DOCTOR: You must decide, Captain.

CAPTAIN: No, no, I can't decide. You must say what you want. I've no feelings in the matter, no feelings at all.

DOCTOR: Yes, but I can't decide——

CAPTAIN: For God's sake, man, say what you want! I've no inclinations in the matter, I couldn't care less what you do! Are you such a nitwit that you don't know what you want? Answer, or I'll get angry!

DOCTOR: If I must decide, then I'll live here!

CAPTAIN: Good! Thank you. Oh—! Forgive me, Doctor, but nothing annoys me so much as to hear people say it's all the same to them!

He rings. The NURSE *enters.*

CAPTAIN: Oh, is it you, Margaret? Tell me, old dear, do you know if the wing is ready for the doctor?

NURSE: Yes, sir, it's all ready.

CAPTAIN: Good. Then I won't keep you, Doctor—I expect you're tired. Good night. I'll look forward to seeing you again tomorrow.

DOCTOR: Good night, Captain.

CAPTAIN: I suppose my wife told you a few things about conditions here, to put you in the picture?

DOCTOR: She did mention one or two details she thought it might be useful for a stranger to know. Good night, Captain.

Scene 7

The CAPTAIN. *The* NURSE.

CAPTAIN: What do you want, old darling? Is something the matter?

NURSE: Now, listen, Mr. Adolf, pet.

CAPTAIN: What is it, Margaret? Speak out, my dear. You're the only one I can listen to without getting spasms.

NURSE: Now, listen, Mr. Adolf. Why don't you go halfway to meet madam about the child? Remember, she's a mother.

CAPTAIN: Remember I'm a father, Margaret.

NURSE: Now, now, now. A father has other things beside his child, but a mother has nothing. She's only got her child.

CAPTAIN: Exactly. She has only one burden, but I have three, including hers. Do you think I'd have stayed a soldier all my life if I hadn't been saddled with her and her child?

NURSE: Oh, I didn't mean that.

CAPTAIN: No, I'm sure you didn't. You're trying to put me in the wrong.

NURSE: Surely you think I want what's best for you, Mr. Adolf?

CAPTAIN: Yes, yes, my dear, I'm sure you do. But you don't know what's best for me. You see, it isn't enough for me to have given the child life. I want to give it my soul too.

NURSE: Well, I don't understand that. But I still think you ought to be able to come to some agreement.

CAPTAIN: You are not my friend, Margaret.

NURSE: I? Why, Mr. Adolf, how can you say such a thing? Do you think I can forget you were my baby when you were little?

CAPTAIN: Have *I* ever forgotten it, my dear? You've been like a mother to me—you've supported me, up to now, when everyone's been against me—but now, when I need you most, now you betray me and go over to the enemy.

NURSE: Enemy!

CAPTAIN: Yes, enemy! You know how things are in this house. You've seen it all, from beginning to end.

NURSE: Yes, I've seen enough. Blessed Jesus, why must two human beings torment the life out of each other? You're both so good and kind —madam's never like that to me or anyone else.

CAPTAIN: Only to me. Yes, I know. But I'm telling you, Margaret—if you betray me now, you are committing a sin. A web is being spun around me here, and that doctor is not my friend.

NURSE: Oh, Mr. Adolf, you think bad of everyone. But that's because you don't follow the true faith. That's the cause of it.

CAPTAIN: And you've found the only true faith, you and your Baptists. Aren't you lucky!

NURSE: Well, I'm luckier than you, Mr. Adolf. And happier. Humble your heart, and you'll see. God will make you happy, and you'll love your neighbour.

CAPTAIN: It's extraordinary—as soon as you start talking about God and love, your voice becomes hard and your eyes fill with hatred. No, Margaret, you haven't found the true faith.

NURSE: Ah, you're proud. All your learning won't get you far at the Day of Judgment.

CAPTAIN: How arrogantly thou speakest, O humble heart! Yes, I know learning means nothing to animals like you.

NURSE: Shame on you! Never mind. Old Margaret loves her big, big boy best of all, and when the storm comes he'll creep back to her like the good child he is.

CAPTAIN: Margaret! Forgive me, but—believe me, there's no one here who loves me except you. Help me. I feel something is going to happen here—I don't know what, but there's something evil threatening—— *(There is a scream from within the house.)* What's that? Who's screaming?

Scene 8

The CAPTAIN. *The* NURSE. BERTHA *enters.*

BERTHA: Father, father! Help me! Save me!

CAPTAIN: What is it, my beloved child? Tell me.

BERTHA: Help me! I think she wants to hurt me!

CAPTAIN: Who wants to hurt you? Tell me. Tell me.

BERTHA: Grandmamma. But it was my fault. I played a trick on her.

CAPTAIN: Tell me about it.

BERTHA: But you mustn't say anything! Promise you won't!

CAPTAIN: Very well. But tell me what it is.

The NURSE *goes.*

BERTHA: Well—in the evenings, she turns down the lamp and sits me down at the table with a pen and paper. And then she says that the spirits are going to write.

CAPTAIN: What! Why haven't you told me about this before?

BERTHA: Forgive me—I didn't dare. Grandmamma says the spirits take their revenge if anyone talks about them. And then the pen writes, but I don't know if it's me. And sometimes it won't move at all. And when I'm tired, nothing comes—but it's *got* to come! And tonight I thought it was writing well, but then grandmamma said I was copying from some old poem and playing a trick on her—and then she became so horribly angry!

CAPTAIN: Do you believe that spirits exist?

BERTHA: I don't know.

CAPTAIN: But I know they do not!

BERTHA: But grandmamma says you don't understand, and that you have much worse things, that can see what's happening on other planets.

CAPTAIN: She says that, does she? What else does she say?

BERTHA: She says you can't work magic.

CAPTAIN: I haven't said I can. You know what meteors are? Yes, stones that fall from other heavenly bodies. I can study them and say whether they contain the same elements as our earth. That's all I can see.

BERTHA: But grandmamma says there are things that she can see but you can't.

CAPTAIN: Well, she's lying.

BERTHA: Grandmamma doesn't tell lies.

CAPTAIN: How do you know?

BERTHA: Then mother would be lying too.

CAPTAIN: Hm!

BERTHA: If you say mother's lying, I'll never believe you again!

CAPTAIN: I haven't said that, and you must believe me when I tell you that your happiness and your whole future depend on your leaving this house. Would you like that? Would you like to go and live in town, and learn something new?

BERTHA: Oh, I'd so love to live in town and get away from here! As long as I can see you sometimes—often! In there everything's so gloomy, so horrible, like a winter night—but when you come, father, it's like throwing open the window on a spring morning!

CAPTAIN: My child! My child!

BERTHA: But, father, you must be nice to mother, do you hear? She cries so often.

CAPTAIN: Hm! So you want to go and live in town?

BERTHA: Yes! Yes!

CAPTAIN: But if your mother doesn't want you to?

BERTHA: But she must!

CAPTAIN: But if she doesn't?

BERTHA: Well, then—I don't know. But she must! She must!

CAPTAIN: Will you ask her?

BERTHA: You must ask her, nicely. She doesn't pay any attention to me.

CAPTAIN: Hm! Well, if you want it and I want it, and she doesn't want it, what shall we do then?

BERTHA: Oh, then everything'll be difficult again. Why can't you both——?

Scene 9

The CAPTAIN. BERTHA. LAURA.

LAURA: Oh, she's here. Now perhaps we can hear her opinion. Since her fate is about to be decided.

CAPTAIN: The child can hardly be expected to hold an informed opinion on what a young girl ought to do with her life. We are at least partly qualified to judge, since we have seen a good many young girls grow up.

LAURA: But since we differ, let Bertha decide.

CAPTAIN: No! I permit no one to usurp my rights—neither woman nor child. Bertha, leave us.

BERTHA *goes.*

LAURA: You were afraid to let her speak, because you knew she'd agree with me.

CAPTAIN: I happen to know she wants to leave home. But I also know that you have the power to alter her will at your pleasure.

LAURA: Oh, am I so powerful?

CAPTAIN: Yes. You have a satanic genius for getting what you want. But that's always the way with people who aren't scrupulous about what

means they use. How, for example, did you get rid of Dr. Norling, and find this new man?

LAURA: Well, how did I?

CAPTAIN: You insulted Norling, so that he went, and got your brother to fix this fellow's appointment.

LAURA: Well, that was very simple, wasn't it? And quite legal. Is Bertha to leave at once?

CAPTAIN: In a fortnight.

LAURA: Is that final?

CAPTAIN: Yes.

LAURA: Have you spoken to Bertha?

CAPTAIN: Yes.

LAURA: Then I shall have to stop it.

CAPTAIN: You can't.

LAURA: Can't? You think I'm prepared to let my daughter live with people who'll tell her that everything I taught her is nonsense, so that she'll despise her mother for the rest of her life?

CAPTAIN: Do you think I am prepared to allow ignorant and conceited women to teach my daughter that her father is a charlatan?

LAURA: That should matter less to you.

CAPTAIN: Why?

LAURA: Because a mother is closer to her child. It has recently been proved that no one can be sure who is a child's father.

CAPTAIN: What has that to do with us?

LAURA: You can't be sure that you are Bertha's father.

CAPTAIN: I—can't be sure——!

LAURA: No. No one can be sure, so you can't.

CAPTAIN: Are you trying to be funny?

LAURA: I'm only repeating what you've said to me. Anyway, how do you know I haven't been unfaithful to you?

CAPTAIN: I could believe almost anything of you, but not that. Besides, if it were true you wouldn't talk about it.

LAURA: Suppose I were prepared for anything—to be driven out, despised, anything—rather than lose my child? Suppose I am telling you the truth now, when I say to you: "Bertha is my child, but not yours!" Suppose——!

CAPTAIN: Stop!

LAURA: Just suppose. Your power over her would be ended.

CAPTAIN: If you could prove I was not the father.

LAURA: That wouldn't be difficult. Would you like me to?

CAPTAIN: Stop it! At once!

LAURA: I'd only need to name the true father, and tell you the time and place. When was Bertha born? Three years after our marriage——

CAPTAIN: Stop it, or——!

LAURA: Or what? All right, I'll stop. But think carefully before you take any decision. And, above all, don't make yourself ridiculous.

CAPTAIN: God—I could almost weep——!

LAURA: Then you *will* be ridiculous.

CAPTAIN: But not you!

LAURA: No. Our relationship to our children is not in question.

CAPTAIN: That is why one cannot fight with you.

LAURA: Why try to fight with an enemy who is so much stronger?

CAPTAIN: Stronger?

LAURA: Yes. It's strange, but I've never been able to look at a man without feeling that I am stronger than him.

CAPTAIN: Well, for once you're going to meet your match. And I'll see you never forget it.

LAURA: That'll be interesting.

NURSE: *(enters)* Dinner's ready. Will you come and eat?

LAURA: Thank you.

The CAPTAIN *hesitates, then sits in a chair by the table, next to the sofa.*

LAURA: Aren't you going to eat?

CAPTAIN: No, thank you. I don't want anything.

LAURA: Are you sulking?

CAPTAIN: No. I'm not hungry.

LAURA: Come along, or there'll be questions asked. Be sensible, now. Oh, very well. If you won't, you'd better go on sitting there. *(Goes.)*

NURSE: Mr. Adolf! What is all this?

CAPTAIN: I don't know. Can you explain to me how it is that you women can treat an old man as though he was a child?

NURSE: Don't ask me. I suppose it's because, whether you're little boys or grown men, you're all born of woman.

CAPTAIN: But no woman is born of man. Yes, but I *am* Bertha's father! Tell me, Margaret! You do believe that? Don't you?

NURSE: Lord, what a child you are! Of course you're your own daughter's father. Come and eat now, and don't sit there sulking. There! There now, come along!

CAPTAIN: *(gets up)* Get out, woman! Back to hell, you witches! *(Goes to the door leading to the hall.)* Svaerd! Svaerd!

BATMAN: *(enters)* Sir?

CAPTAIN: Harness the sleigh! At once!

NURSE: Captain! Now, listen——!

CAPTAIN: Out, woman! At once!

NURSE: Lord help us, what's going to happen now?

CAPTAIN: *(puts on his hat and makes ready to go out)* Don't expect me home before midnight! *(Goes.)*

NURSE: Blessed Jesus preserve us, how's all this going to end?

ACT TWO

[*In Act Two* LAURA *is told that "a person who is* non-compos *loses all his civic and family rights." She pretends not to have known this but increases efforts already suggested in Act One to plant suspicions concerning her husband's mental balance. The new doctor is the special target of her efforts; since he does not know the family he is inclined to take her word about past events.*]

Scene 5

LAURA *enters embarrassed. The* CAPTAIN *sits down at the secretaire.*

CAPTAIN: It's late, but we must talk this matter out. Sit down! *(Pause.)* This evening I went to the post office and collected my letters. It is evident from them that you have been intercepting both my outgoing and my incoming correspondence. The resultant waste of time has virtually destroyed the value of my researches.

LAURA: I was acting from kindness. You were neglecting your duties for this work.

CAPTAIN: You were not acting from kindness. You feared that some day I might win more honour through these researches than through my military career, and you were determined that I should not win any honour, because that would throw into relief your insignificance. Now I have confiscated some letters addressed to you.

LAURA: How noble of you.

CAPTAIN: I'm glad you appreciate my qualities. It is clear from these letters that for some time you have been turning all my former friends against me by spreading a rumour concerning my sanity. And you've succeeded, for now hardly one of them, from my commanding officer to my cook, regards me as sane. The situation regarding my mental condition is as follows. My brain is, as you know, unaffected, since I can perform both my professional duties and my duties as a father. I still have my emotions more or less under control, and my will is, to date, fairly unimpaired, but you have been chipping and chafing at it so that soon the cogs will disengage and the wheels will start whirling backwards. I shall not appeal to your feelings, for you have none—that is your strength. But I appeal to your self-interest.

LAURA: Go on.

CAPTAIN: By your behaviour you have succeeded in filling my mind with doubts, so that soon my judgment will be clouded and my thoughts begin to wander. This is the approaching dementia for which you have been waiting, and which may come at any time. Now you must ask yourself the question: is it not more to your interest that I should be well rather than ill? Think carefully? If I break down, I shall lose my job, and you will be without support. If I die, you will receive the insurance on my life; but if I kill myself, you will get nothing. So it is to your own interest that I should go on living.

LAURA: Is this a trap?

CAPTAIN: Yes. It is up to you whether you go round it or stick your neck in it.

LAURA: You say you'll kill yourself. You won't!

CAPTAIN: Are you sure? Do you think a man can live when he has nothing and no one to live for?

LAURA: Then you capitulate?

CAPTAIN: No. I propose an armistice.

LAURA: And your conditions?

CAPTAIN: That I retain my sanity. Free me from my doubts, and I will abandon the battle.

LAURA: What doubts?

CAPTAIN: About Bertha's parentage.

LAURA: Are there any doubts about that?

CAPTAIN: In my mind there are. You have awoken them.

LAURA: I?

CAPTAIN: Yes. You have dripped them into my ear like poison, and events have fostered their growth. Free me from my uncertainty, tell me straight out: "It is so!" and already I forgive you.

LAURA: How can I confess to a crime I have not committed?

CAPTAIN: What does it matter? You know I shan't reveal it. Do you think a man goes around trumpeting his shame?

LAURA: If I say it isn't true, you won't be sure; but if I say it is, you will be. So you would rather it was true.

CAPTAIN: Yes. It's strange, but I suppose it's because the one cannot be proved, whereas the other can.

LAURA: Have you any grounds for your suspicions?

CAPTAIN: Yes and no.

LAURA: I suppose you'd like me to be guilty so that you could throw me out and keep the child to yourself. But you won't catch me with a trick like that.

CAPTAIN: Do you think I'd want to keep some other man's child if I knew you were guilty?

LAURA: I'm sure you wouldn't. And that's why I realise you were lying just now when you said you already forgave me.

CAPTAIN: *(gets up)* Laura, save me and my sanity. You don't understand what I'm saying. If the child is not mine, I have no rights over her, and want none—and that is all that *you* want. Isn't it? Or do you want something else too? Do you want to retain your power over the child, but to keep me here as breadwinner?

LAURA: Power? Yes. What has this life-and-death struggle been for if not for power?

CAPTAIN: I do not believe in resurrection, and to me this child was my life hereafter. She was my idea of immortality—perhaps the only one that has any roots in reality. If you take her away, you cut short my life.

LAURA: Why didn't we part while there was still time?

CAPTAIN: Because the child bound us together. But the bond became a chain. How did it become that? How? I've never thought about it, but now memories return, accusing, condemning. We had been married for two years, and had no children, you best know why. I fell ill, and lay near to death. In a lucid moment I heard voices from the drawing-room. It was you and the lawyer, talking about my money—I still had some then. He is explaining that you cannot inherit anything because we have no children, and he asks if you are pregnant. I didn't hear your reply. I got better, and we had a child. Who is the father?

LAURA: You!

CAPTAIN: No, it is not I! A crime lies buried here, and it's beginning to come to light. And what a fiendish crime! You women were soft-hearted enough to free your black slaves, but you keep your white ones! I have worked and slaved for you, for your child, your mother, your servants. I have sacrificed my life and my career, I have undergone torture, scourging, sleeplessness, every kind of torment for you, my hair has turned grey, all so that you might live free from care and, when you grow old, enjoy new life through your child. All this I have borne without complaint, because I believed I was the father to this child. This is the most arrant form of theft, the most brutal slavery. I have served seventeen years of hard labour for a crime I did not commit. What can you give me in return?

LAURA: Now you really *are* mad.

CAPTAIN: *(sits)* So you hope. And I have seen how you worked to hide your crime. I pitied you, because I didn't understand why you were sad. I often calmed your evil conscience, supposing that I was driving away some sick thought. I heard you cry aloud in your sleep, though I didn't want to listen. Now I remember—the night before last! It was Bertha's birthday. It was between two and three o'clock in the morning, and I was sitting up, reading. You screamed as though someone was trying to strangle you: "Don't come, don't come!" I banged on the wall because —because I didn't want to hear any

more. I have had my suspicions for a long time, but I didn't dare to hear them confirmed. I have suffered all this for you. What will you do for me?

LAURA: What can I do? I will swear by God and all that is sacred that you are Bertha's father.

CAPTAIN: What good will that do, when you have already said that a mother can and should commit any crime for the sake of her child? I entreat you, by the memory of the past —I beg you, as a wounded man begs for mercy—tell me everything! Don't you see that I am as helpless as a child, can't you hear me crying for pity like a child crying to its mother, can't you forget that I am a man, a soldier who with a word can tame men and beasts? I ask only for the pity you would extend to a sick man, I lay down the insignia of my power and cry for mercy—for my life.

LAURA: *(has approached him and lays her hand on his forehead)* What! Man, you're crying!

CAPTAIN: Yes, I am crying, although I am a man. But has not a man eyes? Has not a man hands, limbs, heart, thoughts, passions? Does he not live by the same food, is he not wounded by the same weapons, warmed and cooled by the same summer and winter as a woman? If you prick us, do we not bleed? If you tickle us, do we not laugh? If you poison us, do we not die? Why should a man be forbidden to complain, or a soldier to weep? Because it is unmanly? Why is it unmanly?

LAURA: Weep, my child. Your mother is here to comfort you. Do you remember, it was as your second mother that I first entered into your life? Your big, strong body was afraid. You were a great child who had come too late into the world, or had come unwanted.

CAPTAIN: Yes, I suppose it was that. Father and mother had me against their will, and so I was born without a will. When you and I became one, I thought I was making myself whole; so I let you rule; and I who, in the barracks, among the soldiers, issued commands, was, with you, the one who obeyed. I grew up at your side, looked up to you as though to a superior being, listened to you as though I was your innocent child.

LAURA: Yes. That's how it was, and I loved you as my child. But, do you know—I suppose you noticed it—every time your feelings towards me changed, and you approached me as my lover, I felt bashful, and your embrace was an ecstasy followed by pangs of conscience, as though my blood was ashamed. The mother became the mistress—ugh!

CAPTAIN: Yes, I saw it, but I didn't understand. I thought you despised my lack of masculinity, and I wanted to win you as a woman by being a man.

LAURA: That was where you made your mistake. The mother was your friend, you see, but the woman was your enemy. Love between man and woman is war. And don't think I gave myself. I didn't give, I took—what I wanted to have. But you had the upper hand. I felt it, and I wanted to make you feel it.

CAPTAIN: No, you were always the one who had the upper hand. You could hypnotise me so that I neither saw nor heard, but only obeyed. You could give me a raw potato and make me think it was a peach, you could force me to admire your stupid ideas as strokes of genius, you could

have driven me to crime, yes, even to vice. For you lacked intelligence, and instead of following my advice you did as *you* wanted. But when, later, I awoke and looked about me and saw that my honour had been sullied, I wanted to wipe out the stain through a noble action, a brave deed, a discovery, or an honourable suicide. I wanted to go to war, but I couldn't. It was then that I turned to science. Now, when I should stretch out my hand to receive the fruits of my labour, you chop off my arm. Now I am without honour, and I cannot go on living, for a man cannot live without honour.

LAURA: But a woman—

CAPTAIN: She has her children, but he has none. Yet you and I and all the other men and women in the world have gone on living, as innocently as children, living on fancies, ideals and illusions. And then we awoke. Yes, we awoke, but with our feet on the pillow, and He Who woke us was Himself a sleepwalker. When women grow old and cease to be women, they get beards on their chins. I wonder what men get when they grow old and cease to be men? We who greeted the dawn were no longer cocks but capons, and the hens answered our false call, so that when the sun should have risen we found ourselves sitting in moonlight among ruins, just like in the good old days. It had only been a fretful slumber, a wild dream. It was no awakening.

LAURA: You know, you ought to have been a poet.

CAPTAIN: Perhaps I ought.

LAURA: Well, I'm sleepy. If you've any more fantasies, keep them until morning.

CAPTAIN: One word more—and

this isn't a fantasy. Do you hate me?

LAURA: Sometimes. When you are a man.

CAPTAIN: This is like racial hatred. If it is true that we are descended from the ape, it must have been from two different species. We aren't of the same blood, are we?

LAURA: What do you mean by all that?

CAPTAIN: I feel that, in this war, one of us must go under.

LAURA: Which one?

CAPTAIN: The weaker, of course.

LAURA: And the stronger is in the right?

CAPTAIN: Always. Because he is the one with power.

LAURA: Then I am in the right.

CAPTAIN: You think you have the power?

LAURA: Yes. And tomorrow I shall have it legally, when I have you certified.

CAPTAIN: Certified—?

LAURA: Yes. And then I shall bring up the child myself, without having to listen to your visions.

CAPTAIN: And who will pay for the child's upbringing, when I am gone?

LAURA: Your pension.

CAPTAIN: *(goes towards her threateningly)* How can you have me certified?

LAURA: *(takes out a letter)* By this letter, an attested copy of which I have deposited with the authorities.

CAPTAIN: What letter?

LAURA: *(moves backwards towards the door)* Yours! The one you wrote to the doctor telling him you were mad. *(The* CAPTAIN *looks at her dumbly.)* You have done your job as a father and a breadwinner. Now you are no longer needed, and you can go. You realise that my intel-ligence is equal to my will, and since you are not prepared to stay and admit it, you can go!

The CAPTAIN *goes to the table, takes the burning lamp and throws it at* LAURA, *who has retreated through the door.*

ACT THREE

[*In Act Three the* PASTOR *expresses to Laura his suspicion that she has engineered the plan to incarcerate her husband in an asylum. "You cannot deny that this fits in very nicely with your wish that you should bring up the child yourself." He promises, however, not to reveal her guilt: "After all, blood is thicker than water." But his motivations are mixed; not only protection of his sister is at stake but hostility towards a "freethinker" like her husband. Bertha has become alienated from her father after hearing of his act of violence but not about the preceding provocation. The doctor brings a straitjacket and the trusted old nurse is persuaded to get the Captain into it, circumventing his intention to shoot himself. At the end the Captain suffers a stroke and is on his deathbed. Bertha and Laura embrace.*]

Scene 7

As before. LAURA *gestures to* NÖJD *to go.*

CAPTAIN: Omphale! Omphale! Now you play with the club while Hercules winds your wool!

LAURA: *(comes over to the sofa)* Adolf! Look at me! Do you think I am your enemy?

CAPTAIN: Yes, I do. I think you are all my enemies. My mother was my enemy. She didn't want to bring me into the world because my birth would cause her pain. She robbed my first embryo of its nourishment, so that I was born half-crippled. My sister was my enemy, when she taught me that I was her inferior. The first woman I kissed was my enemy—she gave me ten years of sickness in return for the love I gave her. My daughter became my enemy, when you forced her to choose between you and me. And you, my wife, you were my mortal enemy, for you didn't let go of me until you had throttled the life out of me.

LAURA: I don't know that I ever planned, or intended, what you think I have done. I may have felt a vague desire to be rid of you, because you were an obstacle in my path; but if you see a plan in the way I have acted, then perhaps there was one, though I wasn't aware of it. I didn't plot any of this—it just glided forward on rails which you laid yourself—and before God and my conscience, I feel that I am innocent, even if I am not. Your presence has been like a stone on my heart, pressing and pressing until my heart rebelled against its suffocating weight. This is the truth, and if I have unintentionally hurt you, I ask your forgiveness.

CAPTAIN: That all sounds plausible. But how does it help me? And who is to blame? Perhaps the idea of marriage is to blame. In the old days, one married a wife; now one forms a company with a woman who goes to work, or moves in to live with a friend. And then one seduces the partner, or defiles the friend. What became of love—healthy, sensuous love? It died, starved. And what is the offspring of this broker's-love, a blank cheque drawn on a bankrupt account? Who will honour it when the crash comes? Who is the bodily father to the spiritual child?

LAURA: Those suspicions of yours about the child are completely unfounded.

CAPTAIN: That's just what's so horrible. If they were real, at least one would have something to grip on, something to cling to. Now there are only shadows, hiding in the bushes and poking out their heads to laugh —it's like fighting with air, a mock battle with blank cartridges. A real betrayal would have acted as a challenge, roused my soul to action. But now my thoughts dissolve in twilight, my brain grinds emptiness until it catches fire! Give me a pillow under my head! And put something over me, I'm cold. I'm so terribly cold!

LAURA *takes her shawl and spreads it over him. The* NURSE *goes out to fetch a pillow.*

LAURA: Give me your hand, friend.

CAPTAIN: My hand! Which you have tied behind my back? Omphale! Omphale! But I feel your soft shawl against my mouth. It's warm and smooth like your arm, and it smells of vanilla, as your hair did when you were young. Laura—when you were young—and we walked in the birch woods among the primroses— and thrushes sang! Beautiful, beautiful! How beautiful life was! And now it has become like this. You didn't want it to be like this, I didn't want it, and yet it happened. Who rules our lives?

LAURA: God alone rules——

CAPTAIN: The God of battle, then! Or the goddess, nowadays! Take away this cat that's lying on me! Take it away; *(The* NURSE *enters with the pillow and removes the shawl.)* Give me my tunic. Put it over me! *(The* NURSE *takes his military tunic from the clothes-hanger and drapes it over him.)* Ah, my brave lion's skin, that you would take from me! Omphale! Omphale! O cunning woman, who so loved peace that you discovered the art of disarming men! Awake, Hercules, before they take your club from you! You would rob us of our armour and have us believe that it is only tinsel. No, it was iron before it became tinsel. In the old days it was the smith who forged the soldier's tunic; now it is the seamstress. Omphale! Omphale! Strength has been vanquished by craft and weakness! Curse you, damned woman, and all your sex! *(Raises himself to spit, but falls back on the couch.)* What kind of a pillow have you given me, Margaret? It's so hard, and so cold, so cold! Come and sit beside me here, on the chair. That's right. May I rest my head in your lap? So. That's warm! Bend over so that I can feel your breast. Oh, it is sweet to sleep at a woman's breast, whether a mother's or a mistress's, but sweetest at a mother's!

LAURA: Do you want to see your child, Adolf? Speak!

CAPTAIN: My child? A man has no children. Only women have children, and so the future belongs to them, while we die childless. Gentle Jesus, meek and mild, look upon this little child—!

NURSE: Listen! He's praying to God!

CAPTAIN: No, to you, to send me to sleep. I'm so tired, so tired. Good night, Margaret. Blessed be thou amongst women——

He raises himself, but falls with a cry in the NURSE's *lap.*

Main Street

Sinclair Lewis

The greatest mystery about a human being is not his reaction to sex or praise, but the manner in which he contrives to put in twenty-four hours a day. It is this which puzzles the longshoreman about the clerk, the Londoner about the bushman. It was this which puzzled Carol in regard to the married Vida. Carol herself had the baby, a larger house to care for, all the telephone calls for Kennicott when he was away; and she read everything, while Vida was satisfied with newspaper headlines.

But after detached brown years in boarding-houses, Vida was hungry for housework, for the most pottering detail of it. She had no maid, nor wanted one. She cooked, baked, swept, washed supper-cloths, with the triumph of a chemist in a new laboratory. To her the hearth was veritably the altar. When she went shopping she hugged the cans of soup, and she bought a mop or a side of bacon as though she were preparing for a reception. She knelt beside a bean sprout and crooned, "I raised this with my own hands— I brought this new life into the world."

"I love her for being so happy," Carol brooded. "I ought to be that way. I worship the baby, but the housework—— Oh, I suppose I'm fortunate; so much better off than farm-women on a new clearing, or people in a slum."

It has not yet been recorded that any human being has gained a very large or permanent contentment from meditation upon the fact that he is better off than others.

In Carol's own twenty-four hours a day she got up, dressed the baby, had breakfast, talked to Oscarina about the day's shopping, put the baby on the porch to play, went to the butcher's to choose between steak and pork chops, bathed the baby, nailed up a shelf, had dinner, put the baby to bed for a nap, paid the iceman, read for an hour, took the baby out for a walk, called on Vida, had supper, put the baby to bed, darned socks, listened to Kennicott's yawning comment on what a fool Dr. McGanum was to try to use that cheap X-ray outfit of his on an epithelioma, repaired a frock, drowsily heard Kennicott stoke the furnace, tried to read a page of Thorstein Veblen—and the day was gone.

Except when Hugh was vigorously naughty, or whiney, or laughing, or saying "I like my chair" with thrilling maturity, she was always enfeebled by loneliness. She no longer felt superior about that misfortune. She would gladly have been converted to Vida's satisfaction in Gopher Prairie and mopping the floor.

* * *

The second evening after the movies she impulsively summoned Vida Sherwin and Guy to the house for pop-corn and cider. In the living-room Vida and Kennicott debated "the value of manual training in grades below the eighth," while Carol sat beside Guy at the dining table, buttering pop-corn. She was quickened by the speculation in his eyes. She murmured:

"Guy, do you want to help me?"

"My dear! How?"

"I don't know!"

He waited.

"I think I want you to help me find out what has made the darkness of the women. Gray darkness and shadowy trees. We're all in it, ten million women, young married

women with good prosperous husbands, and business women in linen collars, and grandmothers that gad out to teas, and wives of underpaid miners, and farmwives who really like to make butter and go to church. What is it we want—and need? Will Kennicott there would say that we need lots of children and hard work. But it isn't that. There's the same discontent in women with eight children and one more coming—always one more coming! And you find it in stenographers and wives who scrub, just as much as in girl college-graduates who wonder how they can escape their kind parents. What do we want?"

"Essentially, I think, you are like myself, Carol; you want to go back to an age of tranquillity and charming manners. You want to enthrone good taste again."

"Just good taste? Fastidious people? Oh—no! I believe all of us want the same things—we're all together, the industrial workers and the women and the farmers and the negro race and the Asiatic colonies, and even a few of the Respectables. It's all the same revolt, in all the classes that have waited and taken advice. I think perhaps we want a more conscious life. We're tired of drudging and sleeping and dying. We're tired of seeing just a few people able to be individualists. We're tired of always deferring hope till the next generation. We're tired of hearing the politicians and priests and cautious reformers (and the husbands!) coax us, 'Be calm! Be patient! Wait! We have the plans for a Utopia already made; just give us a bit more time and we'll produce it; trust us; we're wiser than you.' For ten thousand years they've said that. We want our Utopia *now*—and we're going to try our hands at it. All we want is—everything for all of us! For every housewife and every longshoreman and every Hindu nationalist and every teacher. We want everything. We sha'n't get it. So we sha'n't ever be content——"

John Sloan. *Woman's Work.*

A Start in Life

RUTH SUCKOW

The Switzers were scurrying around to get Daisy ready by the time that Elmer Kruse should get through in town. They had known all week that Elmer might be in for her any day. But they hadn't done a thing until he appeared. "Oh, it was so rainy today, the roads were so muddy, they hadn't thought he'd get in until maybe next week." It would have been the same any other day.

Mrs. Switzer was trying now at the last moment to get all of Daisy's things into the battered telescope that lay on the bed. The bed had not "got made"; and just as soon as Daisy was gone, Mrs. Switzer would have to hurry off to the Woodworths, where she was to wash today. Daisy's things were scattered over the dark brown quilt and the rumpled sheet that were dingy and clammy in this damp weather. So was the whole bed-room with its sloping ceiling, and old-fashioned square-paned windows, the commode that they used for a dresser littered with pin trays, curlers, broken combs, ribbons, smoky lamp, all mixed up together; the door of the closet open, showing the confusion of clothes and shabby shoes. . . . They all slept in this room—Mrs. Switzer and Dwight in the bed, the two girls in the cot against the wall.

"Mama, I can't find the belt to that plaid dress."

"Oh, ain't it somewheres around? Well, I guess you'll have to let it go. If I come across it I can send it out to you. Someone'll be going past there."

She had meant to get Daisy all mended and "fixed up" before she went out to the country. But some-how . . . oh, there was always so much to see to when she came

119

home. Gone all day, washing and cleaning for other people; it didn't leave her much time for her own home.

She was late now. The Woodworths liked to have her get the washing out early so that she could do some cleaning too before she left. But she couldn't help it. She would have to get Daisy off first. She had already had on her wraps ready to go, when Elmer came—her cleaning cap, of a blue faded almost gray, and the ancient black coat with gathered sleeves that she wore over her work dress when she went out to wash.

"What's become of all your underclothes? They ain't all dirty, are they?"

"They are, too. You didn't wash for us last week, mama."

"Well, you'll just have to take along what you've got. Maybe there'll be some way of getting the rest to you."

"Elmers come in every week, don't they?" Daisy demanded.

"Yes, but maybe they won't always be bringing you in."

She jammed what she could into the telescope, thinking with her helpless, anxious fatalism that it would have to do somehow.

"Daisy, you get yourself ready now."

"I am ready, mama, I want to put on my other ribbon."

"Oh, that's 'way down in the telescope somewhere. You needn't be so anxious to fix yourself up. This ain't like going visiting."

Daisy stood at the little mirror preening herself—such a homely child, "all Switzer," skinny, with pale sharp eyes set close together and thin, stringy, reddish hair. But she

had never really learned yet how homely she was. She was the oldest, and she got the pick of what clothes were given to the Switzers. Goldie and Dwight envied her. She was important in her small world. She was proud of her blue coat that had belonged to Alice Brooker, the town lawyer's daughter. It hung unevenly above her bony little knees, and the buttons came down too far. Her mother had tried to make it over for her.

Mrs. Switzer looked at her, troubled, but not knowing how she could tell her all the things she ought to be told. Daisy had never been away before except to go to her Uncle Fred's at Lehigh. She seemed to think that this would be the same. She had so many things to learn. Well, she would find them out soon enough—only too soon. Working for other people—she would learn what that meant. Elmer and Edna Kruse were nice young people. They would mean well enough by Daisy. It was a good chance for her to start in. But it wasn't the same.

Daisy was so proud. She thought it was quite a thing to be "starting in to earn." She thought she could buy herself so much with her dollar and a half a week. The other children stood back watching her, round-eyed and impressed. They wished that they were going away, like Daisy.

They heard a car come splashing through the mud on low. "There he is back! Have you got your things on? Goldie—go out and tell him she's coming."

"No, me tell him, me!" Dwight shouted jealously.

"Well—both of you tell him. Land! . . ."

She tried hastily to put on the

cover of the bulging telescope and to fasten the straps. One of them broke.

"Well, you'll have to take it the way it is."

It was an old thing, hadn't been used since her husband, Mert, had "left off canvassing" before he died. And he had worn it all to pieces.

"Well, I guess you'll have to go now. He won't want to wait. I'll try and send you out what you ain't got with you." She turned to Daisy. Her face was working. There was nothing else to do, as everyone said. Daisy would have to help, and she might as well learn it now. Only, she hated to see Daisy go off, to have her starting in. She knew what it meant. "Well—you try and work good this summer, so they'll want you to stay. I hope they'll bring you in sometimes."

Daisy's homely little face grew pale with awe, suddenly, at the sight of her mother crying, at something that she dimly sensed in the pressure of her mother's thin strong arms. Her vanity in her new importance was somehow shamed and dampened.

Elmer's big new Buick, mud-splashed but imposing, stood tilted on the uneven road. Mud was thick on the wheels. It was a bad day for driving, with the roads a yellow mass, water lying in all the wheel ruts. The little road that led past these few houses on the outskirts of town, and up over the hill, had a cold, rainy loneliness. Elmer sat in the front seat of the Buick, and in the back was a big box of groceries.

"Got any room to sit in there?" he asked genially. "I didn't get out, it's so muddy here."

"No, don't get out," Mrs. Switzer said hastily. "She can put this right on the floor there in the back." She added, with a timid attempt at courtesy, "Ain't the roads pretty bad out that way?"

"Yes, but farmers get so they don't think so much about the roads."

"I s'pose that's so."

He saw the signs of tears on Mrs. Switzer's face, and they made him anxious to get away. She embraced Daisy hastily again. Daisy climbed over the grocery box and scrunched herself into the seat.

"I guess you'll bring her in with you some time when you're coming," Mrs. Switzer hinted.

"Sure. We'll bring her."

He started the engine. It roared, half died down as the wheels of the car spun in the thick wet mud.

In that moment, Daisy had a startled view of home—the small house standing on a rough rise of land, weathered to a dim color that showed dark streaks from the rain; the narrow sloping front porch whose edge had a soaked, gnawed look; the chickens, grayish-black, pecking at the wet ground; their playthings, stones, a wagon, some old pail covers littered about; a soaked, discolored piece of underwear hanging on the line in the back yard. The yard was tussocky and overhung the road with shaggy long grass where the yellow bank was caved in under it. Goldie and Dwight were gazing at her solemnly. She saw her mother's face—a thin, weak, loving face, drawn with neglected weeping, with its reddened eyes and poor teeth . . . in the old coat and heavy shoes and cleaning cap, her work-worn hand with its big knuckles clutching at her coat. She saw the playthings they had used yesterday, and the old swing that hung from one of the trees, the ropes sodden, the seat in crooked. . . .

The car went off, slipping on the wet clay. She waved frantically, suddenly understanding that she was leaving them. They waved at her.

Mrs. Switzer stood there a little while. Then came the harsh rasp of the old black iron pump that stood out under the box elder tree. She was pumping water to leave for the children before she went off to work.

Daisy held on as the car skidded going down the short clay hill. Elmer didn't bother with chains. He was too used to the roads. But her eyes brightened with scared excitement. When they were down, and Elmer slowed up going along the tracks in the deep wet grass that led to the main road, she looked back, holding on her hat with her small scrawny hand.

Just down this little hill—and home was gone. The big car, the feel of her telescope on the floor under her feet, the fact that she was going out to the country, changed the looks of everything. She saw it all now.

Dunkels' house stood on one side of the road. A closed-up white house. The windows stared blank and cold between the old shutters. There was a chair with a broken straw seat under the fruit trees. The Dunkels were old Catholic people who seldom went anywhere. In the front yard was a clump of tall pines, the rough brown trunks wet, the green branches, dark and shining, heavy with rain, the ground underneath mournfully sodden and black.

The pasture on the other side. The green grass, lush, wet and cold, and the outcroppings of limestone

that held little pools of rain water in all the tiny holes. Beyond, the low hills gloomy with timber against the lowering sky.

They slid out onto the main road. They bumped over the small wooden bridge above the swollen creek that came from the pasture. Daisy looked down. She saw the little swirls of foam, the long grass that swished with the water, the old rusted tin cans lodged between the rocks.

She sat up straight and important, her thin, homely little face strained with excitement, her sharp eyes taking in everything. The watery mud holes in the road, the little thickets of plum trees, low and wet, in dark interlacings. She held on fiercely, but made no sound when the car skidded.

She felt the grandeur of having a ride. One wet Sunday, Mr. Brooker had driven them all home from church, she and Goldie and Dwight packed tightly into the back seat of the car, shut in by the side curtains, against which the rain lashed, catching the muddy scent of the roads. Sometimes they could plan to go to town just when Mr. Pattey was going to work in his Ford. Then they would run out and shout eagerly, "Mr. Pattey! Are you going through town?" Sometimes he said, with curt good nature, "Well, pile in"; and they all hopped into the truck back. "He says we can go along with him."

She looked at the black wet fields through which little leaves of bright green corn grew in rows, at showery bushes of sumac along the roadside. A gasoline engine pumping water made a loud desolate sound. There were somber-looking cattle in the wet grass, and lonely, thick-foliaged trees growing here and there in the pastures. She felt her telescope on the floor of the car, the box of groceries beside her. She eyed these with a sharp curiosity. There was a fresh pineapple—something the Switzers didn't often get at home. She wondered if Edna would have it for dinner. Maybe she could hint a little to Edna.

She was out in the country. She could no longer see her house even if she wanted to—standing dingy, streaked with rain, in its rough grass on the little hill. A lump came into her throat. She had looked forward to playing with Edna's children. But Goldie and Dwight would play all morning without her. She was still proud of being the oldest, of going out with Elmer and Edna; but now there was a forlornness in the pride.

She wished she were in the front seat with Elmer. She didn't see why he hadn't put her there. She would have liked to know who all the people were who lived on these farms; how old Elmer's babies were; and if he and Edna always went to the movies when they went into town on Saturday nights. Elmer must have lots of money to buy a car like this. He had a new house on his farm, too, and Mrs. Metzinger had said that it had plumbing. Maybe they would take her to the movies, too. She might hint about that.

When she had to visit Uncle Fred, she had had to go on the train. She liked this better. She hoped they had a long way to go. She called out to Elmer:

"Say, how much farther is your place?"

"What's that?" He turned around. "Oh, just down the road a ways. Scared to drive in the mud?"

"No, I ain't scared. I like to drive most any way."

She looked at Elmer's back, the old felt hat crammed down carelessly on his head, the back of his neck with the golden hair on the sunburned skin above the blue of his shirt collar. Strong and easy and slouched a little over the steering wheel that he handled so masterly. Elmer and Edna were just young folks; but Mrs. Metzinger said that they had more to start with than most young farmers did, and that they were hustlers. Daisy felt that the pride of this belonged to her too, now.

"Here we are!"

"Oh, is this where you folks live?" Daisy cried eagerly.

The house stood back from the road, beyond a space of bare yard with a little scattering of grass just starting—small, modern, painted a bright new white and yellow. The barn was new, too, a big splendid barn of frescoed brick, with a silo of the same. There were no trees. A raw, desolate wind blew across the back yard as they drove up beside the back door.

Edna had come out on the step. Elmer grinned at her as he took out the box of groceries, and she slightly raised her eyebrows. She said kindly enough:

"Well, you brought Daisy. Hello, Daisy, are you going to stay with us this summer?"

"I guess so," Daisy said importantly. But she suddenly felt a little shy and forlorn as she got out of the car and stood on the bare ground in the chilly wind.

"Yes, I brought her along," Elmer said.

"Are the roads very bad?"

"Kind of bad. Why?"

"Well, I'd like to get over to mama's some time today."

"Oh, I guess they aren't too bad for that."

Daisy pricked up her sharp little ears. Another ride. That cheered her.

"Look in the door," Edna said in a low fond voice, motioning with her head.

Two little round, blond heads were pressed tightly against the screen door. There was a clamor of "Daddy, daddy!" Elmer grinned with a bashful pride as he stood with the box of groceries, raising his eyebrows with mock surprise and demanding, "Who's this? What you shoutin' 'daddy' for? You don't think daddy's got anything for you, do you?" He and Edna were going into the kitchen together, until Edna remembered and called back hastily:

"Oh, come in, Daisy!"

Daisy stood, a little left out and solitary, there in the kitchen, as Billy, the older of the babies, climbed frantically over Elmer, demanding candy, and the little one toddled smilingly about. Her eyes took in all of it. She was impressed by the shining blue-and-white linoleum, the range with its nickel and enamel, the bright new woodwork. Edna was laughing and scolding at Elmer and the baby. Billy had made his father produce the candy. Daisy's sharp little eyes looked hungrily at the lemon drops and Edna remembered her.

"Give Daisy a piece of your candy," she said.

He would not go up to Daisy. She had to come forward and take one of the lemon drops herself. She saw where Edna put the sack, in a dish

high in the cupboard. She hoped they would get some more before long.

"My telescope's out there in the car," she reminded them.

"Oh! Elmer, you go and get it and take it up for her," Edna said.

"What?"

"Her valise—or whatever it is—out in the car."

"Oh, sure," Elmer said with a cheerful grin.

"It's kind of an old telescope," Daisy said conversationally. "I guess it's been used a lot. My papa used to have it. The strap broke when Mama was fastening it this morning. We ain't got any suitcase. I had to take this because it was all there was in the house, and mama didn't want to get me a new one."

Edna raised her eyebrows politely. She leaned over and pretended to spat the baby as he came toddling up to her, then rubbed her cheek against his round head with its funny fuzz of hair.

Daisy watched solemnly. "I didn't know both of your children was boys. I thought one of 'em was a girl. That's what there is at home now—one boy and one girl."

"Um-hm," Edna replied absently. "You can go up with Elmer and take off your things, Daisy," she said. "You can stop and unpack your valise now, I guess, if you'd like to. Then you can come down and help me in the kitchen. You know we got you to help me," she reminded.

Daisy, subdued, followed Elmer up the bright new stairs. In the upper hall, two strips of very clean rag were laid over the shining yellow of the floor. Elmer had put her telescope in one of the bedrooms.

"There you are!"

She heard him go clattering down the stairs, and then a kind of murmuring and laughing in the kitchen. The back door slammed. She hurried to the window in time to see Elmer go striding off toward the barn.

She looked about her room with intense curiosity. It, too, had a bright varnished floor. She had a bed all her own—a small, old-fashioned bed, left from some old furnishings, that had been put in this room that had the pipes and the hot water tank. She had to see everything, but she had a stealthy look as she tiptoed about, started to open the drawers of the dresser, looked out of her window. She put her coat and hat on the bed. She would rather be down in the kitchen with Edna than unpack her telescope now.

She guessed she would go down where the rest of them were.

Elmer came into the house for dinner. He brought in a cold, muddy, outdoor breath with him. The range was going, but the bright little kitchen seemed chilly, with the white oilcloth on the table, the baby's varnished high chair and his little fat mottled hands.

Edna made a significant little face at Elmer. Daisy did not see. She was standing back from the stove, where Edna was at work, looking at the baby.

"He can talk pretty good, can't he? Dwight couldn't say anything but 'mama' when he was that little."

Edna's back was turned. She said meaningly:

"Now, Elmer's come in for dinner, Daisy, we'll have to hurry. You must help me get on the dinner. You can cut bread and get things on the table. You must help, you know. That's what you are supposed to do."

Daisy looked startled, a little scared and resentful. "Well, I don't know where you keep your bread."

"Don't you remember where I told you to put it this morning? Right over in the cabinet, in that big box. You must watch, Daisy, and learn where things are."

Elmer, a little embarrassed at the look that Edna gave him, whistled as he began to wash his hands at the sink. "How's daddy's old boy?" he said loudly, gving a poke at the baby's chin.

As Edna passed him, she shook her head and her lips just formed, "Been like that all morning!"

He grinned comprehendingly. Then both their faces became expressionless.

Daisy had not exactly heard, but she looked from one to the other, silent and dimly wondering. The queer ache that had kept starting all through the morning, under her interest in Edna's things and doings, came over her again. She sensed something different in the atmosphere than she had ever known before— some queer difference between the position of herself and of the two babies, a faint notion of what mama had meant when she had said that this would not be visiting.

"I guess I'm going to have the toothache again," she said faintly.

No one seemed to hear her.

Edna whisked off the potatoes, drained the water. . . . "You might bring me a dish, Daisy." Daisy searched a long time while Edna turned impatiently and pointed. Edna put the rest of the things on the table herself. Her young, fresh, capable mouth was tightly closed, and she was making certain resolutions.

Daisy stood hesitating in the middle of the room, a scrawny, unappealing little figure. Billy—fat, blond, in funny, dark blue union-alls—was trotting busily about the kitchen. Daisy swooped down upon him and tried to bring him to the table. He set up a howl. Edna turned, looked astonished, severe.

"I was trying to make him come to the table," Daisy explained weakly.

"You scared him. He isn't used to you. He doesn't like it. Don't cry, Billy. The girl didn't mean anything."

"Here, daddy'll put him in his place," Elmer said hastily.

Billy looked over his father's shoulder at Daisy with suffused, resentful blue eyes. She did not understand it, and felt strangely at a loss. She had been left with Goldie and Dwight so often. She had always made Dwight go to the table. She had been the boss.

Edna said in a cool, held-in voice, "Put these things on the table, Daisy."

They sat down. Daisy and the other children had always felt it a great treat to eat away from home instead of at their own scanty, hastily set table. They had hung around Mrs. Metzinger's house at noon, hoping to be asked to stay, not offended when told that "it was time for them to run off now." Her pinched little face had a hungry look as she stared at the potatoes and fried ham and pie. But they did not watch and urge her to have more, as Mrs. Metzinger did, and Mrs. Brooker when she took pity on the Switzers and had them there. Daisy wanted more pie. But none of them seemed to be taking more, and so she said nothing. She remembered what her mother had

said, with now a faint comprehension. "You must remember you're out working for other folks, and it won't be like it is at home.

After dinner Edna said, "Now you can wash the dishes, Daisy."

She went into the next room with the children. Daisy, as she went hesitatingly about the kitchen alone, could hear Edna's low contented humming as she sat in there rocking, the baby in her lap. The bright kitchen was empty and lonely now. Through the window, Daisy could see the great barn looming up against the rainy sky. She hoped that they would drive to Edna's mother's soon.

She finished as soon as she could and went into the dining room where Edna was sewing on the baby's rompers. Edna went on sewing. Daisy sat down disconsolately. That queer low ache went all through her. She said in a small dismal voice:

"I guess I got the toothache again."

Edna bit off a thread.

"I had it awful hard awhile ago. Mama come pretty near taking me to the dentist."

"That's too bad," Edna murmured politely. But she offered no other condolence. She gave a little secret smile at the baby asleep on a blanket and a pillow in one corner of the shiny leather davenport.

"Is Elmer going to drive into town tomorrow?"

"Tomorrow? I don't suppose so."

"Mama couldn't find the belt of my plaid dress and I thought if he was, maybe I could go along and get it. I'd like to have it."

Daisy's homely mouth drooped at the corners. Her toothache did not seem to matter to anyone. Edna did not seem to want to see that anything was wrong with her. She had expected Edna to be concerned, to mention remedies. But it wasn't toothache, that strange lonesome ache all over her. Maybe she was going to be terribly sick. Mama wouldn't come home for supper to be told about it.

She saw mama's face as in that last glimpse of it—drawn with crying, and yet trying to smile, under the old cleaning cap, her hand holding her coat together . . .

Edna glanced quickly at her. The child was so mortally unattractive, unappealing even in her forlornness. Edna frowned a little, but said kindly:

"Now you might take Billy into the kitchen out of my way, Daisy, and amuse him."

"Well, he cries when I pick him up," Daisy said faintly.

"He won't cry this time. Take him out and help him play with his blocks. You must help me with the children, you know."

"Well, if he'll go with me."

"He'll go with you, won't he, Billy boy? Won't you go with Daisy, sweetheart?"

Billy stared and then nodded. Daisy felt a thrill of comfort as Billy put his little fat hand in hers and trotted into the kitchen beside her. He had the fattest hands, she thought. Edna brought the blocks and put the box down on the floor beside Daisy.

"Now, see if you can amuse him so that I can get my sewing done."

"Shall you and me play blocks, Billy?" Daisy murmured.

He nodded. Then he got hold of the box with one hand, tipped out all the blocks on the floor with a bang and a rattle, and looked at her with a pleased proud smile.

"Oh no, Billy. You mustn't spill out the blocks. Look, you're too little to play with them. No, now—now wait! Let Daisy show you. Daisy'll build you something real nice—shall she?"

He gave a solemn nod of consent.

Daisy set out the blocks on the bright linoleum. She had never had such blocks as these to handle before. Dwight's were only a few old, unmatched broken ones. Her spirit of leadership came back, and she firmly put away that fat hand of Billy's whenever he meddled with her building. She could make something really wonderful with these blocks.

"No, Billy, you mustn't. See, when Daisy's got it all done, then you can see what the lovely building is."

She put the blocks together with great interest. She knew what she was going to make—it was going to be a new house; no, a new church. Just as she got the walls up, in came that little hand again, and then with a delighted grunt Billy swept the blocks pellmell about the floor. At the clatter, he sat back, pursing his mouth to give an ecstatic "Ooh!"

"Oh, Billy—you mustn't, the building wasn't done! Look, you've spoiled it. Now, you've got to sit 'way off here while I try to build it over again."

Billy's look of triumph turned to surprise and then to vociferous protest as Daisy picked him up and firmly transplanted him to another corner of the room. He set up a tremendous howl. He had never been set aside like that before. Edna came hurrying out. Daisy looked at Edna for justification, but instinctively on the defensive.

"Billy knocked over the blocks. He spoiled the building."

"Wah! Wah!" Billy gave loud heartbroken sobs. The tears ran down his fat cheeks and he held out his arms piteously toward his mother.

"I didn't hurt him," Daisy said, scared.

"Never mind, lover," Edna was crooning. "Of course he can play with his blocks. They're Billy's blocks, Daisy," she said. "He doesn't like to sit and see you put up buildings. He wants to play, too. See, you've made him cry now."

"Do' wanna stay here," Billy wailed.

"Well, come in with mother then." She picked him up, wiping his tears.

"I didn't hurt him," Daisy protested.

"Well, never mind now. You can pick up the blocks and then sweep up the floor, Daisy. You didn't do that when you finished the dishes. Never mind," she was saying to Billy. "Pretty soon daddy'll come in and we'll have a nice ride."

Daisy soberly picked up the blocks and got the broom. What had she done to Billy? He had tried to spoil her building. She had always made Dwight keep back until she had finished. Of course it was Daisy, the oldest, who should lead and manage. There had been no one to hear her side. Everything was different. She winked back tears as she swept, poorly and carelessly.

Then she brightened up as Elmer came tramping up on the back porch and then through the kitchen.

"Edna!"

"She's in there," Daisy offered.

"Want to go now? What? Is the baby asleep?" he asked blankly.

Edna gave him a warning look and the door was closed.

Daisy listened hard. She swept very softly. She could catch only a little of what they said—"Kind of hate to go off . . . I know, but if we once start . . . not a thing all day . . . what we got her for . . ." She had no real comprehension of it. She hurried and put away the broom. She wanted to be sure and be ready to go.

Elmer tramped out, straight past her. She saw from the window that he was backing the car out from the shed. She could hear Edna and Billy upstairs, could hear the baby cry a little as he was wakened. Maybe she ought to go out and get her wraps, too.

Elmer honked the horn. A moment later Edna came hurrying downstairs, in her hat and coat, and Billy in a knitted cap and a red sweater crammed over his union-alls, so that he looked like a little brownie. The baby had on his little coat, too.

Edna called out, "Come in and get this boy, daddy." She did not look at Daisy, but said hurriedly, "We're going for a little ride, Daisy. Have you finished the sweeping? Well, then, you can pick up those pieces in the dining room. We won't be gone so very long. When it's a quarter past five, you start the fire, like I showed you this noon, and slice the potatoes that were left, and the meat. And set the table."

The horn was honked again.

"Yes! Well, we'll be back, Daisy. Come, lover, daddy's in a hurry."

Daisy stood looking after them. Billy clamored to sit beside his daddy. Edna took the baby from Elmer and put him beside her on the back seat. There was room—half of the big back seat. There wasn't anything, really, to be done at home. That was the worst of it. They just didn't want to take her. They all belonged together. They didn't want to take anyone else along. She was an outsider. They all—even the baby —had a freshened look of expectancy.

The engine roared—they had started; slipping on the mud of the drive, then forging straight ahead, around the turn, out of sight.

She went forlornly into the dining room. The light from the windows was dim now in the rainy, late afternoon. The pink pieces from the baby's rompers were scattered over the gay rug. She got down on her hands and knees, slowly picking them up, sniffing a little. She heard the Big Ben clock in the kitchen ticking loudly.

That dreadful ache submerged her. No one would ask about it, no one would try to comfort her. Before, there had always been mama coming home, anxious, scolding sometimes, but worried over them if they didn't feel right, caring about them. Mama and Goldie and Dwight cared about her—but she was away out in the country, and they were at home. She didn't want to stay here, where she didn't belong. But mama had told her that she must begin helping this summer.

Her ugly little mouth contorted into a grimace of weeping. But silent weeping, without any tears; because she already had the cold knowledge that no one would notice or comfort it.

Jerome B. Thompson. *A "Pic Nick," Camden, Maine.*

The Dog

CAROL REILLEY

The man said every son should have a dog. It was a part of his training, part of his growing up. The woman said she wasn't sure, after all, the furniture and all, you know. Of course, if it were a little dog . . .

"Little dog!" The man pounded his fist on the table, until the forks bounced, and the coffee splashed dangerously. "A boy needs a big dog; little dogs are for women and sissies. It takes a big dog with a big heart for a boy."

The man glared angrily, and the woman shivered back into her chair. The subject was closed. For a minute there was only the sound of knives and forks against the battered china plates. Then the man began speaking slowly of his day at work.

But the dog still lurked in the shadows. Peter toyed with his food, mixing the horrible spinach and the horrible peas together, and then carefully separating them into two sickening piles. Dogs were not good things. He had never known one very well, but he knew they were not good.

"Can you eat your peas, Peter?" the woman asked. "They're good for you. They'll make you a strong, big man."

"Yes'm," Peter said, and squashed one cruelly with his spoon. He did not dare look at her face. He knew that she wanted him to call her "Mother," but he could not. He felt sorry for her, dimly sensing the longing that lay at the bottom of her weakness. He squashed another pea. He wondered if it felt the pain. He hoped it did. He hated peas.

"Peter," the man said, "clean up your plate, and when you're finished, we'll go into the living room while Mother clears off the table, and we'll talk about the dog. Would you like that, Peter, boy?"

Peter put his spoon down on the plate. He stared at the horrible spinach and the horrible peas. He felt suddenly, softly sick.

"Sir," he mumbled. "I don't think I would care for a dog." He wiggled uneasily, sensing the man's hurt. "They're so expensive and things," he added quickly.

"Nonsense," said the man. "Now clean up your plate!"

"Oh, don't make him, Charles, if he doesn't want to, just this once—"

"Quiet!" shouted the man. "He'll eat that food before he leaves the table. He's only been here a couple of weeks and already you want to spoil him!"

The woman moved from the table and silently cleared the dishes away. Peter knew that her silence was not anger. It had something to do with the hurt in her body, and the hurt in the body of her man. Peter knew that he had put the hurt there.

He ate the spinach and peas quietly. The sickness in his stomach was not from them. There was a dark shadow in the corner by the stove. Peter moved on his chair, away from it. His eyes kept coming back in tingling horror. He was sure he saw the shapeless form of a dog, crouching there.

Bed was a terrible thing, in this new house. It was a frightening thing, there, all alone in the room, high, cold, and white.

Before, bed had always been a little cot, lined up against the wall with four other little cots. Before it had been friendly, with pillow fights, and the light of a street lamp through the window, and sleep, and the quiet breathing coming from the moonlit humps of other boys.

Here there was only loneliness. There was not even a street light. There were only shadows, blending into one frightening shadow in the darkness. Peter hated bed in the new house. He had never known his mother, never cried for her, and inside he had always felt a superiority over these boys who sobbed softly into their pillows in the nighttime. But here, in the new house, he, too, knew the sickening sobs. When he could not sleep with his fear, he lay and cried for his little cot, and the warmth of the Marshall Home.

This night he climbed into the bed and lay trembling beneath the cold sheet. He lay still while the woman bent to kiss him. He had never known such fear. Every shadow was a dog. He could not let the woman go. He clung to her.

"What's wrong, Peter, darling?" she asked softly.

He was ashamed of his fear, "Nothing, ma'am," he said.

She opened the window, and adjusted the shade. She smoothed the covers across his legs. He lay stiff, unspeaking, afraid of her leaving.

"Good night, Peter, honey," she said. She smiled down at him. She turned out the light, and stood there for a moment, her fingers lingering on the switch. "Good night," she said again, and waited for his answer.

"Good night, ma'am," Peter said. She sighed, and then he heard her slowly going down the steps.

He lay stiffly in the darkness. He did not dare to close his eyes. He watched the shadows warily, fearfully, trying to make out the crouched figure of a dog. He lay still, so it would not see him. He pressed his hands against his throat, to protect it from the tearing fangs. He waited. The shadows drew together, came closer. There was a scraping across the floor. He could not breathe. There was a roaring in his ears. A cold breath brushed his cheek. His stomach tightened. He clutched the sheets, and drew himself up. The dog shadows were all around him. All around him . . .

Peter screamed. "Mother!" he screamed. "Father!" He choked, and then screamed again, hysterically.

He could not hear the sounds they made, until they were there, and the light had blinded on, and chased the shadow dog away. He sobbed, and clung to them.

They asked him things he could not bear to answer, until the sobs inside of him had died away.

"Peter, Peter, honey!" the woman said, and her arms were close around him, and her hand was warm in his hair.

"What's wrong, son?" the man said. "Have a nightmare, or something?"

He lay quietly, sensing the new warmth in them. He did not dare to say anything about the dog. He was afraid the hurt would come back to their eyes.

They sat with him a while. The man's arm was warm across Peter's shoulders. "You don't need to be afraid, Peter, we're here. And soon you'll have a dog of your own to sleep here with you . . ."

"On the bed?" the woman started to say, and then she stopped, and smiling, brushed her cheek across the man's chest.

They left him there.

"Should I leave the light on, Peter?" the woman said.

"Yes—Mother," Peter answered carefully, "if you would."

She came over to him, and kissed him again, and held him close. Peter knew her hunger, and slipped his arm around her neck experimentally.

She let him go, and moved away. "Good night," she said, and waited.

"Good night, Mother," said Peter carefully. "Good night, Father."

He heard them laugh. He heard the man say, "You're spoiling him, darling."

The woman answered, "A boy needs a father to give him a dog, and make a man out of him, and a mother to spoil him, just a little."

Peter listened to their happy laughter. It was a new thing to him, and he knew that he had put it there.

He lay for a long time, his eyes still open, staring unseeingly at the books and toys that lined the walls of the lonely room. His mind was dulled from fear. He did not actually think of the dog. He just lay there, dumbly, until sleep came.

The next three days were slow ones. Peter tried very hard to remember to call them "Mother" and "Father." It seemed such a little thing to say, and he could sense the glow inside of them when he said it.

He still could not make himself go up into the room, and play there with all the toys and books they had brought him. He was afraid to be alone in the room, even in the daytime. The fear of that night was still with him.

Bed was sheer terror. He spent those next three days in shadow of the coming night. They left the light on for him now. The first time he asked for it again, the man was angry. Lord, he shouted, what am I raising for a son, a pantywaist. The woman had pleaded with him, clinging to his arm. No, no, he had cried.

"Please, Dad," Peter had said.

"Just for now," the woman begged.

"Dad, please," Peter whispered.

The man had turned to him, hugged him close. "I'm spoiling him, too," he had said. "For a while then, Peter, but remember, if I get you a dog, you'll have to act like a man, and not be afraid of the dark."

Peter could say nothing, only slip back among the pillows, and pray hollowly the man would forget the dog.

So the three days passed. Peter spent the afternoon of the third one in the kitchen with the woman. She was making an apple pie. He stayed close to her, following her wherever she went, not daring to be away from the warmth of her body.

"Be a good boy, Peter, darling," she laughed, "and fetch me the sugar."

"Yes, ma'am," he said, and ran across the floor to the cupboard. He liked the cold sound of his feet against the linoleum. He liked this linoleum. It was red and white and black, not the smooth dull brown of the linoleum on the floors of the Marshall Home. He brought the sugar back to her, put the can into her waiting hands. He saw a kind of pain around her eyes. He did not like it there.

"Here—Mother," he said, and smiled.

The woman laughed and tried to lift him in her arms.

"You're heavy, Peter," she said. "O Peter, Peter, you do like it here, don't you?"

"Yes, ma'am—Mother," he said.

"We love you so," and she was not talking to him any more. "We don't want them ever to take—to, oh, Peter, honey! Miss Rodgers is com-

ing tomorrow, maybe, or Saturday. You know who she is. She brought you here. She'll come quite often at first, once a month, maybe, until she's sure you're going to fit in here. And you will, Peter. We love you so; we'll be good to you. Sometimes, at night . . ." She stopped, and laughed, a little ashamed.

She turned back to the mixing bowl. "Your father will be home soon," she said. "This will be a good night, Peter."

He slipped to the floor, and sat contemplating the good, clean, black and white squares of linoleum. He traced his fingers around a slim red line of circle. The color was warm, but the touch was cool.

The woman moved back and forth beside him, humming and talking. Her legs brushed his back. He did not listen to her words. He began playing a game on the squares, seeing how many he could touch, with his palm pushed down, and his fingers spread out across the cold floor.

The afternoon was gone, and dusk had come. The front door rattled.

"It's your father," the woman said. "You go, Peter."

He gave his truck another shove across the soft livingroom rug, and then made himself stand up, and go to the door. He opened it slowly, and stared into the dusk.

The man was standing there, smiling. "Surprise, Peter," he said, and Peter watched dumbly, while the man pulled gently at the rope in his hand.

"It's a dog, Peter," the man said. "See, it's an Irish setter. Isn't she a beauty? Here, Peter."

Peter stared at the dog. The woman came up behind him. "It is a little big," she said doubtfully, "but it is pretty." She rested a hand on Peter's shoulder. "What's her name?"

"Well, whatever Peter wants to call her, I suppose," the man said.

Peter lost the sound of their words in a dull drone. He stared at the dog. It looked back at him, and whimpered. Peter backed up against the woman.

"Don't be afraid, son," the man said. "Here, pet her. She won't hurt you."

The man stepped inside. The dog moved toward Peter. Its cool nose brushed his hand.

Out of the depths of his mind, a cold fear arose. It came from a cold, sharp, blue place, in the back of his head. It put its fingers around his heart, and pushed into his throat. It choked his breath.

The dog pushed its eager body against him. Peter screamed, and wrenched free. He ran, not knowing where he was going. He thought he could feel the breath of the dog against his neck.

He slammed the kitchen door shut behind him, and held it, screaming hysterically. He swayed, and the linoleum reeled dizzily up to meet him.

Peter lay there, and retched weakly all over the cool black and white squares, and the swimming red circles.

The next day, Miss Rodgers came, and Peter went back with her. He clung to her, screaming, begging when she came, and the man and woman stood there together, stunned, and crying a little.

"It happens this way, sometimes,"

Miss Rodgers explained, "that the child doesn't adjust. Once in a while it comes suddenly like this. Other times, it comes slowly, several months, or even a year. Sometimes it is a difference in temperament not apparent at first, and sometimes, like this, it is a fear buried in the child. I'm sorry for this, but I think you'll understand."

Peter got his coat, and Miss Rodgers helped him pack some of his things in a little box. They left the bedroom, and there was no sadness inside of Peter. They went downstairs to say good-by, and he clung tightly to the familiar warmth of her plump fingers.

Miss Rodgers talked to the man and the woman a little longer, and then she told him it was time to go.

"Say good-by, Peter," she said gently.

The woman sat lost in the heavy chair, and there were tears on her cheeks. The man stood beside her.

The dog scratched on the basement door, and barked hoarsely. Peter did not even shiver. The man swore softly. Peter saw the woman's fingers tighten around his hand. He knew again that there was hurt in their bodies, he knew again that he had put it there.

"Good-by, ma'am," he said softly. "Good-by, sir."

All the way back to the Marshall Home, he sat stiffly next to Miss Rodgers in the car. It was not until they were almost there, that he began to cry, softly, and slowly. He was not sure, inside of him, the reason why.

William Hogarth. *Gin Lane.*

from

Mrs. Warren's Profession

GEORGE BERNARD SHAW

MRS WARREN: *(Again raising her voice angrily)* Do you know who youre speaking to, Miss?

VIVIE: *(Looking across at her without raising her head from her book)* No. Who are you? What are you?

MRS WARREN: *(rising breathless)* You young imp!

VIVIE: Everybody knows my reputation, my social standing, and the profession I intend to pursue. I know nothing about you. What is that way of life which you invite me to share with you and Sir George Crofts, pray?

MRS WARREN: Take care. I shall do something I'll be sorry for after, and you too.

VIVIE: *(putting aside her books with cool decision)* Well, let us drop the subject until you are better able to face it. *(Looking critically at her mother)* You want some good walks and a little lawn tennis to set you up. You are shockingly out of condition: you were not able to manage twenty yards uphill today without stopping to pant; and your wrists are mere rolls of fat. Look at mine. *(She holds out her wrists.)*

MRS WARREN: *(after looking at her helplessly, begins to whimper)* Vivie—

VIVIE: *(springing up sharply)* Now pray dont begin to cry. Anything but that. I really cannot stand whimpering. I will go out of the room if you do.

MRS WARREN: *(piteously)* Oh, my darling, how can you be so hard on me? Have I no rights over you as your mother?

VIVIE: Are you my mother?

MRS WARREN: *(appalled)* Am I your mother! Oh, Vivie!

VIVIE: Then where are our relatives? my father? our family friends? You claim the rights of a mother:

the right to call me fool and child; to speak to me as no woman in authority over me at college dare speak to me; to dictate my way of life; and to force on me the acquaintance of a brute whom anyone can see to be the most vicious sort of London man about town. Before I give myself the trouble to resist such claims, I may as well find out whether they have any real existence.

MRS WARREN: *(distracted, throwing herself on her knees)* Oh no, no. Stop, stop. I am your mother: I swear it. Oh, you cant mean to turn on me—my own child! it's not natural. You believe me, dont you? Say you believe me.

VIVIE: Who was my father?

MRS WARREN: You dont know what youre asking. I cant tell you.

VIVIE: *(determinedly)* Oh yes you can, if you like. I have a right to know; and you know very well that I have that right. You can refuse to tell me, if you please; but if you do, you will see the last of me tomorrow morning.

MRS WARREN: Oh, it's too horrible to hear you talk like that. You wouldnt—you couldnt leave me.

VIVIE: *(ruthlessly)* Yes, without a moment's hesitation, if you trifle with me about this. *(Shivering with disgust)* How can I feel sure that I may not have the contaminated blood of that brutal waster in my veins?

MRS WARREN: No, no. On my oath it's not he, nor any of the rest that you have ever met. I'm certain of that, at least.

Vivie's eyes fasten sternly on her mother as the significance of this flashes on her.

VIVIE: *(slowly)* You are certain of that, at least. Ah! You mean that that is all you are certain of. *(Thoughtfully)* I see. *(Mrs. Warren buries her face in her hands).* Dont do that, mother: you know you dont feel it a bit. *(Mrs. Warren takes down her hands and looks up deplorably at Vivie, who takes out her watch and says)* Well, that is enough for tonight. At what hour would you like breakfast? Is half-past eight too early for you?

MRS WARREN: *(wildly)* My God, what sort of woman are you?

VIVIE: *(coolly)* The sort the world is mostly made of, I should hope. Otherwise I dont understand how it gets its business done. Come *(taking her mother by the wrist, and pulling her up pretty resolutely)*, pull yourself together. Thats right.

MRS WARREN: *(querulously)* Youre very rough with me, Vivie.

VIVE: Nonsense. What about bed? It's past ten.

MRS WARREN: *(passionately)* Whats the use of my going to bed? Do you think I could sleep?

VIVIE: Why not? I shall.

MRS WARREN: You! youve no heart. *(She suddenly breaks out vehemently in her natural tongue—the dialect of a woman of the people—with all her affectations of maternal authority and conventional manners gone, and an overwhelming inspiration of true conviction and scorn in her)* Oh, I won't bear it: I won't put up with the injustice of it. What right have you to set yourself up above me like this? You boast of what you are to me—to me, who gave you the chance of being what you are. What chance had I? Shame on you for a bad daughter and a stuck-up prude!

VIVIE: *(sitting down with a shrug, no longer confident; for her replies,* which have sounded sensible and strong to her so far, now begin to ring rather woodenly and even priggishly against the new tone of her mother) Dont think for a moment I set myself above you in any way. You attacked me with the conventional authority of a mother: I defended myself with the conventional superiority of a respectable woman. Frankly, I am not going to stand any of your nonsense; and when you drop it I shall not expect you to stand any of mine. I shall always respect your right to your own opinions and your own way of life.

MRS WARREN: My own opinions and my own way of life! Listen to her talking! Do you think I was brought up like you? able to pick and choose my own way of life? Do you think I did what I did because I liked it, or thought it right, or wouldnt rather have gone to college and been a lady if I'd had the chance?

VIVIE: Everybody has some choice, mother. The poorest girl alive may not be able to choose between being Queen of England or Principal of Newnham; but she can choose between ragpicking and flowerselling, according to her taste. People are always blaming their circumstances for what they are. I dont believe in circumstances. The people who get on in this world are the people who get up and look for the circumstances they want, and, if they cant find them, make them.

MRS WARREN: Oh, it's easy to talk, very easy, isn't it? Here! would you like to know what my circumstances were?

VIVIE: Yes: you had better tell me. Wont you sit down?

MRS WARREN: Oh, I'll sit down: dont you be afraid. (*She plants her* chair farther forward with brazen energy, and sits down. Vivie is impressed in spite of herself). D'you know what your gran'mother was?

VIVIE: No.

MRS WARREN: No you dont. I do. She called herself a widow and had a fried-fish shop down by the Mint, and kept herself and four daughters out of it. Two of us were sisters: that was me and Liz; and we were both good-looking and well made. I suppose our father was a well-fed man: mother pretended he was a gentleman; but I dont know. The other two were only half sisters: undersized, ugly, starved looking, hard working, honest poor creatures: Liz and I would have half-murdered them if mother hadnt half-murdered us to keep our hands off them. They were the respectable ones. Well, what did they get by their respectability? I'll tell you. One of them worked in a whitelead factory twelve hours a day for nine shillings a week until she died of lead poisoning. She only expected to get her hands a little paralyzed; but she died. The other was always held up to us as a model because she married a Government laborer in the Deptford victualling yard, and kept his room and the three children neat and tidy on eighteen shillings a week—until he took to drink. That was worth being respectable for, wasn't it?

VIVIE: (*now thoughtfully attentive*) Did you and your sister think so?

MRS WARREN: Liz didnt, I can tell you: she had more spirit. We both went to a church school—that was part of the ladylike airs we gave ourselves to be superior to the children that knew nothing and went nowhere—and we stayed there until Liz went out one night and never came back. I know the schoolmistress thought I'd soon follow her example; for the clergyman was always warning me that Lizzie'd end up by jumping off Waterloo Bridge. Poor fool: that was all he knew about it! But I was more afraid of the whitelead factory than I was of the river; and so would you have been in my place. That clergyman got me a situation as scullery maid in a temperance restaurant where they sent out for anything you liked. Then I was waitress; and then I went to the bar at Waterloo station: fourteen hours a day serving drinks and washing glasses for four shillings a week and my board. That was considered a great promotion for me. Well, one cold, wretched night, when I was so tired and I could hardly keep myself awake, who should come up for a half of Scotch but Lizzie, in a long fur cloak, elegant and comfortable, with a lot of sovereigns in her purse.

VIVIE: *(grimly)* My aunt Lizzie!

MRS WARREN: Yes; and a very good aunt to have, too. She's living down at Winchester now, close to the cathedral, one of the most respectable ladies there. Chaperones girls at the country ball, if you please. No river for Liz, thank you! You remind me of Liz a little: she was a first-rate business woman—saved money from the beginning—never let herself look too like what she was—never lost her head or threw away a chance. When she saw I'd grown up good-looking she said to me across the bar "What are you doing there, you little fool? wearing out your health and your appearance for other people's profit!" Liz was saving money then to take a house for herself in Brussels; and she thought we two could save faster than one. So she lent me some money and gave me a start; and I saved

steadily and first paid her back, and then went into business with her as her partner. Why shouldnt I have done it? The house in Brussels was real high class: a much better place for a woman to be in than the factory where Anne Jane got poisoned. None of our girls were ever treated as I was treated in the scullery of that temperance place, or at the Waterloo bar, or at home. Would you have had me stay in them and become a worn out old drudge before I was forty?

VIVIE: *(intensely interested by this time)* No; but why did you choose that business? Saving money and good management will succeed in any business.

MRS WARREN: Yes, saving money. But where can a woman get the money to save in any other business? Could you save out of four shillings a week and keep yourself dressed as well? Not you. Of course, if youre a plain woman and cant earn anything more; or if you have a turn for music, or the stage, or newspaper-writing: thats different. But neither Liz nor I had any turn for such things: all we had was our appearance and our turn for pleasing men. Do you think we were such fools as to let other people trade in our good looks by employing us as shopgirls, or barmaids, or waitresses, when we could trade in them ourselves and get all the profits instead of starvation wages? Not likely.

VIVIE: You were certainly quite justified—from the business point of view.

MRS WARREN: Yes; or any other point of view. What is any respectable girl brought up to do but to catch some rich man's fancy and get the benefit of his money by marrying him?—as if a marriage ceremony could make any difference in the right or wrong of the thing! Oh, the hypocrisy of the world makes me sick! Liz and I had to work and save and calculate just like other people; elseways we should be as poor as any good-for-nothing drunken waster of a woman that thinks her luck will last for ever. *(With great energy)* I despise such people: theyve no character; and if theres a thing I hate in a woman, it's want of character.

VIVIE: Come now, mother: frankly! Isnt it part of what you call character in a woman that she should greatly dislike such a way of making money?

MRS WARREN: Why, of course. Everybody dislikes having to work and make money; but they have to do it all the same. I'm sure Ive often pitied a poor girl, tired out and in low spirits, having to try to please some man that she doesn't care two straws for—some half-drunken fool that thinks he's making himself agreeable when he's teasing and worrying and disgusting a woman so that hardly any money could pay her for putting up with it. But she has to bear with disagreeables and take the rough with the smooth, just like a nurse in a hospital or anyone else. It's not work that any woman would do for pleasure, goodness knows; though to hear the pious people talk you would suppose it was a bed of roses.

VIVIE: Still, you consider it worth while. It pays.

MRS WARREN: Of course it's worth while to a poor girl, if she can resist temptation and is good-looking and well conducted and sensible. It's far better than any other employment open to her. I always thought that oughtnt to be. It cant be right, Vivie, that there shouldnt be better opportunities for women. I stick to that: it's wrong. But it's so, right or wrong; and a girl must make the best of it. But of course it's not worth while for a lady. If you took to it youd be a fool; but I should have been a fool if I'd taken to anything else.

VIVIE: *(more and more deeply moved)* Mother: suppose we were both as poor as you were in those wretched old days, are you quite sure that you wouldnt advise me to try the Waterloo bar, or marry a laborer, or even go into the factory?

MRS WARREN: *(indignantly)* Of course not. What sort of mother do you take me for! How could you keep your self-respect in such starvation and slavery? And whats a woman worth? whats life worth? without self-respect! Why am I independent and able to give my daughter a first-rate education, when other women that had just as good opportunities are in the gutter? Because I always knew how to respect myself and control myself. Why is Liz looked up to in a cathedral town? The same reason. Where would we be now if we'd minded the clergyman's foolishness? Scrubbing floors for one and sixpence a day and nothing to look forward to but the workhouse infirmary. Dont you be led astray by people who dont know the world, my girl. The only way for a woman to provide for herself decently is for her to be good to some man that can afford to be good to her. If she's in his own station of life, let her make him marry her; but if she's far beneath him she cant expect it: why should she? it wouldn't be for her own happiness. Ask any lady in London society that has daughters; and she'll tell you the same, except that I tell you straight and she'll tell you crooked. Thats all the difference.

VIVIE: *(fascinated, gazing at her)* My dear mother: you are a wonderful woman: you are stronger than all England. And are you really and truly not one wee bit doubtful—or—or—ashamed?

MRS WARREN: Well, of course, dearie, it's only good manners to be ashamed of it: it's expected from a woman. Women have to pretend to feel a great deal that they dont feel. Liz used to be angry with me for plumping out the truth about it. She used to say that when every woman could learn enough from what was going on in the world before her eyes, there was no need to talk about it to her. But then Liz was such a perfect lady! She had the true instinct of it; while I was always a bit of a vulgarian. I used to be so pleased when you sent me your photos to see that you were growing up like Liz: youve just her ladylike, determined way. But I cant stand saying one thing when everyone knows I mean another. Whats the use in such hypocrisy? If people arrange the world that way for women, theres no good pretending it's arranged the other way. No: I never was a bit ashamed really. I consider I had a right to be proud of how we managed everything so respectably, and never had a word against us, and how the girls were so well taken care of. Some of them did very well: one of them married an ambassador. But of course now I darent talk about such things: whatever would they think of us! *(She yawns).* Oh dear! I do believe I'm getting sleepy after all. *(She stretches herself lazily, thoroughly relieved by her explosion, and placidly ready for her night's rest.)*

VIVIE: I believe it is I who will not be able to sleep now. *(She goes to the dresser and lights the candle. Then she extinguishes the lamp, darkening the room a good deal).* Better let in some fresh air before locking up. *(She opens the cottage door, and finds that it is broad moonlight). What a beautiful night! Look! (She draws aside the curtains of the window. The landscape is seen bathed in the radiance of the harvest moon rising over Blackdown).*

MRS WARREN: *(with a perfunctory glance at the scene)* Yes, dear; but take care you dont catch your death of cold from the night air.

VIVIE: *(contemptuously)* Nonsense.

MRS WARREN: *(querulously)* Oh yes: everything I say is nonsense, according to you.

VIVIE: *(turning to her quickly)* No: really that is not so, mother. You have got completely the better of me tonight, though I intended it to be the other way. Let us be good friends now.

MRS WARREN: *(shaking her head a little ruefully)* So it has been the other way. But I suppose I must give in to it. I always got the worst of it from Liz; and now I suppose it'll be the same with you.

VIVIE: Well, never mind. Come: goodnight, dear old mother. *(She takes her mother in her arms).*

MRS WARREN: *(fondly)* I brought you up well, didn't I, dearie?

VIVIE: You did.

MRS WARREN: And youll be good to your poor old mother for it, wont you?

VIVIE: I will, dear. *(Kissing her)* Goodnight.

MRS WARREN: *(with unction)* Blessings on my own dearie darling! a mother's blessing!

She embraces her daughter protectively, instinctively looking upward for divine sanction.

from "Down These Mean Streets"

Puerto Rican Paradise

PIRI THOMAS

Poppa didn't talk to me the next day. Soon he didn't talk much to anyone. He lost his night job—I forget why, and probably it was worth forgetting—and went back on home relief. It was 1941, and the Great Hunger called Depression was still down on Harlem.

But there was still the good old WPA. If a man was poor enough, he could dig a ditch for the government. Now Poppa was poor enough again.

The weather turned cold one more time, and so did our apartment. In the summer the cooped-up apartments in Harlem seem to catch all the heat and improve on it. It's the same in the winter. The cold, plastered walls embrace that cold from outside and make it a part of the apartment, till you don't know whether it's better to freeze out in the snow or by the stove, where four jets, wide open, spout futile, blue-yellow flames. It's hard on the rats, too.

Snow was falling. "My *Cristo*," Momma said, "*qué frío*. Doesn't that landlord have any *corazón*? Why don't he give more heat?" I wondered how Pops was making out working a pick and shovel in that falling snow.

Momma picked up a hammer and began to beat the beat-up radiator that's copped a plea from so many beatings. Poor steam radiator, how could it give out heat when it was freezing itself? The hollow sounds Momma beat out of it brought echoes from other freezing people in the building. Everybody picked up the beat and it seemed a crazy, good idea. If everybody took turns beating on the radiators, everybody could keep warm from the exercise.

We drank hot cocoa and talked about summertime. Momma talked

143

about Puerto Rico and how great it was, and how she'd like to go back one day, and how it was warm all the time there and no matter how poor you were over there, you could always live on green bananas, *bacalao,* and rice and beans. *"Dios mío,"* she said, "I don't think I'll ever see my island again."

"Sure you will, Mommie," said Miriam, my kid sister. She was eleven. "Tell us, tell us all about Porto Rico."

"It's not Porto Rico, it's Puerto Rico," said Momma.

"Tell us, Moms," said nine-year-old James, "about Puerto Rico."

"Yeah, Mommie," said six-year-old José.

Even the baby, Paulie, smiled.

Moms copped that wet-eyed look and began to dream-talk about her *isla verde,* Moses' land of milk and honey.

"When I was a little girl," she said, "I remember the getting up in the morning and getting the water from the river and getting the wood for the fire and the quiet of the greenlands and the golden color of the morning sky, the grass wet from the *lluvia . . . Ai, Dios,* the *coquis* and the *pajaritos* making all the *música . . ."*

"Mommie, were you poor?" asked Miriam.

"Sí, muy pobre, but very happy. I remember the hard work and the very little bit we had, but it was a good little bit. It counted very much. Sometimes when you have too much, the good gets lost within and you have to look very hard. But when you have a little, then the good does not have to be looked for so hard."

"Moms," I asked, "did everybody love each other—I mean, like if everybody was worth something, not like if some weren't important because they were poor—you know what I mean?"

"Bueno hijo, you have people everywhere who, because they have more, don't remember those who have very little. But in Puerto Rico those around you share *la pobreza* with you and they love you, because only poor people can understand poor people. I like *los Estados Unidos,* but it's sometimes a cold place to live—not because of the winter and the landlord not giving heat but because of the snow in the hearts of the people."

"Moms, didn't our people have any money or land?" I leaned forward, hoping to hear that my ancestors were noble princes born in Spain.

"Your grandmother and grandfather had a lot of land, but they lost that."

"How come, Moms?"

"Well, in those days there was nothing of what you call *contratos,* and when you bought or sold something, it was on your word and a handshake, and that's the way your *abuelos* bought their land and then lost it."

"Is that why we ain't got nuttin' now?" James asked pointedly.

"Oh, it—"

The door opened and put an end to the kitchen yak. It was Poppa coming home from work. He came into the kitchen and brought all the cold with him. Poor Poppa, he looked so lost in the clothes he had on. A jacket and coat, sweaters on top of sweaters, two pairs of long johns, two pairs of pants, two pairs of socks and a woolen cap. And under all that he was cold. His eyes were cold; his ears were red with pain. He took off his gloves and his fingers were stiff with cold.

"*Cómo está?*" said Momma. "I will make you coffee."

Poppa said nothing. His eyes were running hot frozen tears. He worked his fingers and rubbed his ears, and the pain made him make faces. "Get me some snow, Piri," he said finally.

I ran to the window, opened it, and scraped all the snow on the sill into one big snowball and brought it to him. We all watched in frozen wonder as Poppa took that snow and rubbed it on his ears and hands.

"Gee, Pops, don't it hurt?" I asked.

"*Sí*, but it's good for it. It hurts a little first, but it's good for the frozen parts."

I wondered why.

"How was it today?" Momma asked

"Cold. My God, ice cold."

Gee, I thought, *I'm sorry for you, Pops. You gotta suffer like this.*

"It was not always like this," my father said to the cold walls. "It's all the fault of the damn depression."

"Don't say 'damn,' " Momma said.

"Lola, I say 'damn' because that's what it is—*damn*."

And Momma kept quiet. She knew it was "damn."

My father kept talking to the walls. Some of the words came out loud, others stayed inside. I caught the inside ones—the damn WPA, the damn depression, the damn home relief, the damn poorness, the damn cold, the damn crummy apartments, the damn look on his damn kids, living so damn damned and his not being able to do a damn thing about it.

And Momma looked at Poppa and at us and thought about her Puerto Rico and maybe being there where you didn't have to wear a lot of extra clothes and feel so full of damns, and how when she was a little girl all the green was wet from the *lluvias.*

And Poppa looking at Momma and us, thinking how did he get trapped and why did he love us so much that he dug in damn snow to give us a piece of chance? And why couldn't he make it from home, maybe, and keep running?

And Miriam, James, José, Paulie, and me just looking and thinking about snowballs and Puerto Rico and summertime in the street and whether we were gonna live like this forever and not know enough to be sorry for ourselves.

Frida Kahlo de Rivera. *My Pre-Natal Life.*

from

Birth of a Man

MAXIM GORKY

. . . A low moan in the bushes—a sound of human distress, which always shakes the soul with sympathy.

Making my way through the bushes, I came upon the peasant woman with the yellow kerchief. She was sitting with her back against the trunk of a nut tree. Her head was resting on her shoulder, her mouth was gaping in an ugly way, her eyes were starting out of her head, and there was a crazy look in them. She was clutching her enormous abdomen with her hands and breathing so unnaturally that it moved up and down convulsively, and she was making a muffled, cow-like sound, baring yellow, wolf-like teeth.

"Someone gave you a beating?" I asked her, bending over her. Her bare legs, covered with ashen dust, were jerking like a fly's, and shaking her heavy head, she managed:

"Go away . . . you shameless fellow . . . go . . ."

I understood what it all meant, I had seen it happen before. Of course, I was frightened, jumped away, and the woman let out a long-drawn-out wail. From her eyes, which looked ready to burst, came troubled tears that ran down her purple, strained face.

This brought me back to her. I threw my bundle on the ground, together with the tea-pot and kettle, put her on her back, and tried to bend her knees. She pushed me away, hitting me on the face and chest, turned around, and roaring like a bear and cursing, crawled on all fours further into the bushes:

"You bandit . . . devil . . ." she brought out.

Her arms giving way under her, she fell, her face striking the earth, and again she howled convulsively, stretching out her legs.

In a fever of excitement, and quickly recalling everything I knew about this business, I turned her around and laid her on her back and bent her legs.

"Lie quiet," I said to her, "you will be delivered in no time. . . ."

I ran down to the sea, tucked up my sleeves, washed my hands, returned, and became an accoucheur.

The woman was writhing like birch-bark in the fire, she thrashed about with her hands, and, plucking the faded grass, tried to push it into her mouth. She strewed earth over her terrible, inhuman face with its wild, bloodshot eyes. Already the child's head was showing. I had to keep her legs from writhing, help the child, and see that she did not put grass into her wry, bellowing mouth.

We swore at each other a little, she through her teeth, I too under my breath, she from pain and, perhaps, also from shame, I because I was ill at ease and tormented by pity for her.

"L-lord!" she repeated, bringing out the word with a rattling sound. Her blue lips were bitten and frothy, and from her eyes, which looked as though they had suddenly been faded by the sun, tears kept pouring, the abundant tears of a mother's unbearable suffering, and her body was writhing, breaking, dividing in two. "G-go away, you devil . . ." she kept saying.

With weak, dislocated hands she kept pushing me away, while I repeated persuasively:

"Get through, you fool, get through quickly. . . ."

I was racked by pity for her, it was as though her tears were in my eyes, anguish squeezed my heart, I felt like shouting, and I shouted: "Come on, hurry up!"

At last, a human being was in my hands. Through my tears I saw that he was all red, and already he was discontented with the world. He struggled, carried on, and howled in a thick voice, although he was still tied to his mother. He had blue eyes, his nose was ludicrously crushed against his red, crumpled face, his lips moved, and he screamed: "I . . . I . . ."

He was so slippery that if I had not taken care, he would have slipped away from me. Kneeling, I looked at him and laughed—I was very glad to see him. And I had forgotten what must be done next.

"Cut it . . ." whispered the mother gently. Her eyes were closed, her face relaxed. It was earth-colored, as if she were dead, her blue lips barely moved:

"Cut it . . . with a knife. . . ."

My knife had been stolen in the barracks. I bit through the cord. The baby howled in an Oryol bass, and the mother smiled; her bottomless eyes blossomed out marvelously and burned with a blue fire. Her dark hand fumbled in her skirt, feeling for the pocket, and her bleeding, bitten lips were barely able to produce:

"I haven't . . . strength . . . tape . . . in the pocket . . . to bind . . . the navel. . . ."

I got the tape and bound up the navel. Her smile was even brighter, it was indeed so warm and brilliant that it nearly dazzled me.

"Now set yourself to rights," I said, "and I'll go and give him a wash. . . ."

"But look out," she murmured, uneasily, "go gently. . . ."

This red fellow didn't have to be treated with care, not at all: he clenched his fists and bawled, bawled as if challenging someone to a fight: "I . . . I . . ."

"You . . . you! Assert yourself firmly, brother, or else your fellow men will break your neck for you straight off. . . ."

He gave a particularly loud and earnest yell when he was splashed for the first time by a frothy wave which gaily dashed against us both. Then, when I bathed his chest and back, he screwed up his eyes, struggled violently and screamed piercingly, while the waves kept splashing over him.

"Make a noise, old fellow! Shout at the top of your lungs. . . ."

When I took him back to his mother, she lay with her eyes closed again, biting her lips. She was undergoing the pangs of expelling the after-birth. Nevertheless, through her sighs and groans I heard her faint whisper:

"Give . . . give him to me. . . ."

"He'll wait."

"No . . . give him here."

And with unsteady, trembling hands she unbuttoned her blouse. I helped her to free her breast, prepared by nature for a score of babies, and I placed the obstreperous fellow against her warm body. He grasped the situation at once and grew silent.

"Holy Mother of God, Most Pure Virgin," the woman repeated, shuddering, and rolled her disheveled head from side to side on my bundle.

And suddenly, with a gentle outcry, she grew silent. Then she opened her infinitely beautiful eyes, the hallowed eyes of a mother. Blue, they looked at the blue sky, and there burned and melted in them a grateful, joyous smile. With a heavy hand she was slowly making the sign of the cross over herself and the child. . . .

"Glory be to Thee, Most Pure Mother of God," she repeated; "oh . . . glory . . ."

Her eyes grew tired and sunken. For a long time she was silent, scarcely breathing. And suddenly she said in a matter-of-fact tone, her voice grown firm:

"Untie my sack, lad. . . ."

I did so. She looked at me attentively, smiled weakly, her drawn cheeks and damp forehead flushing slightly.

"Would you mind . . ."

"Don't you do too much. . . ."

"Just leave me. . . ."

I went off into the thicket. Birds were gently singing in my heart and, together with the noise of the sea, this was so wonderful that I thought I could listen to that music for a year on end. . . .

Not far off a stream was babbling: it was as though a girl were telling her friend of her beloved.

Presently above the bushes the woman's head appeared, with the yellow kerchief properly tied.

"Eh, is that you, sister?" I shouted. "It's too soon for you to be stirring about."

Holding on to a bough, she was sitting like a statue, white-faced, with huge blue lakes instead of eyes, and she whispered with emotion:

"Look—how he sleeps. . . ."

He looked well asleep, but as far as I could judge, no better than other babies, and if there was any difference it was due to the surroundings. He lay under a bush, such as

do not grow in the province of Oryol, on a heap of bright autumn
leaves.

"You had better lie down now,
mother . . ." I advised her.

"No," she said, shaking her head,
which seemed to be loosely screwed
to her neck; "I must tidy up and
be off for what-d'ye-call-it. . . ."

"Ochemchiry?"

"That's it. My people must have
gone quite a distance. . . ."

"But can you walk?"

"And what of the Virgin? She will
help. . . ."

Well, if she has the Virgin with
her, there is nothing more to be said!

She looked at the little pouting
face under the bush, and warm rays
of caressing light poured from her
eyes. She licked her lips and passed
her hands slowly over her chest.

I made a fire and set up stones to
put the tea-kettle on.

"Now I am going to treat you to
tea, mother."

"Do . . . my throat is dry. . . ."

"And what about your people?
Have they left you in the lurch?"

"They haven't . . . no. I just
stayed behind. They have had a drop
too much, and . . . it's better, this
way. . . . What would it have been
like with them around?"

Glancing at me, she covered her
face with her elbow; then spat blood,
and smiled bashfully.

"Is this your first one?" I asked.

"The first. . . . And who are
you?"

"A human being, sort of. . . ."

"Of course, a human being! Married?"

"Haven't had the honor. . . ."

"That's not true."

"What do you mean?"

She dropped her eyes, thought
awhile, and said:

"And how is it you know about
these things?"

This time I decided to lie, and I
said:

"I studied these things. I am a
student, understand?"

"Yes, yes. Our priest's eldest son is
also a student. He studied to be a
priest. . . ."

"Yes, that's the kind I am. Well,
I'll fetch some water. . . ."

The woman bent her head in the
direction of the child, listening to
his breathing for a while, then looked
off toward the sea.

"I would like to have a wash, too,"
she said, "but this queer water . . .
What kind is it? It's salt and bitter. . . ."

"You wash yourself with it; it's
good for you!"

"Is it?"

"Sure. Its warmer than the stream;
the streams hereabouts are like
ice. . . ."

"You know best. . . ."

Here an Abkhasian came riding
slowly by, his head drooping sleepily. His small sinewy horse looked
at us out of the corner of its round
black eye, pricking up its ears. Suddenly it snorted and the rider warily
jerked up his head in its shaggy fur
cap, looked in our direction too, and
dropped his head again.

"How queer people are here, and
frightening," said the woman quietly.

I went off. A stream of clear water
as alive as quicksilver flowed over
the stones, and in it autumn leaves
were gaily cavorting. It was wonderful. I washed my hands and face,
filled my tea-kettle, and went back.
Through the bushes I noticed that

the woman was crawling on her knees, casting uneasy glances about her.

"What is it?" I shouted to her.

She turned gray with fright and proceeded to hide something under her skirts. I understood what it was.

"Give it to me," I said; "I'll bury it."

"Oh, dear! But how will you do it? It should be buried in the bath-house entry, under the floor. . . ."

"And how soon do you think they'll build a bath-house here?"

"This is a joke to you, but I am afraid! Maybe, a beast will devour it . . . it must be given back to the earth, you know. . . ."

She turned aside and, handing me a damp, heavy bundle, begged me shamefacedly, under her breath:

"You bury it well, as deep as possible, for Christ's sake . . . Out of pity for my little son, do it well. . . ."

When I came back, she was returning from the beach. Her gait was unsteady and one of her arms was stretched out in front of her; her skirt was wet up to her waist; her face was somewhat flushed and lit by an inner light, as it were. I helped her to walk to the fire, thinking to myself: "What animal strength!" Then we drank tea with honey and she questioned me gently:

"You've given up school?"

"I have."

"After drinking away everything?"

"Yes, I drank away everything, mother, to the last shred!"

"That is the kind of fellow you are! I remember you; I noticed you at Sukhum when you were arguing with the chief over the food; I thought to myself then: he is afraid of nothing, must be a drunk!"

And appreciatively licking the honey off her swollen lips, she kept glancing at the bush under which the latest addition to the population of Oryol was quietly asleep.

"What will his life be like, I wonder?" she said with a sigh, looking at me. "Here you've helped me, and I thank you for it . . . but is it good for him? I don't know. . . ."

She finished her tea and her food, crossed herself, and while I was getting my things together, she was staring at the ground with her faded eyes, swaying sleepily, and thinking. Then she started getting up.

"Are you really going to walk?" I asked her.

"I am."

"Look out, mother!"

"And what of the Virgin? Let me have him!"

"No, I will carry him."

After some argument she gave in, and we set off, shoulder to shoulder.

"I hope I don't drop," she said, smiling guiltily, and laid her hand on my shoulder.

Meanwhile the new inhabitant of the Russian land, a person with an unknown future, was lying in my arms, breathing noisily like a solid citizen. The sea was splashing and swishing, laced with white shavings; the bushes were whispering to each other; the sun, which had already passed the zenith, was shining.

We were walking, slowly. Now and then the mother would stop to draw a deep breath. She would lift up her head and look about, at the sea, the forest, the mountains, and then she would peer into her son's face. Her eyes, thoroughly washed by tears of suffering, were again amazingly clear, again blossoming and burning with the blue fire of inexhaustible love.

Once, as she halted, she said:

"Lord, dear God! How good it all is, how good! I could walk like this, I could walk to the end of the world, and he, my little son, would grow, would grow freely amidst plenty, near his mother's breast, my darling. . . ."

. . . The sea was booming. . . .

Mexican Mother

Gabriela Mistral

Mexican mother: claim for your child, with vigor, what life owes to all beings who are born and never asked to be born. For him you have the right to make great demands. Demand a school that is sunny and clean; demand happy parks; demand a fiesta of images, in books and in educational films; insist on helping make the laws, but when they deal with things that dirty you, or make your life petty, you can demand laws that wash away the shame of the illegitimate child, who is made an outcast from birth, and must live an outcast among other children; and demand laws that regulate your labor and the labor of your children, who work to exhaustion in the brutal routine of the factory.

For that cause you can speak loudly and still be austere; your words will not seem peculiar, they will even be saintly.

You will be heard, sooner or later, Mexican mother. The eyes of honest men will turn to you, for there still are many, because your dignity defeats other kinds of pride; and they will remember the poem of Walt Whitman when they see you: "I tell you there is none greater than the mother of men!"

I have love for you, Mexican mother, sister of my mother, as you embroider exquisitely, and you weave honey-colored mats, as you paint a flowered jar, or cross a field dressed in blue like the women of the Bible, carrying food to a son or husband watering the field of corn.

Our raza will be proven in your sons; in them will we be saved or will we perish. God assigned to them the difficult fate of having the tide from the North break against their chests. That is why, when your sons are fighting or singing, other eyes from farther South look toward here, full of hope and worry at the same time.

Mexican mother; on your knees you rock the entire raza; there is no greater destiny than yours in our time.

I Stand Here Ironing

TILLIE OLSEN

I stand here ironing, and what you asked me moves tormented back and forth with the iron.

"I wish you would manage the time to come in and talk with me about your daughter. I'm sure you can help me understand her. She's a youngster who needs help and whom I'm deeply interested in helping."

"Who needs help." Even if I came, what good would it do? You think because I am her mother I have a key, or that in some way you could use me as a key? She has lived for nineteen years. There is all that life that has happened outside of me, beyond me.

And when is there time to remember, to sift, to weigh, to estimate, to total? I will start and there will be an interruption and I will have to gather it all together again. Or I will become engulfed with all I did or did not do, with what should have been and what cannot be helped.

She was a beautiful baby. The first and only one of our five that was beautiful at birth. You do not guess how new and uneasy her tenancy in her now-loveliness. You did not know her all those years she was thought homely, or see her poring over her baby pictures, making me tell her over and over how beautiful she had been—and would be, I would tell her—and was now, to the seeing eye. But the seeing eyes were few or non-existent. Including mine.

I nursed her. They feel that's important nowadays. I nursed all the children, but with her, with all the fierce rigidity of first motherhood, I did like the books then said. Though her cries battered me to trembling and my breasts ached with swollenness, I waited till the clock decreed.

Why do I put that first? I do not even know if it matters, or if it explains anything.

She was a beautiful baby. She blew

shining bubbles of sound. She loved motion, loved light, loved color and music and textures. She would lie on the floor in her blue overalls patting the surface so hard in ecstasy her hands and feet would blur. She was a miracle to me, but when she was eight months old I had to leave her daytimes with the woman downstairs to whom she was no miracle at all, for I worked or looked for work and for Emily's father, who "could no longer endure" (he wrote in his good-bye note) "sharing want with us."

I was nineteen. It was the pre-relief, pre-WPA world of the depression. I would start running as soon as I got off the streetcar, running up the stairs, the place smelling sour, and awake or asleep to startle awake, when she saw me she would break into a clogged weeping that could not be comforted, a weeping I can hear yet.

After a while I found a job hashing at night so I could be with her days, and it was better. But it came to where I had to bring her to his family and leave her.

It took a long time to raise the money for her fare back. Then she got chicken pox and I had to wait longer. When she finally came, I hardly knew her, walking quick and nervous like her father, looking like her father, thin, and dressed in a shoddy red that yellowed her skin and glared at the pockmarks. All the baby loveliness gone.

She was two. Old enough for nursery school they said, and I did not know then what I know now—the fatigue of the long day, and the lacerations of group life in nurseries that are only parking places for children.

Except that it would have made no difference if I had known. It was the only place there was. It was the only way we could be together, the only way I could hold a job.

And even without knowing, I knew. I knew the teacher that was evil because all these years it has curdled into my memory, the little boy hunched in the corner, her rasp, "why aren't you outside, because Alvin hits you? that's no reason, go out, scaredy." I knew Emily hated it even if she did not clutch and implore "don't go Mommy" like the other children, mornings.

She always had a reason why we should stay home. Momma, you look sick. Momma, I feel sick. Momma, the teachers aren't there today, they're sick. Momma, we can't go, there was a fire there last night. Momma, it's a holiday today, no school, they told me.

But never a direct protest, never rebellion. I think of our others in their three-, four-year-oldness—the explosions, the tempers, the denunciations, the demands—and I feel suddenly ill. I put the iron down. What in me demanded that goodness in her? And what was the cost, the cost to her of such goodness?

The old man living in the back once said in his gentle way: "You should smile at Emily more when you look at her." What *was* in my face when I looked at her? I loved her. There were all the acts of love.

It was only with the others I remembered what he said, and it was the face of joy, and not of care or tightness or worry I turned to them—too late for Emily. She does not smile easily, let alone almost always as her brothers and sisters do. Her face is closed and sombre, but when she wants, how fluid. You must have seen it in her pantomimes, you spoke of her rare gift for comedy on the stage that rouses a laughter out of the audience so dear they applaud and applaud and do not want to let her go.

Where does it come from, that comedy? There was none of it in her when she came back to me that second time, after I had had to send her away again. She had a new daddy now to learn to love, and I think perhaps it was a better time.

Except when we left her alone nights, telling ourselves she was old enough.

"Can't you go some other time, Mommy, like tomorrow?" she would ask. "Will it be just a little while you'll be gone? Do you promise?"

The time we came back, the front door open, the clock on the floor in the hall. She rigid awake. "It wasn't just a little while. I didn't cry. Three times I called you, just three times, and then I ran downstairs to open the door so you could come faster. The clock talked loud. I threw it away, it scared me what it talked."

She said the clock talked loud again that night I went to the hospital to have Susan. She was delirious with the fever that comes before red measles, but she was fully conscious all the week I was gone and the week after we were home when she could not come near the new baby or me.

She did not get well. She stayed skeleton thin, not wanting to eat, and night after night she had nightmares. She would call for me, and I would rouse from exhaustion to sleepily call back: "You're all right, darling, go to sleep, it's just a dream," and if she still called, in a sterner voice, "now go to sleep, Emily, there's nothing to hurt you." Twice, only twice, when I had to get up for Susan anyhow, I went in to sit with her.

Now when it is too late (as if she would let me hold and comfort her like I do the others) I get up and go to her at once at her moan or restless stirring. "Are you awake, Emily? Can I get you something?" And the answer is always the same: "No, I'm all right, go back to sleep, Mother."

They persuaded me at the clinic to send her away to a convalescent home in the country where "she can have the kind of food and care you can't manage for her, and you'll be free to concentrate on the new baby." They still send children to that place. I see pictures on the society page of sleek young women planning affairs to raise money for it, or dancing at the affairs, or decorating Easter eggs or filling Christmas stockings for the children.

They never have a picture of the children so I do not know if the girls still wear those gigantic red bows and the ravaged looks on the every other Sunday when parents can come to visit "unless otherwise notified"—as we were notified the first six weeks.

Oh it is a handsome place, green lawns and tall trees and fluted flower beds. High up on the balconies of each cottage the children stand, the girls in their red bows and white dresses, the boys in white suits and giant red ties. The parents stand below shrieking up to be heard and the children shriek down to be heard, and between them the invisible wall "Not To Be Contaminated by Parental Germs or Physical Affection."

There was a tiny girl who always stood hand in hand with Emily. Her parents never came. One visit she was gone. "They moved her to Rose College," Emily shouted in explanation. "They don't like you to love anybody here."

She wrote once a week, the labored writing of a seven-year-old. "I am fine. How is the baby. If I write my leter nicly I will have a star. Love." There never was a star. We wrote every other day, letters she could never hold or keep but only hear read—once. "We simply do not have room for children to keep any personal possessions," they patiently explained when we pieced one Sunday's shrieking together to plead how much it would mean to Emily, who loved so to keep things, to be allowed to keep her letters and cards.

Each visit she looked frailer. "She isn't eating," they told us.

(They had runny eggs for breakfast or mush with lumps, Emily said later, I'd hold it in my mouth and not swallow. Nothing ever tasted good, just when they had chicken.)

It took us eight months to get her released home, and only the fact that she gained back so little of her seven lost pounds convinced the social worker.

I used to try to hold and love her after she came back, but her body would stay stiff, and after a while she'd push away. She ate little. Food sickened her, and I think much of life too. Oh she had physical lightness and brightness, twinkling by on skates, bouncing like a ball up and down up and down over the jump rope, skimming over the hill; but these were momentary.

She fretted about her appearance, thin and dark and foreign-looking at a time when every little girl was supposed to look or thought she should look a chubby blonde replica of Shirley Temple. The doorbell sometimes rang for her, but no one seemed to come and play in the house or be a best friend. Maybe because we moved so much.

There was a boy she loved painfully through two school semesters. Months later she told me how she had taken pennies from my purse to buy him candy. "Licorice was his favorite and I brought him some every day, but he still liked Jennifer better'n me. Why, Mommy?" The kind of question for which there is no answer.

School was a worry to her. She was not glib or quick in a world where glibness and quickness were easily confused with ability to learn. To her overworked and exasperated teachers she was an overconscientious "slow learner" who kept trying to catch up and was absent entirely too often.

I let her be absent, though sometimes the illness was imaginary. How different from my now-strictness about attendance with the others. I wasn't working. We had a new baby, I was home anyhow. Sometimes, after Susan grew old enough, I would keep her home from school, too, to have them all together.

Mostly Emily had asthma, and her breathing, harsh and labored, would fill the house with a curiously tranquil sound. I would bring the two old dresser mirrors and her boxes of collections to her bed. She would select beads and single earrings, bottle tops and shells, dried flowers and pebbles, old postcards and scraps, all sorts of oddments; then she and Susan would play Kingdom, setting up landscapes and furniture, peopling them with action.

Those were the only times of peaceful companionship between her and Susan. I have edged away from it, that poisonous feeling between them, that terrible balancing of hurts and needs I had to do between the two, and did so badly, those earlier years.

Oh there are conflicts between the others too, each one human, needing, demanding, hurting, taking—but only between Emily and Susan, no, Emily toward Susan that corroding resentment. It seems so obvious on the surface, yet it is not obvious. Susan, the second child, Susan, golden- and curly-haired and chubby, quick and articulate and assured, everything in appearance and manner Emily was not; Susan, not able to resist Emily's precious things, losing or sometimes clumsily breaking them; Susan telling jokes and riddles to company for applause while Emily sat silent (to say to me later: that was *my* riddle, Mother, I told it to Susan); Susan, who for all the five years' difference in age was just a year behind Emily in developing physically.

I am glad for that slow physical development that widened the difference between her and her contemporaries, though she suffered over it. She was too vulnerable for that terrible world of youthful competition, of preening and parading, of constant measuring of yourself against every other, of envy, "If I had that copper hair," "If I had that skin. . . ." She tormented herself enough about not looking like the others, there was enough of the unsureness, the having to be conscious of words before you speak, the con-

stant caring—what are they thinking of me? without having it all magnified by the merciless physical drives.

Ronnie is calling. He is wet and I change him. It is rare there is such a cry now. That time of motherhood is almost behind me when the ear is not one's own but must always be racked and listening for the child cry, the child call. We sit for a while and I hold him, looking out over the city spread in charcoal with its soft aisles of light. "*Shoogily,*" he breathes and curls closer. I carry him back to bed, asleep. *Shoogily.* A funny word, a family word, inherited from Emily, invented by her to say: *comfort.*

In this and other ways she leaves her seal, I say aloud. And startle at my saying it. What do I mean? What did I start to gather together, to try and make cóherent? I was at the terrible, growing years. War years. I do not remember them well. I was working, there were four smaller ones now, there was not time for her. She had to help be a mother, and housekeeper, and shopper. She had to set her seal. Mornings of crisis and near hysteria trying to get lunches packed, hair combed, coats and shoes found, everyone to school or Child Care on time, the baby ready for transportation. And always the paper scribbled on by a smaller one, the book looked at by Susan then mislaid, the homework not done. Running out to that huge school where she was one, she was lost, she was a drop; suffering over the unpreparedness, stammering and unsure in her classes.

There was so little time left at night after the kids were bedded down. She would struggle over books, always eating (it was in those years she developed her enormous appetite that is legendary in our family) and I would be ironing, or preparing food for the next day, or writing V-mail to Bill, or tending the baby. Sometimes, to make me laugh, or out of her despair, she would imitate happenings or types at school.

I think I said once: "Why don't you do something like this in the school amateur show?" One morning she phoned me at work, hardly understandable through the weeping: "Mother, I did it. I won, I won; they gave me first prize; they clapped and clapped and wouldn't let me go."

Now suddenly she was Somebody, and as imprisoned in her difference as she had been in anonymity.

She began to be asked to perform at other high schools, even in colleges, then at city and statewide affairs. The first one we went to, I only recognized her that first moment when thin, shy, she almost drowned herself into the curtains. Then: Was this Emily? The control, the command, the convulsing and deadly clowning, the spell, then the roaring, stamping audience, unwilling to let this rare and precious laughter out of their lives.

Afterwards: You ought to do something about her with a gift like that —but without money or knowing how, what does one do? We have left it all to her, and the gift has as often eddied inside, clogged and clotted, as been used and growing.

She is coming. She runs up the stairs two at a time with her light graceful step, and I know she is happy tonight. Whatever it was that occasioned your call did not happen today.

"Aren't you ever going to finish the ironing, Mother? Whistler painted his mother in a rocker. I'd have to paint mine standing over an ironing board." This is one of her communicative nights and she tells me everything and nothing as she fixes herself a plate of food out of the icebox.

She is so lovely. Why did you want me to come in at all? Why were you concerned? She will find her way.

She starts up the stairs to bed. "Don't get me up with the rest in the morning." "But I thought you were having midterms." "Oh, those," she comes back in, kisses me, and says quite lightly, "in a couple of years when we'll all be atom-dead they won't matter a bit."

She has said it before. She *believes* it. But because I have been dredging the past, and all that compounds a human being is so heavy and meaningful in me, I cannot endure it tonight.

I will never total it all. I will never come in to say: She was a child seldom smiled at. Her father left me before she was a year old. I had to work her first six years when there was work, or I sent her home and to his relatives. There were years she had care she hated. She was dark and thin and foreign-looking in a world where the prestige went to blondeness and curly hair and dimples, she was slow where glibness was prized. She was a child of anxious, not proud, love. We were poor and could not afford for her the soil of easy growth. I was a young mother, I was a distracted mother. There were the other children pushing up, demanding. Her younger sister seemed all that she was not. There were years she did not want me to touch her. She kept too much in herself, her life was such she had to keep too much in herself. My wisdom came too late. She has much to her and probably nothing will come of it. She is a child of her age, of depression, of war, of fear.

Let her be. So all that is in her will not bloom—but in how many does it? There is still enough left to live by. Only help her to know—help make it so there is cause for her to know—that she is more than this dress on the ironing board, helpless before the iron.

from

Medea

EURIPIDES

CHARACTERS

NURSE OF MEDEA
ATTENDANT ON HER CHILDREN
MEDEA
CHORUS OF CORINTHIAN WOMEN
CREON, *King of Corinth*
JASON
AEGEUS, *King of Athens*
MESSENGER
THE TWO SONS OF JASON AND MEDEA

Scene:—Before MEDEA'S *house in Corinth, near the palace of* CREON. *The* NURSE *enters from the house.*

NURSE: . . . and in all things of her own accord was she at one with Jason, the greatest safeguard this when wife and husband do agree; but now their love is all turned to hate, and tenderest ties are weak. For Jason hath betrayed his own children and my mistress dear for the love of a royal bride, for he hath wedded the daughter of Creon, lord of this land. While Medea, his hapless wife, thus scorned, appeals to the oaths he swore, recalls the strong pledge his right hand gave, and bids heaven be witness what requital she is finding from Jason. And here she lies fasting, yielding her body to her grief, wasting away in tears ever since she learnt that she was wronged by her husband, never lifting her eye nor raising her face from off the ground; and she lends as deaf an ear to her friend's warning as if she were a rock or ocean billow, save when she turns her snow-white neck aside and softly to herself bemoans her father dear, her country and her home, which she gave up to come hither with the man who now holds her in dishonour. She, poor lady, hath by sad experience learnt how good a thing it is never to quit one's native land. And she hates her children now and feels no joy at seeing them; I fear she may contrive some untoward scheme; for her mood is dangerous nor will she brook her cruel treatment; full well I know her, and I much do dread that she will plunge the keen sword through their hearts, stealing without a word into the chamber where their marriage couch is spread, or else that she will slay the prince and bridegroom too, and so find some calamity still more grievous than the present; for dreadful is her wrath; verily the man that doth incur her hate will have no easy task to raise o'er her a song of triumph. Lo! where her sons come hither from their childish sports; little they reck of their mother's woes, for the soul of the young is no friend to sorrow.

The ATTENDANT *leads in* MEDEA's *children.*

ATTENDANT: Why dost thou, so long my lady's own handmaid, stand here at the gate alone, loudly lamenting to thyself the piteous tale? how comes it that Medea will have thee leave her to herself?

NURSE: Old man, attendant on the sons of Jason, our masters' fortunes when they go awry make good slaves grieve and touch their hearts. Oh! I have come to such a pitch of grief that there stole a yearning wish upon me to come forth hither and proclaim to heaven and earth my mistress's hard fate.

ATTENDANT: What! has not the poor lady ceased yet from her lamentation?

NURSE: Would I were as thou art! the mischief is but now beginning; it has not reached its climax yet.

ATTENDANT: O foolish one, if I may call my mistress such a name; how little she recks of evils yet more recent!

NURSE: What mean'st, old man? grudge not to tell me.

ATTENDANT: 'Tis naught; I do repent me even of the words I have spoken.

NURSE: Nay, by thy beard I conjure thee, hide it not from thy fellow-slave; I will be silent, if need be, on that text.

ATTENDANT: I heard one say, pretending not to listen as I approached the place where our greybeards sit playing draughts near Pirene's sacred spring, that Creon, the ruler of this land, is bent on driving these children and their mother from the boundaries of Corinth; but I know not whether the news is to be relied upon, and would fain it were not.

NURSE: What! will Jason brook such treatment of his sons, even though he be at variance with their mother?

ATTENDANT: Old ties give way to new; he bears no longer any love to this family.

NURSE: Undone, it seems, are we, if to old woes fresh ones we add, ere we have drained the former to the dregs.

ATTENDANT: Hold thou thy peace, say not a word of this; 'tis no time for our mistress to learn hereof.

NURSE: O children, do ye hear how your father feels towards you? Perdition catch him, but no! he is my master still; yet is he proved a very traitor to his nearest and dearest.

ATTENDANT: And who 'mongst men is not? Art learning only now, that every single man cares for himself more than for his neighbour, some from honest motives, others for

mere gain's·sake? seeing that to indulge his passion their father has
ceased to love these children.

NURSE: Go, children, within the
house; all will be well. Do thou keep
them as far away as may be, and
bring them not near their mother in
her evil hour. For ere this have I seen
her eyeing them savagely, as though
she were minded to do them some
hurt, and well I know she will not
cease from her fury till she have
pounced on some victim. At least
may she turn her hand against her
foes, and not against her friends.

MEDEA: *(chanting within)* Ah, me!
a wretched suffering woman I! O
would that I could die!

NURSE: *(chanting)* 'Tis as I said,
my dear children; wild fancies stir
your mother's heart, wild fury goads
her on. Into the house without delay,
come not near her eye, approach her
not, beware her savage mood, the fell
tempest of her reckless heart. In, in
with what speed ye may. For 'tis
plain she will soon redouble her
fury; that cry is but the herald of the
gathering storm-cloud whose lightning soon will flash; what will her
proud restless soul, in the anguish of
despair, be guilty of?

The ATTENDANT *takes the children
into the house.*

MEDEA: *(chanting within)* Ah, me!
the agony I have suffered, deep
enough to call for these laments!
Curse you and your father too, ye
children damned, sons of a doomed
mother! Ruin seize the whole family!

NURSE: *(chanting)* Ah me! ah me!
the pity of it! Why, pray, do thy children share their father's crime? Why
hatest thou them? Woe is you, poor
children, how do I grieve for you lest
ye suffer some outrage! Strange are

the tempers of princes, and maybe because they seldom have to obey, and
mostly lord it over others, change
they their moods with difficulty. 'Tis
better then to have been trained to
live on equal terms. Be it mine to
reach old age, not in proud pomp,
but in security! Moderation wins the
day first as a better word for men to
use, and likewise it is far the best
course for them to pursue; but great-
ness that doth o'erreach itself, brings
no blessing to mortal men; but pays
a penalty of greater ruin whenever
fortune is wroth with a family.

The CHORUS *enters. The following
lines between the* NURSE, CHORUS, *and*
MEDEA *are sung.*

CHORUS: I heard the voice, uplifted loud, of our poor Colchian
lady, nor yet is she quiet; speak, aged
dame, for as I stood by the house
with double gates I heard a voice of
weeping from within, and I do
grieve, lady, for the sorrows of this
house, for it hath won my love.

NURSE: 'Tis a house no more; all
that is passed away long since; a
royal bride keeps Jason at her side,
while our mistress pines away in her
bower, finding no comfort for her
soul in aught her friends can say.

MEDEA: *(within)* Oh, oh! Would
that Heaven's levin bolt would
cleave this head in twain! What gain
is life to me? Woe, woe is me! O, to
die and win release, quitting this
loathed existence!

CHORUS: Didst hear, O Zeus, thou
earth, and thou, O light, the piteous
note of woe the hapless wife is uttering? How shall a yearning for that
insatiate resting-place ever hasten for
thee, poor reckless one, the end that
death alone can bring? Never pray
for that. And if thy lord prefers a

fresh love, be not angered with him for that; Zeus will judge 'twixt thee and him herein. Then mourn not for thy husband's loss too much, nor waste thyself away.

MEDEA: *(within)* Great Themis, and husband of Themis, behold what I am suffering now, though I did bind that accursed one, my husband, by strong oaths to me! O, to see him and his bride some day brought to utter destruction, they and their house with them, for that they presume to wrong me thus unprovoked. O my father, my country, that I have left to my shame, after slaying my own brother.

NURSE: Do ye hear her words, how loudly she adjures Themis, oft invoked, and Zeus, whom men regard as keeper of their oaths? On no mere trifle surely will our mistress spend her rage.

CHORUS: Would that she would come forth for us to see, and listen to the words of counsel we might give, if haply she might lay aside the fierce fury of her wrath, and her temper stern. Never be my zeal at any rate denied my friends! But go thou and bring her hither outside the house, and tell her this our friendly thought; haste thee ere she do some mischief to those inside the house, for this sorrow of hers is mounting high.

* * *

As the CHORUS *finishes its song,* MEDEA *enters from the house.*

MEDEA: From the house I have come forth, Corinthian ladies, for fear lest you be blaming me . . . on me hath fallen this unforeseen disaster, and sapped my life; ruined I am, and long to resign the boon of existence, kind friends, and die. For he who was all the world to me, as well thou knowest, hath turned out the worst of men, my own husband. Of all things that have life and sense we women are the most hapless creatures; first must we buy a husband at a great price, and o'er ourselves a tyrant set which is an evil worse than the first; and herein lies the most important issue, whether our choice be good or bad. For divorce is not honourable to women, nor can we disown our lords. Next must the wife, coming as she does to ways and customs new, since she hath not learnt the lesson in her home, have a diviner's eye to see how best to treat the partner of her life. If haply we perform these tasks with thoroughness and tact, and the husband live with us, without resenting the yoke, our life is a happy one; if not, 'twere best to die. But when a man is vexed with what he finds indoors, he goeth forth and rids his soul of its disgust, betaking him to some friend or comrade of like age; whilst we must needs regard his single self.

And yet they say we live secure at home, while they are at the wars, with their sorry reasoning, for I would gladly take my stand in battle array three times o'er, than once give birth. But enough! this language suits not thee as it does me; thou hast a city here, a father's house, some joy in life, and friends to share thy thoughts, but I am destitute, without a city, and therefore scorned by my husband, a captive I from a foreign shore, with no mother, brother, or kinsman in whom to find a new haven of refuge from this calamity. Wherefore this one boon and only this I wish to win from thee,—

thy silence, if haply I can some way or means devise to avenge me on my husband for this cruel treatment, and on the man who gave to him his daughter, and on her who is his wife. For though a woman be timorous enough in all else, and as regards courage, a coward at the mere sight of steel, yet in the moment she finds her honour wronged, no heart is filled with deadlier thoughts than hers.

LEADER OF THE CHORUS: This will I do; for thou wilt be taking a just vengeance on thy husband, Medea. That thou shouldst mourn thy lot surprises me not. But lo! I see Creon, king of this land coming hither, to announce some new resolve.

CREON *enters, with his retinue.*

CREON: Hark thee, Medea, I bid take those sullen looks and angry thoughts against thy husband forth from this land in exile, and with thee take both thy children and that without delay, for I am judge in this sentence, and I will not return unto my house till I banish thee beyond the borders of the land.

MEDEA: Ah, me! now is utter destruction come upon me, unhappy that I am! For my enemies are bearing down on me full sail, nor have I any landing-place to come at in my trouble. Yet for all my wretched plight I will ask thee, Creon, wherefore dost thou drive me from the land?

CREON: I fear thee,—no longer need I veil my dread 'neath words,— lest thou devise against my child some cureless ill. Many things contribute to this fear of mine; thou art a witch by nature, expert in countless sorceries, and thou art chafing for the loss of thy husband's affec-

tion. I hear, too, so they tell me, that thou dost threaten the father of the bride, her husband, and herself with some mischief; wherefore I will take precautions ere our troubles come. For 'tis better for me to incur thy hatred now, lady, than to soften my heart and bitterly repent it hereafter.

MEDEA: Alas! this is not now the first time, but oft before, O Creon, hath my reputation injured me and caused sore mischief. . . . Fear me not, Creon, my position scarce is such that I should seek to quarrel with princes. Why should I, for how hast thou injured me? Thou has betrothed thy daughter where thy fancy prompted thee. No, 'tis my husband I hate, though I doubt not thou hast acted wisely herein. And now I grudge not thy prosperity; betroth thy child, good luck to thee, but let me abide in this land, for though I have been wronged I will be still and yield to my superiors.

CREON: Thy words are soft to hear, but much I dread lest thou art devising some mischief in thy heart, and less than ever do I trust thee now; for a cunning woman, and man likewise, is easier to guard against when quick-tempered than when taciturn. Nay, begone at once! speak me no speeches, for this is decreed, nor hast thou any art whereby thou shalt abide amongst us, since thou hatest me.

MEDEA: O, say not so! by thy knees and by thy daughter newly-wed, I do implore!

CREON: Thou wastest words; thou wilt never persuade me.

MEDEA: What, wilt thou banish me, and to my prayers no pity yield?

CREON: I will, for I love not thee above my own family.

MEDEA: O my country! what fond

memories I have of thee in this hour!

CREON: Yea, for I myself love my city best of all things save my children.

MEDEA: Ah me! ah me! to mortal man how dread a scourge is love!

CREON: That, I deem, is according to the turn our fortunes take.

MEDEA: O Zeus! let not the author of these my troubles escape thee.

CREON: Begone, thou silly woman, and free me from my toil.

MEDEA: The toil is mine, no lack of it.

CREON: Soon wilt thou be thrust out forcibly by the hand of servants.

MEDEA: Not that, not that, I do entreat thee, Creon!

CREON: Thou wilt cause disturbance yet, it seems.

MEDEA: I will begone; I ask thee now this boon to grant.

CREON: Why then this violence? why dost thou not depart?

MEDEA: Suffer me to abide this single day and devise some plan for the manner of my exile, and means of living for my children, since their father cares not to provide his babes therewith. Then pity them; thou too hast children of thine own; thou needs must have a kindly heart. For my own lot I care naught, though I an exile am, but for those babes I weep, that they should learn what sorrow means.

CREON: Mine is a nature anything but harsh; full oft by showing pity have I suffered shipwreck; and now albeit I clearly see my error, yet shalt thou gain this request, lady; but I do forewarn thee, if to-morrow's rising sun shall find thee and thy children within the borders of this land, thou diest; my word is spoken and it will not lie. So now, if abide thou must, stay this one day only, for in it thou canst not do any of the fearful deeds I dread. (CREON *and his retinue go out.*)

CHORUS: *(chanting)* Ah! poor lady, woe is thee! Alas, for thy sorrows! Whither wilt thou turn? What protection, what home or country to save thee from thy troubles wilt thou find? O Medea, in what a hopeless sea of misery heaven hath plunged thee!

MEDEA: On all sides sorrow pens me in. Who shall gainsay this? But all is not yet lost! think not so. Still are there troubles in store for the new bride, and for her bridegroom no light toil. Dost think I would ever have fawned on yonder man, unless to gain some end or form some scheme? Nay, I would not so much as have spoken to him or touched him with my hand. But he has in folly so far stepped in that, though he might have checked my plot by banishing me from the land, he hath allowed me to abide this day, in which I will lay low in death three of my enemies—a father and his daughter and my husband too. Now, though I have many ways to compass their death, I am not sure, friends, which I am to try first. Shall I set fire to the bridal mansion, or plunge the whetted sword through their hearts, softly stealing into the chamber where their couch is spread? One thing stands in my way. If I am caught making my way into the chamber, intent on my design, I shall be put to death and cause my foes to mock. 'Twere best to take the shortest way—the way we women are most skilled in —by poison to destroy them. Well, suppose them dead; what city will receive me? What friendly host will give me a shelter in his land, a home secure, and save my soul alive? None. So I will wait yet a little while in case some tower of defence rise up for me; then will I proceed to this bloody deed in crafty silence; but if some unexpected mischance drive me forth, I will with mine own hand seize the sword, e'en though I die for it, and slay them, and go forth on my bold path of daring. By that dread queen whom I revere before all others and have chosen to share my task, by Hecate who dwells within my inmost chamber, not one of them shall wound my heart and rue it not. Bitter and sad will I make their marriage for them; bitter shall be the wooing of it, bitter my exile from the land. Up, then, Medea, spare not the secrets of thy art in plotting and devising; on to the danger. Now comes a struggle needing courage. Dost see what thou art suffering? 'Tis not for thee to be a laughing-stock to the race of Sisyphus by reason of this wedding of Jason, sprung, as thou art, from a noble sire, and of the Sun-god's race. Thou hast cunning; and, more than this, we women, though by nature little apt for virtuous deeds, are most expert to fashion any mischief.

*   *   *

JASON *enters, alone.* MEDEA *comes out of the house.*

JASON: It is not now I first remark, but oft ere this, how unruly a pest is a harsh temper. For instance, thou, hadst thou but patiently endured the will of thy superiors, mightest have remained here in this land and house, but now for thy idle words wilt thou be banished. Thy words are naught to me. Cease not to call Jason basest of men; but for those words thou hast spoken against our rulers, count it all gain that exile is thy only punishment. I ever tried to check the outbursts of the angry monarch, and would have had thee stay, but thou wouldst not forego thy silly rage, always reviling our rulers, and so thou wilt be banished. Yet even after all this I weary not of my goodwill, but am come with thus much forethought, lady, that thou mayst not be destitute nor want for aught, when, with thy sons, thou art cast out. Many an evil doth exile bring in its train with it; for even though thou hatest me, never will I harbour hard thoughts of thee.

MEDEA: Thou craven villain (for that is the only name my tongue can find for thee, a foul reproach on thy unmanliness), comest thou to me, thou, most hated foe of gods, of me, and of all mankind? 'Tis no proof of courage or hardihood to confront thy friends after injuring them, but that worst of all human diseases—loss of shame. Yet hast thou done well to come; for I shall ease my soul by reviling thee, and thou wilt be vexed at my recital. I will begin at the very beginning. I saved thy life . . . I slew the dragon which guarded the golden fleece . . . Father and home of my free will I left and came with thee to Iolcos, 'neath Pelion's hills, for my love was stronger than my prudence. . . . All this have I done for thee, thou traitor! and thou hast cast me over, taking to thyself another wife, though children have been born to us. Hadst thou been childless still, I could have pardoned thy desire for this new union. Gone is now the trust I put in oaths. I cannot even understand whether thou thinkest that the gods of old no longer rule, or that fresh decrees are now in vogue amongst mankind, for thy conscience must tell thee thou hast not kept faith with me. Ah! poor right hand, which thou didst often grasp. These knees thou didst embrace! All in vain, I suffered a traitor to touch me! How short of my hopes I am fallen! But come, I will deal with thee as though thou wert my friend. Yet what kindness can I expect from one so base as thee? But yet I will do it, for my questioning will show thee yet more base. Whither can I turn me now? to my father's house, to my own country, which I for thee deserted to come hither? to the hapless daughters of Pelias? A glad welcome, I trow, would they give me in their home, whose father's death I compassed! My case stands even thus; I am become the bitter foe to those of mine own home, and those whom I need ne'er have wronged I have made mine enemies to pleasure thee. Wherefore to reward me for this thou has made me doubly blest in the eyes of many a wife in Hellas; and in thee I own a peerless, trusty lord. O woe is me, if indeed I am to be cast forth an exile from the land, without one friend; one lone woman with her babes forlorn! Yea, a fine reproach to thee in thy bridal hour, that thy children and the wife who saved thy life are beggars and vagabonds! O Zeus! why hast thou granted unto man clear signs to know the sham in gold, while on man's brow no brand is stamped whereby to gauge the villain's heart?

LEADER OF THE CHORUS: There is a something terrible and past all cure, when quarrels arise 'twixt those who are near and dear.

JASON: Needs must I now, it seems, turn orator, and, like a good helmsman on a ship with close-reefed sails, weather that wearisome tongue of thine. Now, I believe, since thou wilt exaggerate thy favours, that to Cypris alone of gods or men I owe the safety of my voyage. Thou hast a subtle wit enough; yet were it a hateful thing for me to say that the Love-god constrained thee by his resistless shaft to save my life. However, I will not reckon this too nicely; 'twas kindly done, however thou didst serve me. Yet for my safety hast thou received more than ever thou gavest, as I will show. First, thou dwellest in Hellas, instead of thy barbarian land, and hast learnt what justice means and how to live by law, not by the dictates of brute force; and all the Hellenes recognize thy cleverness, and thou hast gained a name; whereas, if thou hadst dwelt upon the confines of the earth, no tongue had mentioned thee. . . . As for the taunts thou urgest against my marriage with the princess, I will prove to thee, first, that I am prudent herein, next chastened in my love, and last a powerful friend to thee and to thy sons; only hold thy peace. Since I have here withdrawn from Iolcos with many a hopeless trouble at my back, what happier device could I, an exile, frame than marriage with the daughter of the king? 'Tis not because I loathe thee for my wife—the thought that rankles in thy heart; 'tis not because I am smitten with desire for a new bride, nor yet that I am eager to vie with others in begetting many children, for those we have are quite enough, and I do not complain. Nay, 'tis that we—and this is most important—may dwell in comfort, instead of suffering want (for well I know that every whilom friend avoids the poor), and that I might rear my sons as doth befit my house; further, that I might be the father of brothers for the children thou hast borne, and raise these to the same high rank, uniting the family in one,—to my lasting bliss. Thou, indeed, hast no need of more children, but me it profits to help my present family by that which is to be. Have I miscarried here? Not even thou wouldest say so unless a rival's charms rankled in thy bosom. No, but you women have such strange ideas, that you think all is well so long as your married life runs smooth; but if some mischance occur to ruffle your love, all that was good and lovely erst you reckon as your foes. Yea, men should have begotten children from some other source, no female race existing; thus would no evil ever have fallen on mankind.

LEADER: This speech, O Jason, hast thou with specious art arranged; but yet I think—albeit in speaking I am indiscreet—that thou hast sinned in thy betrayal of thy wife.

MEDEA: No doubt I differ from the mass of men on many points; for, to my mind, whoso hath skill to fence with words in an unjust cause, incurs the heaviest penalty; for such an one, confident that he can cast a decent veil of words o'er his injustice, dares to practise it; and yet he is not so very clever after all. So do not thou put forth thy specious pleas and clever words to me now, for one word of mine will lay thee low. Hadst thou not had a villain's heart, thou shouldst have gained my consent, then made this match, instead of hiding it from those who loved thee.

JASON: Thou wouldest have lent me ready aid, no doubt, in this proposal, if I had told thee of my marriage, seeing that not even now canst thou restrain thy soul's hot fury.

MEDEA: This was not what restrained thee; but thine eye was turned towards old age, and a foreign wife began to appear a shame to thee.

JASON: Be well assured of this: 'twas not for the woman's sake I wedded the king's daughter, my present wife; but, as I have already told thee, I wished to insure thy safety and to be the father of royal sons bound by blood to my own children—a bulwark to our house.

MEDEA: May that prosperity, whose end is woe, ne'er be mine, nor such wealth as would ever sting my heart!

JASON: Change that prayer as I will teach thee, and thou wilt show more wisdom. Never let happiness appear in sorrow's guise, nor, when thy fortune smiles, pretend she frowns!

MEDEA: Mock on; thou hast a place of refuge; I am alone, an exile soon to be.

JASON: Thy own free choice was this; blame no one else.

MEDEA: What did I do? Marry, then betray thee?

JASON: Against the king thou didst invoke an impious curse.

MEDEA: On thy house too maybe I bring the curse.

JASON: Know this, I will no further dispute this point with thee. But if thou wilt of my fortune somewhat take for the children or thyself to help thy exile, say on; for I am ready to grant it with ungrudging hand, yea and to send tokens to my friends elsewhere who shall treat thee well. If thou refuse this offer, thou wilt do a foolish deed, but if thou cease from anger the greater will be thy gain.

MEDEA: I will have naught to do with friends of thine, naught will I receive of thee, offer it not to me; a villain's gifts can bring no blessing.

JASON: At least I call the gods to witness, that I am ready in all things to serve thee and thy children, but thou dost scorn my favours and thrustest thy friends stubbornly away; wherefore thy lot will be more bitter still.

MEDEA: Away! By love for thy young bride entrapped, too long thou lingerest outside her chamber; go wed, for, if God will, thou shalt have such a marriage as thou wouldst fain refuse. (JASON *goes out.*)

* * *

MEDEA *has been seated in despair on her door-step during the choral song.* AEGEUS *and his attendants enter.*

AEGEUS: All hail, Medea! no man knoweth fairer prelude to the greeting of friends than this.

MEDEA: All hail to thee likewise, Aegeus, son of wise Pandion. Whence comest thou to this land?

AEGEUS: From Phoebus' ancient oracle.

MEDEA: What took thee on thy travels to the prophetic centre of the earth?

AEGEUS: The wish to ask how I might raise up seed unto myself.

MEDEA: Pray tell me, hast thou till now dragged on a childless life?

AEGEUS: I have no child owing to the visitation of some god.

MEDEA: Hast thou a wife, or hast thou never known the married state?

AEGEUS: I have a wife joined to me in wedlock's bond. . . . But why that downcast eye, that wasted cheek?

MEDEA: O Aegeus, my husband has proved most evil.

AEGEUS: What meanest thou? explain to me clearly the cause of thy despondency.

MEDEA: Jason is wronging me though I have given him no cause.

AEGEUS: What hath he done? tell me more clearly.

MEDEA: He is taking another wife to succeed me as mistress of his house.

AEGEUS: Can he have brought himself to such a dastard deed?

MEDEA: Be assured thereof; I, whom he loved of yore, am in dishonour now.

AEGEUS: Hath he found a new love? or does he loathe thy bed?

MEDEA: Much in love is he! A traitor to his friend is he become.

AEGEUS: Enough! if he is a villain as thou sayest.

MEDEA: The alliance he is so much enamoured of is with a princess.

AEGEUS: Who gives his daughter to him? go on, I pray.

MEDEA: Creon, who is lord of this land of Corinth.

AEGEUS: Lady, I can well pardon thy grief.

MEDEA: I am undone, and more than that, am banished from the land.

AEGEUS: By whom? fresh woe this word of thine unfolds.

MEDEA: Creon drives me forth in exile from Corinth.

AEGEUS: Doth Jason allow it? This too I blame him for.

MEDEA: Not in words, but he will not stand out against it. O, I implore thee by this beard and by thy knees, in suppliant posture, pity, O pity my sorrows; do not see me cast forth forlorn, but receive me in thy country, to a seat within thy halls. So may thy wish by heaven's grace be crowned with a full harvest of offspring, and may thy life close in happiness! Thou knowest not the rare good luck thou findest here, for I will make thy childlessness to cease and cause thee to beget fair issue; so potent are the spells I know.

AEGEUS: Lady, on many grounds I am most fain to grant thee this thy boon, first for the gods' sake, next for the children whom thou dost promise I shall beget; for in respect of this I am completely lost. 'Tis thus with me; if e'er thou reach my land, I will attempt to champion thee as I am bound to do. Only one warning I do give thee first, lady; I will not from this land bear thee away, yet if of thyself thou reach my halls, there shalt thou bide in safety and I will never yield thee up to any man. But from this land escape without my aid, for I have no wish to incur the blame of my allies as well.

MEDEA: It shall be even so; but wouldst thou pledge thy word to this, I should in all be well content with thee.

AEGEUS: Surely thou dost trust me? or is there aught that troubles thee?

* * *

MEDEA: Go in peace; all is well, and I with what speed I may, will to thy city come, when I have wrought my purpose and obtained my wish. (AEGEUS *and his retinue depart.*)

CHORUS: *(chanting)* May Maia's princely son go with thee on thy way to bring thee to thy home, and mayest thou attain that on which thy soul is set so firmly, for to my mind thou seemest a generous man, O Aegeus.

MEDEA: . . . But now will I explain to thee my plans in full; do not expect to hear a pleasant tale. A servant of mine will I to Jason send and crave an interview; then when he comes I will address him with soft words, say, "this pleases me," and, "that is well," even the marriage with the princess, which my treacherous lord is celebrating, and add "it suits us both, 'twas well thought out"; then will I entreat that here my children may abide, not that I mean to leave them in a hostile land for foes to flout, but that I may slay the king's daughter by guile. For I will send them with gifts in their hands, carrying them unto the bride to save them from banishment, a robe of finest wool and a chaplet of gold. And if these ornaments she take and put them on, miserably shall she die, and likewise everyone who touches her; with such fell poisons will I smear my gifts. And here I quit this theme; but I shudder at the deed I must do next; for I will slay the children I have borne; there is none shall take them from my toils; and when I have utterly confounded Jason's house I will leave the land, escaping punishment for my dear children's murder, after my most unholy deed. . . . Never shall he see again alive the children I bore to him, nor from his new bride shall he beget issue, for she must die a hideous death, slain by my drugs. Let no one deem me a poor weak woman who sits with folded hands, but of another mould, dangerous to foes and well-disposed to friends; for they win the fairest fame who live their life like me.

LEADER OF THE CHORUS: Since thou hast imparted this design to me, I bid thee hold thy hand, both from a wish to serve thee and because I would uphold the laws men make.

MEDEA: It cannot but be so; thy words I pardon since thou art not in the same sorry plight that I am.

LEADER: O lady, wilt thou steel thyself to slay thy children twain?

MEDEA: I will, for that will stab my husband to the heart.

LEADER: It may, but thou wilt be the saddest wife alive.

MEDEA: No matter; wasted is every word that comes 'twixt now and then. Ho! *(The* NURSE *enters in answer to her call.)* Thou, go call me Jason hither, for thee I do employ on every mission of trust. No word divulge of all my purpose, as thou art to thy mistress loyal and likewise of my sex. *(The* NURSE *goes out.)*

JASON *enters.*

JASON: I am come at thy bidding, for e'en though thy hate for me is bitter thou shalt not fail in this small boon, but I will hear what new request thou hast to make of me, lady.

MEDEA: Jason, I crave thy pardon for the words I spoke, and well thou mayest brook my burst of passion, for ere now we twain have shared much love. For I have reasoned with my soul and railed upon me thus, "Ah! poor heart! why am I thus dis-traught, why so angered 'gainst all good advice, why have I come to hate the rulers of the land, my husband too, who does the best for me he can, in wedding with a princess and rearing for my children noble brothers? Shall I not cease to fret? What possesses me, when heaven its best doth offer? Have I not my children to consider? do I forget that we are fugitives, in need of friends?" When I had thought all this I saw how foolish I had been, how senselessly enraged. So now I do commend thee and think thee most wise in forming this connection for us; but I was mad, I who should have shared in these designs, helped on thy plans, and lent my aid to bring about the match, only too pleased to wait upon thy bride. But what we are, we are, we women, evil I will not say; wherefore thou shouldst not sink to our sorry level nor with our weapons meet our childishness.

I yield and do confess that I was wrong then, but now have I come to a better mind. Come hither, my children, come, leave the house, step forth, and with me greet and bid farewell to your father, be reconciled from all past bitterness unto your friends, as now your mother is; for we have made a truce and anger is no more. *(The* ATTENDANT *comes out of the house with the children.)*

Take his right hand; ah me; my sad fate! when I reflect, as now, upon the hidden future. O my children, since there awaits you even thus a long, long life, stretch forth the hand to take a fond farewell. Ah me! how new to tears am I, how full of fear! For now that I have at last released me from my quarrel with your father, I let the tear-drops stream adown my tender cheek.

LEADER OF THE CHORUS: From my eyes too bursts forth the copious tear; O, may no greater ill than the present e'er befall!

JASON: Lady, I praise this conduct, not that I blame what is past; for it is but natural to the female sex to vent their spleen against a husband when he trafficks in other marriages besides his own. But thy heart is changed to wiser schemes and thou art determined on the better course, late though it be; this is acting like a woman of sober sense. And for you, my sons, hath your father provided with all good heed a sure refuge, by God's grace; for ye, I trow, shall with your brothers share hereafter the foremost rank in this Corinthian realm. Only grow up, for all the rest your sire and whoso of the gods is kind to us is bringing to pass. May I see you reach man's full estate, high o'er the heads of those I hate! But thou, lady, why with fresh tears dost thou thine eyelids wet, turning away thy wan cheek, with no welcome for these my happy tidings?

MEDEA: 'Tis naught; upon these children my thoughts were turned.

JASON: Then take heart; for I will see that it is well with them.

MEDEA: I will do so; nor will I doubt thy word; woman is a weak creature, ever given to tears.

JASON: Why prithee, unhappy one, dost moan o'er these children?

MEDEA: I gave them birth; and when thou didst pray long life for them, pity entered into my soul to think that these things must be. But the reason of thy coming hither to speak with me is partly told, the rest will I now mention. Since it is the pleasure of the rulers of the land to banish me, and well I know 'twere best for me to stand not in the way of thee or of the rulers by dwelling here, enemy as I am thought unto their house, forth from this land in exile am I going, but these children, —that they may know thy fostering hand, beg Creon to remit their banishment.

JASON: I doubt whether I can persuade him, yet must I attempt it.

MEDEA: At least do thou bid thy wife ask her sire this boon, to remit the exile of the children from this land.

JASON: Yea, that I will; and her methinks I shall persuade, since she is a woman like the rest.

MEDEA: I too will aid thee in this task, for by the children's hand I will send to her gifts that far surpass in beauty, I well know, aught that now is seen 'mongst men, a robe of finest tissue and a chaplet of chased gold. But one of my attendants must haste and bring the ornaments hither. *(A servant goes into the house.)* Happy shall she be not once alone but ten thousandfold, for in thee she wins the noblest soul to share her love, and gets these gifts as well which on a day my father's sire, the Sun-god, bestowed on his descendants. *(The servant returns and hands the gifts to the children.)* My children, take in your hands these wedding gifts, and bear them as an offering to the royal maid, the happy bride; for verily the gifts she shall receive are not to be scorned.

JASON: But why so rashly rob thyself of these gifts? Dost think a royal palace wants for robes or gold? Keep them, nor give them to another. For well I know that if my lady hold me in esteem, she will set my price above all wealth.

MEDEA: Say not so; 'tis said that

gifts tempt even gods; and o'er men's minds gold holds more potent sway than countless words. Fortune smiles upon thy bride, and heaven now doth swell her triumph; youth is hers and princely power; yet to save my children from exile I would barter life, not dross alone. Children, when we are come to the rich palace, pray your father's new bride, my mistress, with suppliant voice to save you from exile, offering her these ornaments the while; for it is most needful that she receive the gifts in her own hand. Now go and linger not; may ye succeed and to your mother bring back the glad tidings she fain would hear! (JASON, *the* ATTENDANT, *and the children go out together.*)

CHORUS: *(singing. Strophe I.)* Gone, gone is every hope I had that the children yet might live; forth to their doom they now proceed. The hapless bride will take, ay, take the golden crown that is to be her ruin; with her own hand will she lift and place upon her golden locks the garniture of death.

(Antistrophe 1.) Its grace and sheen divine will tempt her to put on the robe and crown of gold, and in that act will she deck herself to be a bride amid the dead. Such is the snare whereinto she will fall, such is the deadly doom that waits the hapless maid, nor shall she from the curse escape.

(Strophe 2.) And thou, poor wretch, who to thy sorrow art wedding a king's daughter, little thinkest of the doom thou art bringing on thy children's life, or of the cruel death that waits thy bride. Woe is thee! how art thou fallen from thy high estate!

(Antistrophe 2.) Next do I bewail thy sorrows, O mother hapless in thy children, thou who wilt slay thy babes because thou hast a rival, the babes thy husband hath deserted impiously to join him to another bride.

The ATTENDANT *enters with the children.*

ATTENDANT: Thy children, lady, are from exile freed, and gladly did the royal bride accept thy gifts in her own hands, and so thy children made their peace with her.

MEDEA: Ah!

ATTENDANT: Why art so disquieted in thy prosperous hour? Why turnest thou thy cheek away, and hast no welcome for my glad news?

MEDEA: Ah me!

ATTENDANT: These groans but ill accord with the news I bring.

MEDEA: Ah me! once more I say.

ATTENDANT: Have I unwittingly announced some evil tidings? Have I erred in thinking my news was good?

MEDEA: Thy news is as it is; I blame thee not.

ATTENDANT: Then why this downcast eye, these floods of tears?

MEDEA: Old friend, needs must I weep; for the gods and I with fell intent devised these schemes.

ATTENDANT: Be of good cheer; thou too of a surety shalt by thy sons be brought home again.

MEDEA: Ere that shall I bring others to their home, ah! woe is me!

ATTENDANT: Thou art not the only mother from thy children reft. Bear patiently thy troubles as a mortal must.

MEDEA: I will obey; go thou within the house and make the day's provision for the children. *(The* ATTENDANT *enters the house.* MEDEA *turns to the children.)* O my babes, my babes, ye have still a city and a home, where far from me and my sad lot you will live your lives, reft of your mother for ever; while I must to another land in banishment, or ever I have had my joy of you, or lived to see you happy, or ever I have graced your marriage couch, your bride, your bridal bower, or lifted high the wedding torch. Ah me! a victim of my own self-will. So it was all in vain I reared you, O my sons; in vain did suffer, racked with anguish, enduring the cruel pangs of childbirth. 'Fore Heaven I once had hope, poor me! high hope of ye that you would nurse me in my age and deck my corpse with loving hands, a boon we mortals covet; but now is my sweet fancy dead and gone; for I must lose you both and in bitterness and sorrow drag through life. And ye shall never with fond eyes see your mother more, for o'er your life there comes a change. Ah me! ah me! why do ye look at me so, my children? why smile that last sweet smile? Ah me! what am I to do? My heart gives way when I behold my children's laughing eyes. O, I cannot; farewell to all my former schemes; I will take the children from the land, the babes I bore. Why should I wound their sire by wounding them, and get me a twofold measure of sorrow? No, no, I will not do it. Farewell my scheming! And yet what possesses me? Can I consent to let those foes of mine escape from punishment, and incur their mockery? I must face this deed. Out upon my craven heart! to think that I should even have let the soft words escape my soul. Into the house, children! *(The children go into the house.)* And whoso feels he must not be

present at my sacrifice, must see to it himself; I will not spoil my handiwork. Ah! ah! do not, my heart, O do not do this deed! Let the children go, unhappy one, spare the babes! For if they live, they will cheer thee in our exile there. Nay, by the fiends of hell's abyss, never, never will I hand my children over to their foes to mock and flout. Die they must in any case, and since 'tis so, why I, the mother who bore them, will give the fatal blow. In any case their doom is fixed and there is no escape. Already the crown is on her head, the robe is round her, and she is dying, the royal bride; that do I know full well. But now since I have a piteous path to tread, and yet more piteous still the path I send my children on, fain would I say farewell to them. *(The children come out at her call. She takes them in her arms.)* O my babes, my babes, let your mother kiss your hands. Ah! hands I love so well, O lips most dear to me! O noble form and features of my children, I wish ye joy, but in that other land, for here your father robs you of your home. O the sweet embrace, the soft young cheek, the fragrant breath! my children! Go, leave me; I cannot bear to longer look upon ye; my sorrow wins the day. At last I understand the awful deed I am to do; but passion, that cause of direst woes to mortal man, hath triumphed o'er my sober thoughts. *(She goes into the house with the children.)*

CHORUS: *(chanting)* Oft ere now have I pursued subtler themes and have faced graver issues than woman's sex should seek to probe; but then e'en we aspire to culture, which dwells with us to teach us wisdom; I say not all; for small is the class amongst women— (one maybe shalt thou find 'mid many) —that is not incapable of wisdom. And amongst mortals I do assert that they who are wholly without experience and have never had children far surpass in happiness those who are parents. The childless, because they have never proved whether children grow up to be a blessing or curse to men are removed from all share in many troubles; whilst those who have a sweet race of children growing up in their houses do wear away, as I perceive, their whole life through; first with the thought how they may train them up in virtue, next how they shall leave their sons the means to live; and after all this 'tis far from clear whether on good or bad children they bestow their toil. But one last crowning woe for every mortal man I now will name; suppose that they have found sufficient means to live, and seen their children grow to man's estate and walk in virtue's path, still if fortune so befall, comes Death and bears the children's bodies off to Hades. Can it be any profit to the gods to heap upon us mortal men beside our other woes this further grief for children lost, a grief surpassing all?

MEDEA *comes out of the house.*

MEDEA: Kind friends, long have I waited expectantly to know how things would at the palace chance. And lo! I see one of Jason's servants coming hither, whose hurried gasps for breath proclaim him the bearer of some fresh tidings.

A MESSENGER *rushes in.*

MESSENGER: Fly, fly, Medea! who hast wrought an awful deed, transgressing every law; nor leave behind or sea-borne bark or car that scours the plain.

MEDEA: Why, what hath chanced that calls for such a flight of mine?

MESSENGER: The princess is dead, a moment gone, and Creon too, her sire, slain by those drugs of thine.

MEDEA: Tidings most fair are thine! Henceforth shalt thou be ranked amongst my friends and benefactors.

MESSENGER: Ha! What? Art sane? Art not distraught, lady, who hearest with joy the outrage to our royal house done, and art not at the horrid tale afraid?

MEDEA: Somewhat have I, too, to say in answer to thy words. Be not so hasty, friend, but tell the manner of their death, for thou wouldst give me double joy, if so they perished miserably.

MESSENGER: When the children twain whom thou didst bear came with their father and entered the palace of the bride, right glad were we thralls who had shared thy griefs, for instantly from ear to ear a rumour spread that thou and thy lord had made up your former quarrel. One kissed thy children's hands, another their golden hair, while I for very joy went with them in person to the women's chambers. Our mistress, whom now we do revere in thy room, cast a longing glance at Jason, ere she saw thy children twain; but then she veiled her eyes and turned her blanching cheek away, disgusted at their coming; but thy husband tried to check his young bride's angry humour with these words: "O, be not angered 'gainst thy friends; cease from wrath and turn once more thy face this way, counting as friends whomso thy husband counts, and accept these gifts, and for my sake crave thy sire to remit these children's exile." Soon as she saw the ornaments, no longer she held out, but yielded to her lord in all; and ere the father and his sons were far from the palace gone, she took the broidered robe and put it on, and set the golden crown about her tresses, arranging her hair at her bright mirror, with many a happy smile at her breathless counterfeit. Then rising from her seat she passed across the chamber, tripping lightly on her fair white foot, exulting in the gift, with many a glance at her uplifted ankle. When lo! a scene of awful horror did ensue. In a moment she turned pale, reeled backwards, trembling in every limb, and sinks upon a seat scarce soon enough to save herself from falling to the ground. An aged dame, one of her company, thinking belike it was a fit from Pan or some god sent, raised a cry of prayer, till from her mouth she saw the foam-flakes issue, her eyeballs rolling in their sockets, and all the blood her face desert; then did she raise a loud scream far different from her former cry. Forthwith one handmaid rushed to her father's house, another to her new bridegroom to tell his bride's sad fate, and the whole house echoed with their running to and fro. By this time would a quick walker have made the turn in a course of six plethra and reached the goal, when she with one awful shriek awoke, poor sufferer, from her speechless trance and oped her closed eyes, for against her a twofold anguish was warring. The chaplet of gold about her head was sending forth a wondrous stream of ravening flame, while the fine raiment, thy children's gift, was preying on the hapless maiden's fair white flesh; and she starts from her seat in a blaze and seeks to fly, shaking her hair and head this way and that, to cast the crown therefrom; but the gold held firm to its fastenings, and the flame, as she shook her locks, blazed forth the more with double fury. Then to the earth she sinks, by the cruel blow o'ercome; past all recognition now save to a father's eye; for her eyes had lost their tranquil gaze, her face no more its natural look preserved, and from the crown of her head blood and fire in mingled stream ran down; and from her bones the flesh kept peeling off beneath the gnawing of those secret drugs, e'en as when the pine-tree weeps its tears of pitch, a fearsome sight to see. And all were afraid to touch the corpse, for we were warned by what had chanced. Anon came her hapless father unto the house, all unwitting of her doom, and stumbles o'er the dead, and loud he cried, and folding his arms about her kissed her, with words like these the while, "O my poor, poor child, which of the gods hath destroyed.thee thus foully? Who is robbing me of thee, old as I am and ripe for death? O my child, alas! would I come die with thee!" He ceased his sad lament, and would have raised his aged frame, but found himself held fast by the fine-spun robe as ivy that clings to the branches of the bay, and then ensued a fearful struggle. He strove to rise, but she still held him back; and if ever he pulled with all his might, from off his bones his aged flesh he tore. At last he gave it up, and breathed forth his soul in awful suffering; for he could no longer master the pain. So there they lie, daughter and aged sire, dead side by side, a grievous sight that calls for tears.

MEDEA: My friends, I am resolved upon the deed; at once will I slay my children and then leave this land, without delaying long enough to hand them over to some more savage hand to butcher. Needs must they die in any case; and since they must, I will slay them—I, the mother that bare them. O heart of mine, steel thyself! Why do I hesitate to do the awful deed that must be done? Come, take the sword, thou wretched hand of mine! Take it, and advance to the post whence starts thy life of sorrow! Away with cowardice! Give not one thought to thy babes, how dear they are or how thou art their mother. This one brief day forget thy children dear, and after that lament; for though thou wilt slay them yet they were thy darlings still, and I am a lady of sorrows. (MEDEA *enters the house.*)

* * *

FIRST SON: *(within)* Ah, me; what can I do? Whither fly to escape my mother's blows?

SECOND SON: *(within)* I know not, sweet brother mine; we are lost.

CHORUS: *(chanting)* Didst hear, didst hear the children's cry? O lady, born to sorrow, victim of an evil fate! Shall I enter the house? For the children's sake I am resolved to ward off the murder.

FIRST SON: *(within)* Yea, by heaven I adjure you; help, your aid is needed.

SECOND SON: *(within)* Even now the toils of the sword are closing round us.

CHORUS: *(chanting)* O hapless mother, surely thou hast a heart of stone or steel to slay the offspring of thy womb by such a murderous doom. Of all the wives of yore I know but one who laid her hand upon her children dear, even Ino, whom the gods did madden in the day that the wife of Zeus drove her wandering from her home. But she, poor sufferer, flung herself into the sea because of the foul murder of her children, leaping o'er the wave-beat cliff, and in her death was she united to her children twain. Can there be any deed of horror left to follow this? Woe for the wooing of women fraught with disaster! What sorrows hast thou caused for men ere now!

JASON *and his attendants enter.*

JASON: Ladies, stationed near this house, pray tell me is the author of these hideous deeds, Medea, still within, or hath she fled from hence? For she must hide beneath the earth or soar on wings towards heaven's vault, if she would avoid the vengeance of the royal house. Is she so sure she will escape herself unpunished from this house, when she hath slain the rulers of the land? But enough of this! I am forgetting her children. As for her, those whom she hath wronged will do the like by her; but I am come to save the children's life, lest the victim's kin visit their wrath on me, in vengeance for the murder foul, wrought by my children's mother.

LEADER OF THE CHORUS: Unhappy man, thou knowest not the full extent of thy misery, else had thou never said those words.

JASON: How now? Can she want to kill me too?

LEADER: Thy sons are dead; slain by their own mother's hand.

JASON: O God; what sayest thou? Woman, thou hast sealed my doom.

LEADER: Thy children are no more; be sure of this.

JASON: Where slew she them; within the palace or outside?

LEADER: Throw wide the doors and see thy children's murdered corpses.

JASON: Haste, ye slaves, loose the bolts, undo the fastenings, that I may see the sight of twofold woe, my murdered sons and her, whose blood in vengeance I will shed.

MEDEA *appears above the house, on a chariot drawn by dragons; the children's corpses are beside her.*

MEDEA: Why shake those doors and attempt to loose their bolts, in quest of the dead and me their murderess? From such toil desist. If thou wouldst aught with me, say on, if so thou wilt; but never shalt thou lay hand on me, so swift the steeds the sun, my father's sire, to me doth give to save me from the hand of my foes.

JASON: Accursed woman! by gods, by me and all mankind abhorred as never woman was, who hadst the heart to stab thy babes, thou their mother, leaving me undone and childless; this hast thou done and still dost gaze upon the sun and earth after this deed most impious. Curses on thee! I now perceive what then I missed in the day I brought thee, fraught with doom, from thy home in a barbarian land to dwell in Hellas, traitress to thy sire and to the land that nurtured thee. On me the gods have hurled the curse that dogged thy steps, for thou didst slay thy brother at his hearth ere thou cam'st aboard our fair ship, Argo. Such was the outset of thy life of crime; then didst thou wed with me, and having borne me sons to glut thy passion's lust, thou now hast slain them. Not one amongst the wives of Hellas e'er had dared this deed; yet before them all I chose thee for my wife, wedding a foe to be my doom, no woman, but a lioness fiercer than Tyrrhene Scylla in nature. But with reproaches heaped a thousandfold I cannot wound thee, so brazen is thy nature. Perish, vile sorceress, murderess of thy babes! Whilst I must mourn my luckless fate, for I shall ne'er enjoy my new-found bride, nor shall I have the children, whom I bred and reared, alive to say the last farewell to me; nay, I have lost them.

MEDEA: To this thy speech I could have made a long reply, but Father Zeus knows well all I have done for thee, and the treatment thou hast given me. Yet thou wert not ordained to scorn my love and lead a life of joy in mockery of me, nor was thy royal bride nor Creon, who gave thee a second wife, to thrust me from this land and rue it not. Wherefore, if thou wilt, call me e'en a lioness, and Scylla, whose home is in the Tyrrhene land; for I in turn have wrung thy heart, as well I might.

JASON: Thou, too, art grieved thyself, and sharest in my sorrow.

MEDEA: Be well assured I am; but it relieves my pain to know thou canst not mock at me.

JASON: O my children, how vile a mother ye have found!

MEDEA: My sons, your father's feeble lust has been your ruin!

JASON: 'Twas not my hand, at any rate, that slew them.

MEDEA: No, but thy foul treatment of me, and thy new marriage.

JASON: Didst think that marriage cause enough to murder them?

MEDEA: Dost think a woman counts this a trifling injury?

JASON: So she be self-restrained; but in thy eyes all is evil.

MEDEA: Thy sons are dead and gone. That will stab thy heart.

JASON: They live, methinks, to bring a curse upon thy head.

MEDEA: The gods know, whoso of them began this troublous coil.

JASON: Indeed, they know that hateful heart of thine.

MEDEA: Thou art as hateful. I am aweary of thy bitter tongue.

JASON: And I likewise of thine. But parting is easy.

MEDEA: Say how; what am I to do? for I am fain as thou to go.

JASON: Give up to me those dead, to bury and lament.

MEDEA: No, never! I will bury them myself, bearing them to Hera's sacred field, who watches o'er the Cape, that none of their foes may insult them by pulling down their tombs; and in this land of Sisyphus I will ordain hereafter a solemn feast and mystic rites to atone for this impious murder. Myself will now to the land of Erechtheus, to dwell with Aegeus, Pandion's son. But thou, as well thou mayst, shalt die a caitiff's death, thy head crushed 'neath a shattered relic of Argo, when thou hast seen the bitter ending of my marriage.

JASON: The curse of our sons' avenging spirit and of Justice, that calls for blood, be on thee!

MEDEA: What god or power divine hears thee, breaker of oaths and every law of hospitality?

JASON: Fie upon thee! cursed witch! child-murderess!

MEDEA: To thy house! go, bury thy wife.

JASON: I go, bereft of both my sons.

MEDEA: Thy grief is yet to come; wait till old age is with thee too.

JASON: O my dear, dear children!

MEDEA: Dear to their mother, not to thee.

JASON: And yet thou didst slay them?

MEDEA: Yea, to vex thy heart.

JASON: One last fond kiss, ah me! I fain would on their lips imprint.

MEDEA: Embraces now, and fond farewells for them; but then a cold repulse!

JASON: By heaven I do adjure thee, let me touch their tender skins.

MEDEA: No, no! in vain this word has sped its flight.

JASON: O Zeus, dost hear how I am driven hence; dost mark the treatment I receive from this she-lion, fell murderess of her young? Yet so far as I may and can, I raise for them a dirge, and do adjure the gods to witness how thou hast slain my sons, and wilt not suffer me to embrace or bury their dead bodies. Would I had never begotten them to see thee slay them after all! (*The chariot carries* MEDEA *away.*)

La Llorona in Mexico

Maria del Refugio

According to our ancestors, there existed a story called la llorona. This was during colonial times, or rather during the conquest, when Mexico was still called New Spain, when Hernan Cortés. . . .

During that time, for his work among the Indians, Cortés had as an interpreter a beautiful Indian woman. Her real name was Marina but she was nicknamed La Malinche. As time passed, Cortés began to notice her since she was so beautiful. She attracted him so much that love blossomed between them and soon they had a son, who was baptised with the name of Hernando. Time passed, the child grew, and when Cortés had no further business here, he wanted to return to Spain, but he didn't want to abandon his son who had caused him so many difficulties. He insisted so much that La Malinche (as she was called) got ill from worrying that her son would be taken to Spain. She became insane and was put away in a well-guarded dungeon. Since her malady was only temporary, she soon regained her health and was set free. Cortés continued with his plan but he was never successful because he didn't want her for a wife; she was always his mistress.

He wanted his son, knowing that Spanish blood ran in his veins. Cortés couldn't take Marina with him since she was an Indian. When their son was seven years old, Cortés couldn't wait any longer. He said that he would take the child through agreement or by force. Marina not knowing what to do, took their child in her arms and brought him to the balcony. From her breast she withdrew an obsidian knife and told

Cortés: "If you want your son, here he is, but you shall not take him alive." And she plunged the sharp knife into the child's breast, piercing his heart. After making sure that her son was dead, she stabbed herself, mother and child lying in a heap on the floor. Afterwards her soul escaped from her body issuing a lament . . . "¡Aaayyy!"

According to history, from that time, her spirit roams everywhere at night, catching people's attention by her painful lament. People call her "la llorona."

part **3**

Extending...

Jerome Myers. *An Interlude.*

The Family Meadow

JOHN UPDIKE

The family always reconvenes in the meadow. For generations it has been traditional, this particular New Jersey meadow, with its great walnut tree making shade for the tables and its slow little creek where the children can push themselves about in a rowboat and nibble watercress and pretend to fish. Early this morning, Uncle Jesse came down from the stone house that his father's father's brother had built and drove the stakes, with their carefully tied rag flags, that would tell the cars where to park. The air was still, inert with the postdawn laziness that foretells the effort of a hot day, and between blows of his hammer Jesse heard the breakfast dishes clinking beneath the kitchen window and the younger collie barking behind the house. A mild man, Jesse moved scrupulously, mildly through the wet grass that he had scythed yesterday. The legs of

his gray workman's pants slowly grew soaked with dew and milkweed spittle. When the stakes were planted, he walked out the lane with the REUNION signs, past the houses. He avoided looking at the houses, as if glancing into their wide dead windows would wake them.

By nine o'clock Henry has come up from Camden with a carful—Eva, Mary, Fritz, Fred, the twins, and, incredibly, Aunt Eula. It is incredible she is still alive, after seven strokes. Her shrivelled head munches irritably and her arms twitch, trying to shake off assistance, as if she intends to dance. They settle her in an aluminum chair beneath the walnut tree. She faces the creek, and the helpless waggle of her old skull seems to establish itself in sympathy with the oscillating shimmer of the sunlight on the slow water. The men, working in silent pairs whose unison is as profound as blood, carry down the tables from the barn, where they are stacked from one year to the next. In truth, it has been three summers since the last reunion, and it was feared that there might never be another. Aunt Jocelyn, her gray hair done up in braids, comes out of her kitchen to say hello on the dirt drive. Behind her lingers her granddaughter, Karen, in white Levis and bare feet, with something shadowy and doubtful about her dark eyes, as if she had been intensely watching television. The girl's father—not here; he is working in Philadelphia—is Italian, and as she matures an alien beauty estranges her, so that during her annual visits to her grandparents' place, which when she was a child had seemed to her a green island, it is now she herself, at thirteen, who seems the island. She feels surrounded by the past, cut off from the images—a luncheonette, a civic swimming pool, an auditorium festooned with crêpe paper—that represent life to her, the present, her youth. The air around her feels brown, as in old photographs. These men greeting her seem to have stepped from an album. The men, remembering their original prejudice against her mother's marrying a Catholic, are especially cordial to her, so jovially attentive that Jocelyn suddenly puts her arm around the girl, expressing a strange multitude of things; that she loves her, that she is one of them, that she needs to be shielded, suddenly, from the pronged kidding of men.

By ten-thirty Horace's crowd has come down from Trenton, and the Oranges clan is arriving, in several cars. The first car says it dropped Cousin Claude in downtown Burlington because he was sure that the second car, which had faded out of sight behind them, needed to be told the way. The second car, with a whoop of hilarity, says it took the bypass and never saw him. He arrives in a third car, driven by Jimmy and Ethel Thompson from Morristown, who say they saw his forlorn figure standing along Route 130 trying to thumb a ride and as they were passing him Ethel cried, "Why, I think that's Claude." Zealous and reckless, a true believer in good deeds, Claude is always getting into scrapes like this, and enjoying it. He stands surrounded by laughing women, a typical man of this family, tall, with a tribal boyishness, a stubborn refusal to look his age, to lose his hair. Though his face is pitted and gouged by melancholy, Claude looks closer to forty than the sixty he is, and, though he works in Newark, he still speaks with the rural softness and slide of middle New Jersey. He has the gift—the privilege—of making these women laugh; the women uniformly run to fat and their laughter has a sameness, a quality both naïve and merciless, as if laughter meant too much to them. Jimmy and Ethel Thompson, whose name is not the family name, stand off to one side, in the unscythed grass, a fragile elderly couple whose links to family have all died away but who come because they received a mimeographed postcard inviting them. They are like those isolated corners of interjections and foreign syllables in a poorly planned crossword puzzle.

The twins bring down from the barn the horseshoes and the quoits. Uncle Jesse drives the stakes and pegs in the places that, after three summers, still show as spots of depressed sparseness in the grass. The sun, reaching toward noon, domineers over the meadow; the shade of the walnut tree grows smaller and more noticeably cool. By noon, all have arrived, including the Dodge station wagon from central Pennsylvania, the young pregnant Wilmington cousin who married an airline pilot, and the White Plains people, who climb from their car looking like clowns, wearing red-striped shorts and rhinestone-studded sunglasses. Handshakes are exchanged that feel to one man like a knobbed wood carving and to the other like a cow's slippery, unresisting teat. Women kiss, kiss stickily, with little overlapping patches of adhesive cheek and clicking conflicts of spectacle rims, under the white unslanting sun. The very insects shrink toward the shade. The eating begins. Clams steam, corn steams, salad wilts, butter runs, hot dogs turn, torn chicken shines in the savage light. Iced tea, brewed in forty-quart milk cans, chuckles when sloshed. Paper plates buckle on broad laps. Plastic butter knives, asked to cut cold ham, refuse. Children underfoot in the pleased frenzy eat only potato chips. Somehow, as the first wave of appetite subsides, the long tables turn musical, and a murmur rises to the blank sky, a cackle rendered harmonious by a remote singleness of ancestor; a kind of fabric is woven and hung, a tapestry of the family fortunes, the threads of which include milkmen, ministers, mailmen, bankruptcy, death by war, death by automobile, insanity—a strangely prevalent thread, the thread of insanity. Never far from a farm or the memory of a farm, the family has

hovered in honorable obscurity, between poverty and wealth, between jail and high office. Real-estate dealers, schoolteachers, veterinarians are its noblemen; butchers, electricians, door-to-door salesmen its yeomen. Protestant, teetotalling, and undaring, ironically virtuous and mildly proud, it has added to America's statistics without altering their meaning. Whence, then, this strange joy?

Watermelons smelling of childhood cellars are produced and massively sliced. The sun passes noon and the shadows relax in the intimate grass of this antique meadow. To the music of reminiscence is added the rhythmic chunking of thrown quoits. They are held curiously, between a straight thumb and four fingers curled as a unit, close to the chest, and thrown with a soft constrained motion that implies realms of unused strength. The twins and the children, as if superstitiously, have yielded the game to the older men, Fritz and Ed, Fred and Jesse, who, in pairs, after due estimation and measurement of the fall, pick up their four quoits, clink them together to clean them, and alternately send them back through the air on a high arc, floating with a spin-held slant like that of gyroscopes. The other pair measures, decides, and stoops. When they tap their quoits together, decades fall away. Even their competitive crowing has something measured about it, something patient, like the studied way their shirtsleeves are rolled up above their elbows. The backs of their shirts are ageless. Generations have sweated in just this style, under the arms, across the shoulder blades, and wherever the suspenders rub. The younger men and the teen-age girls play a softball game along the base paths that Jesse has scythed. The children discover the rowboat and, using the oars as poles, bump from bank to bank. When they dip their hands into the calm brown water, where no fish lives, a mother watching from beneath the walnut tree shrieks, "Keep your hands inside the boat! Uncle Jesse says the creek's polluted!"

And there is a stagnant fragrance the lengthening afternoon strains from the happy meadow. Aunt Eula nods herself asleep, and her false teeth slip down, so her face seems mummified and the children giggle in terror. Flies, an exploding population, discover the remains of the picnic and skate giddily on its odors. The softball game grows boring, except to the airline pilot, a rather fancy gloveman excited by the admiration of Cousin Karen in her tight white Levis. The Pennsylvania and New York people begin to pack their cars. The time has come for the photograph. Their history is kept by these photographs of timeless people in changing costumes standing linked and flushed in a moment of midsummer heat. All line up, from resurrected Aunt Eula, twitching and snapping like a mud turtle, to the unborn baby in the belly of the Delaware cousin. To get them all in, Jesse has to squat, but in doing so he brings the houses into his viewfinder. He does not want them in the picture, he does not want them there at all. They surround his meadow on three sides, raw ranch shacks built from one bastard design but painted in a patchwork of pastel shades. Their back yards, each nurturing an aluminum clothes tree, come right to the far bank of the creek, polluting it, and though a tall link fence holds back the children who have gathered in these yards to watch the picnic as if it were a circus or a zoo, the stare of the houses—mismatched kitchen windows squinting above the gaping cement mouth of a garage—cannot be held back. Not only do they stare, they speak, so that Jesse can hear them even at night. *Sell,* they say. *Sell.*

from "Going to Meet the Man"

Sonny's Blues

JAMES BALDWIN

Mama tried to tell me something about this, just after Daddy died. I was home on leave from the army.

This was the last time I ever saw my mother alive. Just the same, this picture gets all mixed up in my mind with pictures I have of her when she was younger. The way I always see her is the way she used to be on a Sunday afternoon, say, when the old folks were talking after the big Sunday dinner. I always see her wearing pale blue. She'd be sitting on the sofa. And my father would be sitting in the easy chair, not far from her. And the living room would be full of church folks and relatives. There they sit, in chairs all around the living room, and the night is creeping up outside, but nobody knows it yet. You can see the darkness growing against the windowpanes and you hear the street noises every now and again, or maybe the jangling beat of a tambourine from one of the churches close by, but it's real quiet in the room. For a moment nobody's talking, but every face looks darkening, like the sky outside. And my mother rocks a little from the waist, and my father's eyes are closed. Everyone is looking at something a child can't see. For a minute they've forgotten the children. Maybe a kid is lying on the rug, half asleep. Maybe somebody's got a kid in his lap and is absent-mindedly stroking the kid's head. Maybe there's a kid, quiet and big-eyed, curled up in a big chair in the corner. The silence, the darkness coming, and the darkness in the faces frightens the child obscurely. He hopes that the hand which strokes his forehead will never stop—will never die. He hopes that there will never come a time when the old folks won't be sitting around the living room, talking about where they've come from, and what they've seen, and what's happened to them and their kinfolk.

But something deep and watchful in the child knows that this is bound to end, is already ending. In a moment someone will get up and turn on the light. Then the old folks will remember the children and they won't talk any more that day. And when light fills the room, the child is filled with darkness. He knows that every time this happens he's moved just a little closer to that darkness outside. The darkness outside is what the old folks have been talking about. It's what they've come from. It's what they endure. The child knows that they won't talk any more because if he knows too much about what's happened to *them,* he'll know too much too soon, about what's going to happen to *him.*

The last time I talked to my mother, I remember I was restless, I wanted to get out and see Isabel. We weren't married then and we had a lot to straighten out between us.

There Mama sat, in black, by the window. She was humming an old church song, *Lord, you brought me from a long ways off.* Sonny was out somewhere. Mama kept watching the streets.

"I don't know," she said, "if I'll ever see you again, after you go off from here. But I hope you'll remember the things I tried to teach you."

"Don't talk like that," I said, and smiled. "You'll be here a long time yet."

She smiled, too, but she said nothing. She was quiet for a long time. And I said, "Mama, don't you worry about nothing. I'll be writing all

the time, and you be getting the checks. . . ."

"I want to talk to you about your brother," she said, suddenly. "If anything happens to me he ain't going to have nobody to look out for him."

"Mama," I said, "ain't nothing going to happen to you *or* Sonny. Sonny's all right. He's a good boy and he's got good sense."

"It ain't a question of his being a good boy," Mama said, "nor of his having good sense. It ain't only the bad ones, nor yet the dumb ones that gets sucked under." She stopped, looking at me. "Your Daddy once had a brother," she said, and she smiled in a way that made me feel she was in pain. "You didn't never know that, did you?"

"No," I said, "I never knew that," and I watched her face.

"Oh, yes," she said, "your Daddy had a brother." She looked out of the window again. "I know you never saw your Daddy cry. But *I* did —many a time, through all these years."

I asked her, "What happened to his brother? How come nobody's ever talked about him?"

This was the first time I ever saw my mother look old.

"His brother got killed," she said, "when he was just a little younger than you are now. I knew him. He was a fine boy. He was maybe a little full of the devil, but he didn't mean nobody no harm."

Then she stopped and the room was silent, exactly as it had sometimes been on those Sunday afternoons. Mama kept looking out into the streets.

"He used to have a job in the mill," she said, "and, like all young folks, he just liked to perform on

Saturday nights. Saturday nights, him and your father would drift around to different places, go to dances and things like that, or just sit around with people they knew, and your father's brother would sing, he had a fine voice, and play along with himself on his guitar. Well, this particular Saturday night, him and your father was coming home from some place, and they were both a little drunk and there was a moon that night, it was bright like day. Your father's brother was feeling kind of good, and he was whistling to himself, and he had his guitar slung over his shoulder. They was coming down a hill and beneath them was a road that turned off from the highway. Well, your father's brother, being always kind of frisky, decided to run down this hill, and he did, with that guitar banging and clanging behind him, and he ran across the road, and he was making water behind a tree. And your father was sort of amused at him and he was still coming down the hill, kind of slow. Then he heard a car motor and that same minute his brother stepped from behind the tree, into the road, in the moonlight. And he started to cross the road. And your father started to run down the hill, he says he don't know why. This car was full of white men. They was all drunk, and when they seen your father's brother they let out a great whoop and holler and they aimed the car straight at him. They was having fun, they just wanted to scare him, the way they do sometimes, you know. But they was drunk. And I guess the boy, being drunk, too, and scared, kind of lost his head. By the time he jumped it was too late. Your father says he heard his brother scream when the car rolled over

him, and he heard the wood of that guitar when it give, and he heard them strings go flying, and he heard them white men shouting, and the car kept on a-going and it ain't stopped till this day. And, time your father got down the hill, his brother weren't nothing but blood and pulp."

Tears were gleaming on my mother's face. There wasn't anything I could say.

"He never mentioned it," she said, "because I never let him mention it before you children. Your Daddy was like a crazy man that night and for many a night thereafter. He says he never in his life seen anything as dark as that road after the lights of that car had gone away. Weren't nothing, weren't nobody on that road, just your Daddy and his brother and that busted guitar. Oh, yes. Your Daddy never did really get right again. Till the day he died he weren't sure but that every white man he saw was the man that killed his brother."

She stopped and took out her handkerchief and dried her eyes and looked at me.

"I ain't telling you all this," she said, "to make you scared or bitter or to make you hate nobody. I'm telling you this because you got a brother. And the world ain't changed."

I guess I didn't want to believe this. I guess she saw this in my face. She turned away from me, toward the window again, searching those streets.

"But I praise my Redeemer," she said at last, that He called your Daddy home before me. I ain't saying it to throw no flowers at myself, but, I declare, it keeps me from feeling too cast down to know

I helped your father get safely through this world. Your father always acted like he was the roughest, strongest man on earth. And everybody took him to be like that. But if he hadn't had *me* there—to see his tears!"

She was crying again. Still, I couldn't move. I said, "Lord, Lord, Mama, I didn't know it was like that."

"Oh, honey," she said, "there's a lot that you don't know. But you are going to find it out." She stood up from the window and came over to me. "You got to hold on to your brother," she said, "and don't let him fall, no matter what it looks like is happening to him and no matter how evil you gets with him. You going to be evil with him many a time. But don't you forget what I told you, you hear?"

"I won't forget," I said. "Don't you worry, I won't forget. I won't let nothing happen to Sonny."

My mother smiled as though she were amused at something she saw in my face. Then, "You may not be able to stop nothing from happening. But you got to let him know you's *there.*"

Two days later I was married, and then I was gone. And I had a lot of things on my mind and I pretty well forgot my promise to Mama until I got shipped home on a special furlough for her funeral.

And, after the funeral, with just Sonny and me alone in the empty kitchen, I tried to find out something about him.

"What do you want to do?" I asked him.

"I'm going to be a musician," he said.

For he had graduated, in the time I had been away, from dancing to the juke box to finding out who was playing what, and what they were doing with it, and he had bought himself a set of drums.

"You mean, you want to be a drummer?" I somehow had the feeling that being a drummer might be all right for other people but not for my brother Sonny.

"I don't think," he said, looking at me very gravely, "that I'll ever be a good drummer. But I think I can play a piano."

I frowned. I'd never played the role of the older brother quite so seriously before, had scarcely ever, in fact, *asked* Sonny a damn thing. I sensed myself in the presence of something I didn't really know how to handle, didn't understand. So I made my frown a little deeper as I asked: "What kind of musician do you want to be?"

He grinned. "How many kinds do you think there are?"

"Be *serious,*" I said.

He laughed, throwing his head back, and then looked at me. "I *am* serious."

"Well, then, for Christ's sake, stop kidding around and answer a serious question. I mean, do you want to be a concert pianist, you want to play classical music and all that, or—or what?" Long before I finished he was laughing again. "For Christ's *sake,* Sonny!"

He sobered, but with difficulty. "I'm sorry. But you sound so— *scared!*" and he was off again.

"Well, you may think it's funny now, baby, but it's not going to be so funny when you have to make your living at it, let me tell you *that.*" I was furious because I knew he was laughing at me and I didn't know why.

Mary Mulcahy

CHRISTOPHER LA FARGE

Lying quite still on her bed, so that the pain in her back was only a warning and not an actuality, Mary Mulcahy considered her problems. There were the matters of her back, of the broken novena, of Kathleen's hundred dollars, and, most difficult to face, of this room. Whatever else she thought about, she was aware always of the room. Even while she lay thus, with her eyes closed, she could see its comforting detail, be aware of the fine light that swept in at the east window, of the huge ailanthus that grew in the next back yard. You might look and look in all that district of Manhattan and not find another room like this one, so convenient, so cheap, so clean and pretty, its own running water in the bright washbasin; a quiet room, in a jutting ell to an old house, that looked out down the long reach of the back yards to trees and to privet bushes; a room into which the morning sun poured like a blessing from the dear Virgin.

Oh, but cheap as it was, it was the wasteful extravagance, surely! All the long summer, four solid months, she could, by living in one of the servants' rooms at the Park Avenue apartment where she worked, save the rent she was now paying. And so, without touching the savings that now mounted so slowly again toward security, she could set aside the money for Kathleen's wedding present. Hadn't she given a hundred dollars to the other nieces when they married? To Rosie and to Agnes? But that was before the bad days, before the savings of a lifetime melted suddenly away. If you had asked her in 1929, she would have said she could stop work in another five years, living secure and content in such a room as this —if another such existed. Sixty years would have been the good age for stopping.

But there could be no stopping now. There was the present for Kathleen, and the savings that were so small, and always a little bit here and a little bit there for the blessed Church. You could not say no to a good man like Father Elliott, or to the sweet Sisters who knocked at your door, their lives in their faces, their smiles the reflection of the light of the Blessed Mother herself. She was lucky indeed to have so good a job, for taking care of the apartment in summer was within her powers and the half-days in winter were not that hard to do that she could not do them a while yet—if her back grew no worse. Last night it was so bad after she came home from the apartment that she had to lie down, as she had today, and, for all that her soul troubled her, she had stayed in her bed and so the novena was broken, and it in its eighth day. Surely Father Elliott would understand and be kind, there would be never a reproach to her, but within herself there was the knowledge that what she had set out to do for the glory of God and for the salvation of her own soul was left but partly done. It was like a warning, a portent. What if Mr. Gore, or indeed him and his wife too, should come to New York while she was like this now? How would she stand to fix their tea and to give them breakfast? Mrs. Gore was a beautiful, sweet woman and Mr. Gore the fine, big, strong man. It was a pleasure to make him comfortable and have things nice for him. They would be kind, they would understand, but wouldn't they say to themselves, "Mary Mulcahy is too old now"? Wouldn't they begin casting about for a young one who would not be ill to fail them?

I keep it clean and lovely, she thought, the lovely apartment, but if I am not there for them when they come, what good is the cleanness of it? It's well that I never married, indeed, for a fine wife I'd have been to the old husband in these the last years of his age.

So she lay on the bed, conscious of the warmth of the room, of its high ceiling, of the light beyond the window, the soft, glowing, eastern light of evening, and she felt the problems grow heavy in her breast. The smallness of the savings weighed and weighed on her, the present for Kathleen was a deadly weight, her back warned her, distantly but clearly, of the weight of her years. And behind it all, behind even the distress of the broken novena, lay the knowledge that if she gave up the room, lived in at the apartment, it would be all solved, all simple, and that once she let this room go, it would not be twenty-four hours before another would have it, and then, as sure as God was in His heaven, it would be gone forever. Weren't there people waiting for it now? Hadn't she waited for three years herself to get it? Hadn't she waited till Ellen Smith had died? Ellen had wanted never to move from it, and only death in the end had moved her. You live in house after house, room after room—life was a series of rooms in a long series of houses. Surely when a room at last said home to you, you had the good right to cling to it?

It was past six o'clock when she heard the knocking at her door. Something told her that the knocking had been going on for some time and that only now had she become fully conscious of it, and it frightened her.

"Come in," she called out. The door rattled.

"It's locked," said the mild voice of Mrs. Gustafson, her landlady.

It was necessary to rise, then, thought Mary Mulcahy. Slowly she began the process, easing herself up till she sat on the edge of her bed, pushing herself upright from there, one hand on her hip, the other hand clinging to the head of her bed. Her back ached painfully. Her left hand still pressed to her hip, she limped to the door, unlatched it, and opened it.

"It's a telegram," said Mrs Gustafson. "I'm sorry to get you up."

"A telegram!" cried Mary. "Oh!" She took the yellow envelope from Mrs. Gustafson. "Oh God!" she said.

"I hope it ain't some bad news," Mrs. Gustafson said, patting her silver hair in the dusk of the hallway. "It's maybe nothing at all. Sometimes—"

"Aah," Mary said, "them things! You never know." She opened the envelope and read the message on the yellow slip aloud and slowly: " 'ARRIVED UNEXPECTEDLY AND SHALL SPEND THE NIGHT PLEASE BRING KEYS TO SILVER CLOSET.' It's himself. It's Mr. Gore."

"And you yust home to rest!" said Mrs. Gustafson. "It's a shame!"

"No, no," Mary said. "He's a fine man. Sometimes he don't have a chance to send word before he has to come. I must go right around." She turned away.

"Can I help you?"

"Oh, no, I thank you," Mary said.

"You look sick, Miss Mulcahy."

"I'm all right," Mary said. "I'm all right. It's only I have the ache in my back. It's the weather, is all it is. It'll pass. It'll pass. It is nothing at all. I thank you for bringing up the telegram, Mrs. Gustafson."

"That's all right," Mrs. Gustafson said. "Let me know if I can help you. It's a shame when you were home already and lying down." She shut the door softly.

"Home already," Mary repeated as she put on her old black hat. Then she slipped her feet into a rather battered pair of white sneakers and left the room.

It was only six blocks to the apartment, so Mary walked it, as always. The beginning of the journey was worse than the end. She stopped at John O'Meara's and bought oranges and lemons and a pint of cream and a loaf of the Italian bread that Mr. Gore liked so well. Mr. Gore was waiting for her at the apartment when she let herself in at the kitchen entrance.

"Well, Mary," he said. "How are you?"

"Fine, fine," said Mary. "And how is yourself and the family?"

"All blooming," he said. "Sorry to rout you out like this. I'd no time to telegraph, even, till I arrived."

"Aah," she said, "that's all right. That's all right. Will you have some tea now?"

"It's rather late," he said, looking at his watch.

"It'll do you good," she said.

"Well—it would. If it's not too much bother?"

"Not a bit of bother in the world," said Mary. "I'll have it for you in a minyute. In just a minyute."

She bustled back to the kitchen,

took off her hat and put on an apron, and made tea and toast. It was swiftly and competently done. She took the tray in to him, shuffling in her sneakers, the shuffling walk she had found was easier these days.

"And how is your niece, Kathleen?" he asked her.

"Fine, fine," said Mary. "She's to be wed in October."

"Indeed?" he said. "So soon? I could have wished she'd stayed on. Mrs. Gore liked her so much. A first-rate girl."

"Aah," said Mary, "them young ones today! They don't know when they're well off. All they think of is the men and the marrying."

"Natural enough," said Mr. Gore.

"To be sure," Mary said. "Natural enough. But there's many a one lives to regret it. She had a lovely place here, with lovely people, and she should have stayed in it. Then you know where you are. Aah, them young girls! They're all alike. What time will you be wanting breakfast, sir?"

"Call me at seven-thirty," he said. "Breakfast at eight. I'm off for home again tomorrow. I'll try to give you warning next time I come up. Leave the key of the silver closet on my dresser, will you? I'll give it back in the morning."

"I will," she said. "I will. Is there anything else now?"

"No," he said. "Good night, Mary. You toddle along home. I'll be out of here in half an hour."

"Good night, sir," she said.

She went to his bedroom then and turned down the bed for him, unpacked his bag, and laid out his toilet things and his pajamas. She drew him a warm bath and checked to be sure there were plenty of clean towels. She was pleased now that she'd put the fine, embroidered sheets on his bed, but it was lucky Mrs. Gore wasn't here too, for there'd only been the one pair home from the wash and on her bed were the plain ones. Then Mary went to the kitchen, made herself some tea, and sat down to wait for him to go. It was a long wait. She sat in a wooden chair instead of the upholstered wicker one, because it was easier to get up from, with her back and all, if he should ring for her.

When he had gone out at last she went to his room and hung up his suit, treed his shoes, and laid his underclothes neatly, ready for the morning. Then she cleaned the bathroom and straightened it up and rinsed out the tub, though that was painful, so painful it was frightening. But she was too busy to be really frightened. Then she collected the tea things and washed them and put ice in the ice carafe. She set out a tray with glasses and whiskey and the carafe in the living room. She emptied his ashtray and plumped the cushions. When everything was done, when the apartment was as neat as a pin, proper and comfortable for his return, she looked around her and felt a sort of panic. Her occupations were over. There was nothing to divert her mind from her troubles now.

I'll stay here, she thought. I'll never get home and back. There's Cook's bed I can lie upon. That's the best.

Here she could lie down, lie down indeed. When she got up again, for all the pain, she would be here, where she had to be. There would be no stairs up, no stairs down, no six blocks to walk. She got a blanket from a bureau drawer and unrolled the mattress on the bed in the little, dark, viewless bedroom. Then she took off her sneakers, lay down on the bed, and pulled the one blanket over her. Lying there, she said her prayers twice through and begged the good God and His lovely Son and the merciful Mother of Christ to forgive her that she hadn't the strength left in her to kneel to Their worship. When she lay still, her back was only a portent of pain.

In the morning, before Mr. Gore had left, Mary Mulcahy came to her decision. She told him merely that she was going to give up her room and sleep at the apartment. "Till you and Madam come back in October," she said.

"That's good," he said. "I should think that much more convenient and economical."

"It is, it is," she said. That was all. There was no use bothering him with your own little problems and him probably bowed down with his great ones. And how could you tell a man in his lovely ten-room apartment what a room in a rooming house might mean to you?

When she had everything clean and in order, she put on her hat and left the apartment. On the way home she went by way of St. Anthony's, entered the church, and, going to the Lady Chapel, she knelt there and said many prayers and many of them over. She prayed the Virgin for strength to continue and for forgiveness of her sins. If she had seen Father Elliott, she would have told him about her back and the novena, but there was no one around but the old sacristan, dusting and dusting. All he did, she thought, was move the blessed dust from one place to another. Then she left and walked home.

When she had packed her two old, battered suitcases, she found Mrs. Gustafson in her kitchen.

"I've got to go," she said. "I've got to give up the room. I must live in till October."

"Oh!" cried Mrs. Gustafson. "Why, Miss Mulcahy, I thought you would be here for *always* now."

"It's a lovely room," Mary said. "A lovely, lovely room. But it's my niece is to be married and so on and so on. And my people come often to the apartment and all."

"Well," said Mrs. Gustafson. "If you must, you must. I hate to have you going, Miss Mulcahy. Till October, you say?"

"Till October," Mary said. "That's the truth. I'll want a room then, if you've a vacancy, Mrs. Gustafson. Here's what I owe you." She held out a small wad of crumpled bills.

Mrs. Gustafson took it and counted it. "That's right," she said, and smiled. "We'll be waiting for you. Goodbye. Good luck."

"Goodbye," Mary said.

She said the same thing in a whisper to her room as she left it. The leavetaking was not prolonged. That would have been sinful, and useless. She picked up the two bags and looked at the room around her and said, in a whisper, "Goodbye." That was all. Then she began the painful descent of the stairs. The pain reassured her now. It was so right to go. She opened the front door and stepped out onto the high stoop. She was just going to close the door behind her when Mrs. Gustafson appeared.

"See," said Mrs. Gustafson awkwardly. "We are not young now. Life ain't all money, money. Isn't it? When you come back to me, when you come back in October, the little room will be for you. So."

"Aah," Mary said. "My room! God will remember you, Mrs. Gustafson. It's a lovely thing, it is. A lovely thing. You put the heart back into me."

"It is nothing," said Mrs. Gustafson. "Till October, when you come home. We will expect you, Miss Mulcahy. So long." She closed the door then, rather abruptly.

Mary picked up the suitcases again and descended the stoop. There were tears in her eyes and she had to stop when she reached the sidewalk, put down the cases, and wipe her eyes. "The kind woman," she said aloud. "The lovely, kind woman."

John O'Meara's delivery boy, Julian, pushing his cart, stopped beside her.

"You off to the apartment, Miss Mulcahy?" he said.

"I am," she said. "I am. Good morning, Julian."

"I'll put your bags in and take them up," he said.

"Aah," she said, "it's out of your way, then."

"No," he said. "Never a bit." He took the two suitcases and put them in his cart and together they began walking toward Park Avenue, he in his thick shoes, lightly, as a young man; she in her white sneakers, shuffling.

"The Lord is my shepherd," said Mary Mulcahy.

from *"A Puerto Rican in New York"*

Little Things Are Big

Jesús Colón

It was very late at night on the eve of Memorial Day. She came into the subway at the 34th Street Pennsylvania Station. I am still trying to remember how she managed to push herself in with a baby on her right arm, a valise in her left hand and two children, a boy and girl about three and five years old, trailing after her. She was a nice looking white lady in her early twenties.

At Nevins Street, Brooklyn, we saw her preparing to get off at the next station—Atlantic Avenue—which happened to be the place where I too had to get off. Just as it was a problem for her to get on, it was going to be a problem for her to get off the subway with two small children to be taken care of, a baby on her right arm and a medium sized valise in her left hand.

And there I was, also preparing to get off at Atlantic Avenue, with no bundles to take care of—not even the customary book under my arm without which I feel that I am not completely dressed.

As the train was entering the Atlantic Avenue station, some white man stood up from his seat and helped her out, placing the children on the long, deserted platform. There were only two adult persons on the long platform some time after midnight on the eve of last Memorial Day.

I could perceive the steep, long concrete stairs going down to the Long Island Railroad or into the street. Should I offer my help as the American white man did at the subway door placing the two children outside the subway car? Should I take care of the girl and the boy, take them by their hands until they

202

reached the end of the steep long concrete stairs of the Atlantic Avenue station?

Courtesy is a characteristic of the Puerto Rican. And here I was—a Puerto Rican—hours past midnight, a valise, two white children and a white lady with a baby on her arm palpably needing somebody to help her at least until she descended the long concrete stairs.

But how could I, a Negro and a Puerto Rican approach this white lady who very likely might have preconceived prejudices against Negroes and everybody with foreign accents, in a deserted subway station very late at night?

What would she say? What would be the first reaction of this white American woman, perhaps coming from a small town, with a valise, two children and a baby on her right arm? Would she say: Yes, of course, you may help me. Or would she think that I was just trying to get too familiar? Or would she think worse than that perhaps? What would I do if she let out a scream as I went toward her to offer my help?

Was I misjudging her? So many slanders are written every day in the daily press against the Negroes and Puerto Ricans. I hesitated for a long, long minute. The ancestral manners that the most illiterate Puerto Rican passes on from father to son were struggling inside me. Here was I, way past midnight, face to face with a situation that could very well explode into an outburst of prejudices and chauvinistic conditioning of the "divide and rule" policy of present day society.

It was a long minute. I passed on by her as if I saw nothing. As if I was insensitive to her need. Like a rude animal walking on two legs, I just moved on half running by the long subway platform leaving the children and the valise and her with the baby on her arm. I took the steps of the long concrete stairs in twos until I reached the street above and the cold air slapped my warm face.

This is what racism and prejudice and chauvinism and official artificial divisions can do to people and to a nation!

Perhaps the lady was not prejudiced after all. Or not prejudiced enough to scream at the coming of a Negro toward her in a solitary subway station a few hours past midnight.

If you were not that prejudiced, I failed you, dear lady. I know that there is a chance in a million that you will read these lines. I am willing to take that millionth chance. If you were not that prejudiced, I failed you, lady, I failed you, children. I failed myself to myself.

I buried my courtesy early on Memorial Day morning. But here is a promise that I make to myself here and now; if I am ever faced with an occasion like that again, I am going to offer my help regardless of how the offer is going to be received.

Then I will have my courtesy with me again.

from

Main Street

SINCLAIR LEWIS

In reading popular stories and seeing plays, asserted Carol, she had found only two traditions of the American small town. The first tradition, repeated in scores of magazines every month, is that the American village remains the one sure abode of friendship, honesty, and clean sweet marriageable girls. Therefore all men who succeed in painting Paris or in finance in New York at last become weary of smart women, return to their native towns, assert that cities are vicious, marry their childhood sweethearts and, presumably, joyously abide in those towns until death.

The other tradition is that the significant features of all villages are whiskers, iron dogs upon lawns, gold bricks, checkers, jars of gilded cattails, and shrewd comic old men who are known as "hicks" and who ejaculate "Waal I swan." This altogether admirable tradition rules the vaudeville stage, facetious illustrators, and syndicated newspaper humor, but out of actual life it passed forty years ago. Carol's small town thinks not in hoss-swapping but in cheap motor cars, telephones, ready-made clothes, silos, alfalfa, kodaks, phonographs, leather-upholstered Morris chairs, bridge-prizes, oil-stocks, motion-pictures, land-deals, unread sets of Mark Twain, and a chaste version of national politics.

With such a small-town life a Kennicott or a Champ Perry is content, but there are also hundreds of thousands, particularly women and young men, who are not at all content. The more intelligent young people (and the fortunate widows!) flee to the cities with agility and, despite the fictional tradition, resolutely stay there, seldom returning even for holidays. The most protesting patriots of the towns leave them in old age, if they can afford it, and go to live in California or in the cities.

The reason, Carol insisted, is not a whiskered rusticity. It is nothing so amusing!

It is an unimaginatively standardized background, a sluggishness of speech and manners, a rigid ruling of the spirit by the desire to appear respectable. It is contentment . . . the contentment of the quiet dead, who are scornful of the living for their restless walking. It is negation canonized as the one positive virtue. It is the prohibition of happiness. It is slavery self-sought and self-defended. It is dullness made God.

A savorless people, gulping tasteless food, and sitting afterward, coatless and thoughtless, in rocking-chairs prickly with inane decorations, listening to mechanical music, saying mechanical things about the excellence of Ford automobiles, and viewing themselves as the greatest race in the world.

*   *   *

Vida was indignant; Carol was apologetic; they talked for another hour, the eternal Mary and Martha —an immoralist Mary and a reformist Martha. It was Vida who conquered.

The fact that she had been left out of the campaign for the new school-building disconcerted Carol. She laid her dreams of perfection aside. When Vida asked her to take charge of a group of Camp Fire Girls, she obeyed, and had definite pleasure out of the Indian dances and ritual and costumes. She went more regularly to the Thanatopsis. With Vida as lieutenant and unofficial commander she campaigned for a village nurse to attend poor families, raised the fund herself, saw to it that the nurse was young and strong and amiable and intelligent.

Yet all the while she beheld the burly cynical Frenchman and the diaphanous dancers as clearly as the child sees its air-born playmates; she relished the Camp Fire Girls not because, in Vida's words, "this Scout training will help so much to make them Good Wives," but because she hoped that the Sioux dances would bring subversive color into their dinginess.

She helped Ella Stowbody to set out plants in the tiny triangular park at the railroad station; she squatted in the dirt, with a small curved trowel and the most decorous of gardening gauntlets; she talked to Ella about the public-spiritedness of fuchsias and cannas; and she felt that she was scrubbing a temple deserted by the gods and empty even of incense and the sound of chanting. Passengers looking from trains saw her as a village woman of fading prettiness, incorruptible virtue, and no abnormalities; the baggageman heard her say, "Oh yes, I do think it will be a good example for the children"; and all the while she saw herself running garlanded through the streets of Babylon.

Planting led her to botanizing. She never got much farther than recognizing the tiger lily and the wild rose, but she rediscovered Hugh. "What does the buttercup say, mummy?" he cried, his hand full of straggly grasses, his cheek gilded with pollen. She knelt to embrace him; she affirmed that he made life more than full; she was altogether reconciled . . . for an hour.

But she awoke at night to hovering death. She crept away from the hump of bedding that was Kennicott; tiptoed into the bathroom and, by the mirror in the door of the medicine-cabinet, examined her pallid face.

Wasn't she growing visibly older in ratio as Vida grew plumper and younger? Wasn't her nose sharper? Wasn't her neck granulated? She stared and choked. She was only thirty. But the five years since her marriage—had they not gone by as hastily and stupidly as though she had been under ether; would time not slink past till death? She pounded her fist on the cool enameled rim of the bathtub and raged mutely against the indifferent gods:

"I don't care! I won't endure it! They lie so—Vida and Will and Aunt Bessie—they tell me I ought to be satisfied with Hugh and a good home and planting seven nasturtiums in a station garden! I am I! When I die the world will be annihilated, as far as I'm concerned. I am I! I'm not content to leave the sea and the ivory towers to others. I want them for me! Damn Vida! Damn all of them! Do they think they can make me believe that a display of potatoes at Howland & Gould's is enough beauty and strangeness?"

Winter Night

Kay Boyle

There is a time of apprehension which begins with the beginning of darkness, and to which only the speech of love can lend security. It is there, in abeyance, at the end of every day, not urgent enough to be given the name of fear but rather of concern for how the hours are to be reprieved from fear, and those who have forgotten how it was when they were children can remember nothing of this. It may begin around five o'clock on a winter afternoon when the light outside is dying in the windows. At that hour the New York apartment in which Felicia lived was filled with shadows, and the little girl would wait alone in the living room, looking out at the winter-stripped trees that stood black in the park against the isolated ovals of unclean snow. Now it was January, and the day had been a cold one; the water of the artificial lake was frozen fast, but because of the cold and the coming darkness, the skaters had ceased to move across its surface. The street that lay between the park and the apartment house was wide, and the two-way streams of cars and buses, some with their headlamps already shining, advanced and halted and poured swiftly on to the tempo of the traffic signals' altering lights. The time of apprehension had set in, and Felicia, who was seven, stood at the window in the evening and waited before she asked the question. When the signals below would change from red to green again, or when the double-decker bus would turn the corner below, she would ask it. The words of it were already there, tentative in her mouth, when the answer came from the far end of the hall.

'Your mother," said the voice among the sound of kitchen things, "she telephoned up before you came in from nursery school. She won't be back in time for supper. I was to tell you a sitter was coming in from the sitting parents' place."

Felicia turned back from the window into the obscurity of the living room and she looked toward the open door, and into the hall beyond it where the light from the kitchen fell in a clear yellow angle across the wall and onto the strip of carpet. Her hands were cold, and she put them in her jacket pockets as she walked carefully across the living-room rug and stopped at the edge of light.

"Will she be home late?" she said.

For a moment there was the sound of water running in the kitchen, a long way away, and then the sound of the water ceased, and the high, Southern voice went on:

'She'll come home when she gets ready to come home. That's all I have to say. If she wants to spend two dollars and fifty cents and ten cents' carfare on top of that three or four nights out of the week for a sitting parent to come in here and sit, it's her own business. It certainly ain't nothing to do with you or me. She makes her money, just like the rest of us does. She works all day down there in the office, or whatever it is, just like the rest of us works, and she's entitled to spend her money like she wants to spend it. There's no law in the world against buying your own freedom, that's all we're doing. And we're not doing nobody no harm."

"Do you know who she's having supper with?" said Felicia from the edge of dark. There was one more step to take, and then she would be standing in the light that fell on the strip of carpet, but she did not take the step.

"Do I know who she's having supper with?" the voice cried out in what might have been derision, and there was the sound of dishes striking the metal ribs of the drainboard by the sink. "Maybe it's Mr. Van Johnson, or Mr. Frank Sinatra, or maybe it's just the Duke of Wincers for the evening. All I know is you're having softboiled egg and spinach and applesauce for supper, and you're going to have it quick now because the time is getting away."

The voice from the kitchen had no name. It was as variable as the faces and figures of the women who came and sat in the evenings. Month by month the voice in the kitchen altered to another voice, and the sitting parents were no more than lonely aunts of an evening or two who sometimes returned and sometimes did not to this apartment in which they had sat before. Nobody stayed anywhere very long any more, Felicia's mother told her. It was part of the time in which you lived, and part of the life of the city, but when the fathers came back, all this would be miraculously changed. Perhaps you would live in a house again, a small one, with fir trees on either side of the short brick walk, and Father would drive up every night from the station just after darkness set in. When Felicia thought of this, she stepped quickly into the clear angle of light, and she left the dark of the living room behind her and ran softly down the hall.

The drop-leaf table stood in the kitchen between the refrigerator and the sink, and Felicia sat down at the place that was set. The voice at the sink was speaking still, and while Felicia ate it did not cease to speak until the bell of the front door rang abruptly. The girl walked around the table and went down the hall, wiping her dark palms in her apron, and, from the drop-leaf table, Felicia watched her step from the angle of light into darkness and open the door.

"You put in an early appearance," the girl said, and the woman who had rung the bell came into the hall. The door closed behind her, and the girl showed her into the living room, and lit the lamp on the bookcase, and the shadows were suddenly bleached away. But when the girl turned, the woman turned from the living room too and followed her, humbly and in silence, to the threshold of the kitchen. "Sometimes they keep me standing around waiting after it's time for me to be getting on home, the sitting parents do," the girl said, and she picked up the last two dishes from the table and put them in the sink. The woman who stood in the doorway was a small woman, and when she undid the white silk scarf from around her head, Felicia saw that her hair was black. She wore it parted in the middle, and it had not been cut, but was drawn back loosely into a knot behind her head. She had very clean white gloves on, and her face was pale, and there was a look of sorrow in her soft black eyes. "Sometimes I have to stand out there in the hall with my hat and coat on, waiting for the sitting parents to turn up," the girl said, and, as she turned on the water in the sink, the contempt she had for them hung on the kitchen air. "But you're ahead of

time," she said, and she held the dishes, first one and then the other, under the flow of steaming water.

The woman in the doorway wore a neat black coat, not a new-looking coat, and it had no fur on it, but it had a smooth velvet collar and velvet lapels. She did not move, or smile, and she gave no sign that she had heard the girl speaking above the sound of water at the sink. She simply stood looking at Felicia, who sat at the table with the milk in her glass not finished yet.

"Are you the child?" she said at last, and her voice was low, and the pronunciation of the words a little strange.

"Yes, this here's Felicia," the girl said, and the dark hands dried the dishes and put them away. "You drink up your milk quick now, Felicia, so's I can rinse your glass."

"I will wash the glass," said the woman. "I would like to wash the glass for her," and Felicia sat looking across the table at the face in the doorway that was filled with such unspoken grief. "I will wash the glass for her and clean off the table," the woman was saying quietly. "When the child is finished, she will show me where her night things are."

"The others, they wouldn't do anything like that," the girl said, and she hung the dishcloth over the rack. "They wouldn't put their hand to housework, the sitting parents. That's where they got the name for them," she said.

Whenever the front door closed behind the girl in the evening, it would usually be that the sitting parent who was there would take up a book of fairy stories and read aloud for a while to Felicia; or else would settle herself in the big chair in the living room and begin to tell the words of a story in drowsiness to her, while Felicia took off her clothes in the bedroom, and folded them, and put her pajamas on, and brushed her teeth, and did her hair. But this time, that was not the way it happened. Instead, the woman sat down on the other chair at the kitchen table, and she began at once to speak, not of good fairies, or bad, or of animals endowed with human speech, but to speak quietly, in spite of the eagerness behind her words, of a thing that seemed of singular importance to her.

"It is strange that I should have been sent here tonight," she said, her eyes moving slowly from feature to feature of Felicia's face, "for you look like a child that I knew once, and this is the anniversary of that child."

"Did she have hair like mine?" Felicia asked quickly, and she did not keep her eyes fixed on the unfinished glass of milk in shyness any more.

"Yes, she did. She had hair like yours," said the woman, and her glance paused for a moment on the locks which fell straight and thick on the shoulders of Felicia's dress. It may have been that she thought to stretch out her hand and touch the ends of Felicia's hair, for her fingers stirred as they lay clasped together on the table, and then they relapsed into passivity again. "But it is not the hair alone, it is the delicacy of your face, too, and your eyes the same, filled with the same spring lilac color," the woman said, pronouncing the words carefully. "She had little coats of golden fur on her arms and legs," she said, "and when

we were closed up there, the lot of us in the cold, I used to make her laugh when I told her that the fur was so pretty, like a little fawn's skin on her arms, would always help to keep her warm."

"And did it keep her warm?" asked Felicia, and she gave a little jerk of laughter as she looked down at her own legs hanging under the table, with the bare calves thin and covered with a down of hair.

"It did not keep her warm enough," the woman said, and now the mask of grief had come back upon her face. "So we used to take everything we could spare from ourselves, and we would sew them into cloaks and other kinds of garments for her and for the other children. . . ."

"Was it a school?" said Felicia when the woman's voice had ceased to speak.

"No," said the woman softly, "it was not a school, but still there were a lot of children there. It was a camp—that was the name the place had; it was a camp. It was a place where they put people until they could decide what was to be done with them." She sat with her hands clasped, silent a moment, looking at Felicia. "That little dress you have on," she said, not saying the words to anybody, scarcely saying them aloud. "Oh, she would have liked that little dress, the little buttons shaped like hearts, and the white collar——"

"I have four school dresses," Felicia said. "I'll show them to you. How many dresses did she have?"

"Well, there, you see, there in the camp," said the woman, "she did not have any dresses except the little skirt and the pullover. That was all she had. She had brought just a handkerchief of her belongings with her, like everybody else—just enough for three days away from home was what they told us, so she did not have enough to last the winter. But she had her ballet slippers," the woman said, and her clasped fingers did not move. "She had brought them because she thought during her three days away from home she would have the time to practice her ballet."

"I've been to the ballet," Felicia said suddenly, and she said it so eagerly that she stuttered a little as the words came out of her mouth. She slipped quickly down from the chair and went around the table to where the woman sat. Then she took one of the woman's hands away from the other that held it fast, and she pulled her toward the door. "Come into the living room and I'll do a pirouette for you," she said, and then she stopped speaking, her eyes halted on the woman's face. "Did she—did the little girl—could she do a pirouette very well?" she said.

"Yes, she could. At first she could," said the woman, and Felicia felt uneasy now at the sound of sorrow in her words. "But after that she was hungry. She was hungry all winter," she said in a low voice. "We were all hungry, but the children were the hungriest. Even now," she said, and her voice went suddenly savage, "when I see milk like that, clean, fresh milk standing in a glass, I want to cry out loud, I want to beat my hands on the table, because it did not have to be . . ." She had drawn her fingers abruptly away from Felicia now, and Felicia stood before her, cast off, forlorn, alone again in the time of apprehension. "That was

three years ago," the woman was saying, and one hand was lifted, as in weariness, to shade her face. "It was somewhere else, it was in another country," she said, and behind her hand her eyes were turned upon the substance of a world in which Felicia had played no part.

"Did—did the little girl cry when she was hungry?" Felicia asked, and the woman shook her head.

"Sometimes she cried," she said, "but not very much. She was very quiet. One night when she heard the other children crying, she said to me, 'You know, they are not crying because they want something to eat. They are crying because their mothers have gone away.'"

"Did the mothers have to go out to supper?" Felicia asked, and she watched the woman's face for the answer.

"No," said the woman. She stood up from her chair, and now that she put her hand on the little girl's shoulder, Felicia was taken into the sphere of love and intimacy again. "Shall we go into the other room, and you will do your pirouette for me?" the woman said, and they went from the kitchen and down the strip of carpet on which the clear light fell. In the front room, they paused hand in hand in the glow of the shaded lamp, and the woman looked about her, at the books, the low tables with the magazines and ash trays on them, the vase of roses on the piano, looking with dark, scarcely seeing eyes at these things that had no reality at all. It was only when she saw the little white clock on the mantelpiece that she gave any sign, and then she said quickly: "What time does your mother put you to bed?"

Felicia waited a moment, and in the interval of waiting the woman lifted one hand and, as if in reverence, touched Felicia's hair.

"What time did the little girl you knew in the other place go to bed?" Felicia asked.

"Ah, God, I do not know, I do not remember," the woman said.

"Was she your little girl?" said Felicia softly, stubbornly.

"No," said the woman. "She was not mine. At least, at first she was not mine. She had a mother, a real mother, but the mother had to go away."

"Did she come back late?" asked Felicia.

"No, ah, no, she could not come back, she never came back," the woman said, and now she turned, her arm around Felicia's shoulders, and she sat down in the low soft chair. "Why am I saying all this to you, why am I doing it?" she cried out in grief, and she held Felicia close against her. "I had thought to speak of the anniversary to you, and that was all, and now I am saying these others things to you. Three years ago today, exactly, the little girl became my little girl because her mother went away. That is all there is to it. There is nothing more."

Felicia waited another moment, held close against the woman, and listening to the swift, strong heart-beats in the woman's breast.

"But the mother," she said then in a small, persistent voice, "did she take a taxi when she went?"

"This is the way it used to happen," said the woman, speaking in hopelessness and bitterness in the softly lighted room. "Every week they used to come into the place where we were and they would read a list of names out. Sometimes it

would be the names of children they would read out, and then a little later they would have to go away. And sometimes it would be the grown people's names, the names of the mothers or big sisters, or other women's names. The men were not with us. The fathers were somewhere else, in another place."

"Yes," Felicia said. "I know."

"We had been there only a little while, maybe ten days or maybe not so long," the woman went on, holding Felicia against her still, "when they read the name of the little girl's mother out, and that afternoon they took her away."

"What did the little girl do?" Felicia said.

"She wanted to think up the best way of getting out so that she could go find her mother," said the woman, "but she could not think of anything good enough until the third or fourth day. And then she tied her ballet slippers up in the handkerchief again, and she went up to the guard standing at the door." The woman's voice was gentle, controlled now. "She asked the guard please to open the door so that she could go out. 'This is Thursday,' she said, 'and every Tuesday and Thursday I have my ballet lessons. If I miss a ballet lesson, they do not count the money off, so my mother would be just paying for nothing, and she cannot afford to pay for nothing. I missed my ballet lesson on Tuesday,' she said to the guard, 'and I must not miss it again today.'"

Felicia lifted her head from the woman's shoulder, and she shook her hair back and looked in question and wonder at the woman's face.

"And did the man let her go?" she said.

"No, he did not. He could not do

that," said the woman. "He was a soldier and he had to do what he was told. So every evening after her mother went, I used to brush the little girl's hair for her," the woman went on saying. "And while I brushed it, I used to tell her the stories of the ballets. Sometimes I would begin with *Narcissus*," the woman said, and she parted Felicia's locks with her fingers, "so if you go and get your brush now, I will tell it while I brush your hair."

"Oh, yes," said Felicia, and she made two whirls as she went quickly to the bedroom. On the way back, she stopped and held on to the piano with the fingers of one hand while she went up on her toes. "Did you see me? Did you see me standing on my toes?" she called to the woman, and the woman sat smiling in love and contentment at her.

"Yes, wonderful, really wonderful," she said. "I am sure I have never seen anyone do it so well." Felicia came spinning toward her, whirling in pirouette after pirouette, and she flung herself down in the chair close to her, with her thin bones pressed against the woman's soft, wide hip. The woman took the silver-backed, monogrammed brush and the tortoise-shell comb in her hands, and now she began to brush Felicia's hair. "We did not have any soap at all and not very much water to wash in, so I never could fix her as nicely and prettily as I wanted to," she said, and the brush stroked regularly, carefully down, caressing the shape of Felicia's head.

"If there wasn't very much water, then how did she do her teeth?" Felicia said.

"She did not do her teeth," said the woman, and she drew the comb through Felicia's hair. "There were

not any toothbrushes or tooth paste, or anything like that."

Felicia waited a moment, constructing the unfamiliar scene of it in silence, and then she asked the tentative question.

"Do I have to do my teeth tonight? she said.

"No," said the woman, and she was thinking of something else, "you do not have to do your teeth."

"If I am your little girl tonight, can I pretend there isn't enough water to wash?" said Felicia.

"Yes," said the woman, "you can pretend that if you like. You do not have to wash," she said, and the comb passed lightly through Felicia's hair.

"Will you tell me the story of the ballet?" said Felicia, and the rhythm of the brushing was like the soft, slow rocking of sleep.

"Yes," said the woman. "In the first one, the place is a forest glade with little pale birches growing in it, and they have green veils over their faces and green veils drifting from their fingers, because it is the springtime. There is the music of a flute," said the woman's voice softly, softly, "and creatures of the wood are dancing——"

"But the mother," Felicia said as suddenly as if she had been awakened from sleep. "What did the little girl's mother say when she didn't do her teeth and didn't wash at night?"

"The mother was not there, you remember," said the woman, and the brush moved steadily in her hand. "But she did send one little letter back. Sometimes the people who went away were able to do that. The mother wrote it in a train, standing up in a car that had no seats," she said, and she might have been telling the story of the ballet still, for her voice was gentle and the brush did not falter on Felicia's hair. "There were perhaps a great many other people standing up in the train with her, perhaps all trying to write their little letters on the bits of papers they had managed to hide on them, or that they had found in forgotten corners as they traveled. When they had written their letters, then they must try to slip them out through the boards of the car in which they journeyed, standing up," said the woman, "and these letters fell down on the tracks under the train, or they were blown into the fields or onto the country roads, and if it was a kind person who picked them up, he would seal them in envelopes and send them to where they were addressed to go. So a letter came back like this from the little girl's mother," the woman said, and the brush followed the comb, the comb the brush in steady pursuit through Felicia's hair. "It said good-by to the little girl, and it said please to take care of her. It said: 'Whoever reads this letter in the camp, please take good care of my little girl for me, and please have her tonsils looked at by a doctor if this is possible to do.' "

"And then," said Felicia softly, persistently, "what happened to the little girl?"

"I do not know. I cannot say," the woman said. But now the brush and comb had ceased to move, and in the silence Felicia turned her thin, small body on the chair, and she and the woman suddenly put their arms around each other. "They must all be asleep now, all of them," the woman said, and in the silence that fell on them again, they held each other closer. "They must be quietly asleep somewhere, and not crying all night because they are hungry and because they are cold. For three years I have been saying 'They must all be asleep, and the cold and the hunger and the seasons or night or day or nothing matters to them——' "

It was after midnight when Felicia's mother put her key in the lock of the front door, and pushed it open, and stepped into the hallway. She walked quickly to the living room, and just across the threshold she slipped the three blue foxskins from her shoulders and dropped them, with her little velvet bag, upon the chair. The room was quiet, so quiet that she could hear the sound of breathing in it, and no one spoke to her in greeting as she crossed toward the bedroom door. And then, as startling as a slap across her delicately tinted face, she saw the woman lying sleeping on the divan, and Felicia, in her school dress still, asleep within the woman's arms.

from "Down These Mean Streets"

If You Ain't Got Heart,
You Ain't Got Nada

Piri Thomas

We were moving—our new pad was back in Spanish Harlem—to 104th Street between Lex and Park Avenue.

Moving into a new block is a big jump for a Harlem kid. You're torn up from your hard-won turf and brought into an "I don't know you" block where every kid is some kind of enemy. Even when the block belongs to your own people, you are still an outsider who has to prove himself a down stud with heart.

As the moving van rolled to a stop in front of our new building, number 109, we were all standing there, waiting for it—Momma, Poppa, Sis, Paulie, James, José, and myself. I made out like I didn't notice the cats looking us over, especially me—I was gang age. I read their faces and found no trust, plenty of suspicion, and a glint of rising hate. I said to myself, *These cats don't mean noth-*

in'. They're just nosy. But I remembered what had happened to me in my old block, and that it had ended with me in the hospital.

This was a tough-looking block. That was good, that was cool; but my old turf had been tough, too. *I'm tough enough.* A voice within said. *I hope I'm tough enough. I am tough enough. I've got* mucho corazón, *I'm king wherever I go. I'm a killer to my heart. I not only* can *live, I* will *live, no punk out, no die out, walk bad; be down, cool breeze, smooth.* My mind raced, and thoughts crashed against each other, trying to reassemble themselves into a pattern of rep. I turned slowly and with eyelids half-closed I looked at the rulers of this new world and with a cool shrug of my shoulders I followed the movers into the hallway of number 109 and dismissed the coming war from my mind.

The next morning I went to my new school, called Patrick Henry, and strange, mean eyes followed me.

"Say, pops," said a voice belonging to a guy I later came to know as Waneko, "where's your territory?"

In the same tone of voice Waneko had used, I answered, "I'm on it, dad, what's shaking?"

"Bad, huh?" He half-smiled.

"No, not all the way. Good when I'm cool breeze and bad when I'm down."

"What's your name, kid?"

"That depends. 'Piri' when I'm smooth and 'Johnny Gringo' when stomping time's around."

"What's your name now?" he pushed.

"You name me, man," I answered, playing my role like a champ.

He looked around, and with no kind of words, his boys cruised in. Guys I would come to know, to fight, to hate, to love, to take care of. Little Red, Waneko, Little Louie, Indio, Carlito, Alfredo, Crip, and plenty more. I stiffened and said to myself, *Stomping time, Piri boy, go with heart.*

I fingered the garbage-can handle in my pocket—my homemade brass knuckles. They were great for breaking down large odds into small, chopped-up ones.

Waneko, secure in his grandstand, said, "We'll name you later, *panín.*"

I didn't answer. Scared, yeah, but wooden-faced to the end, I thought, *Chévere, panín.*

It wasn't long in coming. Three days later, at about 6 p.m., Waneko and his boys were sitting around the stoop at number 115. I was cut off from my number 109. For an instant I thought, *Make a break for it down the basement steps and through the*

back yards—get away in one piece! Then I thought, *Caramba! Live punk, dead hero. I'm no punk kid. I'm not copping any pleas.* I kept walking, hell's a-burning, hell's a-churning, rolling with cheer. *Walk on, baby man, roll on without fear. What's he going to call?*

"Whatta ya say, Mr. Johnny Gringo?" drawled Waneko.

Think, man, I told myself, *think your way out of a stomping. Make it good.* "I hear you 104th Street coolies are supposed to have heart," I said. "I don't know this for sure. You know there's a lot of streets where a whole 'click' is made out of punks who can't fight one guy unless they all jump him for the stomp." I hoped this would push Waneko into giving me a fair one. His expression didn't change.

"Maybe we don't look at it that way."

Crazy, man. I cheer inwardly, the cabrón is falling into my setup. We'll see who gets messed up first, baby! "I wasn't talking to you," I said. "Where I come from, the pres is president 'cause he got heart when it comes to dealing."

Waneko was starting to look uneasy. He had bit on my worm and felt like a sucker fish. His boys were now light on me. They were no longer so much interested in stomping me as in seeing the outcome between Waneko and me. "Yeah," was his reply.

I smiled at him. "You trying to dig where I'm at and now you got me interested in you. I'd like to see where you're at."

Waneko hesitated a tiny little second before replying, "Yeah."

I knew I'd won. Sure, I'd have to fight; but one guy, not ten or fifteen.

If I lost I might still get stomped, and if I won I might get stomped. I took care of this with my next sentence. "I don't know you or your boys," I said, "but they look cool to me. They don't feature as punks."

I had left him out purposely when I said "they." Now his boys were in a separate class. I had cut him off. He would have to fight me on his own, to prove his heart to himself, to his boys, and most important, to his turf. He got away from the stoop and asked, "Fair one, Gringo?"

"Uh-uh," I said, "roll all the way— anything goes." I thought, *I've got to beat him bad and yet not bad enough to take his prestige all away.* He had *corazón.* He came on me. *Let him draw first blood,* I thought, *it's his block.* Smish, my nose began to bleed. His boys cheered, his heart cheered, his turf cheered. "Waste this chump," somebody shouted.

Okay, baby, now it's my turn. He swung. I grabbed innocently, and my forehead smashed his nose. His eyes crossed. His fingernails went for my eye and landed in my mouth— crunch, I bit hard. I punched him in the mouth as he pulled away from me, and he slammed his foot into my chest.

We broke, my nose running red, my chest throbbing, his finger—well, that was his worry. I tied up with body punching and slugging. We rolled onto the street. I wrestled for acceptance, he for rejection or, worse yet, acceptance on his terms. It was time to start peace talks. I smiled at him. "You got heart, baby," I said.

He answered with a punch to my head. I grunted and hit back, harder now. I had to back up my overtures of peace with strength. I hit him in the ribs, I rubbed my knuckles in his ear as we clinched. I tried again. "You deal good," I said.

"You too," he muttered, pressuring out. And just like that, the fight was over. No more words. We just separated, hands half up, half down. My heart pumped out, *You've established your rep. Move over, 104th Street. Lift your wings, I'm one of your baby chicks now.*

Five seconds later my spurs were given to me in the form of introductions to streetdom's elite. There were no looks of blankness now; I was accepted by heart.

"What's your other name, Johnny Gringo?"

"Piri."

"Okay, Pete, you wanna join my fellows?"

"Sure, why not?"

Winslow Homer. *Snap the Whip.*

A House Divided

JEANNE A. TAYLOR

Asa Brown hated his counterparts who lived on the Eastside. Afro-Americans, the delta immigrants called themselves now. They deserved to stay right where they were. The majority of them were too stupid and too lazy to get out anyway. Real black men were found across town in the tracts of two- and three-bedroom stucco houses. Where the lawns are impeccably cut, fertilized, and watered; early morning traffic is the new Ford and Chevrolet car pools of husbands who work in the high rise civil service buildings downtown or in the suburban aircraft plants; brown-skinned "white collar" mothers take their children to the nursery or the Eastside-stranded grandparent baby-sitters each morning; Saturday nights, couples play whist or listen to Willie Bobo and Cannonball Adderly at a Sunset Strip or pier front club; holidays are spent barbecuing and drinking

Scotch; and the only police visits are the result of the disturbing the peace complaints filled by the aged whitey who is asking $30,000 for his $16,000 house and finally has it in escrow.

These are Asa Brown's people.

Asa Brown Investments was located on the northwest side of Sample Street. The heart of Los Angeles' Eastside. The office was sandwiched between Stein's Second Hand Store and an unoccupied store front whose former tenants had been The Lasting Church of God in Christ. The Lasting Church hadn't lasted very long. Six months after the store front was leased, it had become vacant again. The rumor had circulated up and down Sample Street that the Reverend Mister Smallwood, the Lasting Church's Jehovah, had disappeared with his honey-hued pianist and all of his congregation's $2500.

"No better for 'em," Asa said,

when Dock the bartender at JoJo's told Asa about it. Asa emptied the shot glass of Jim Beam, tapped the bar with his manicured forefinger, and smiled as Dock refilled the glass. "No better for 'em. If they can't see nothing but hand clapping and tambourine banging. No better for 'em."

Asa had grown up on Mott Avenue. Three blocks from where Sample Street crossed Fifth Place. His widowed mother still lived in the thirty-five-year-old whitewashed frame, surrounded by a steel mesh fence. Asa pleaded, nagged, and finally played the buffoon in his attempt to get the old woman to move west of Main Street.

"Listen, Mattie Brown," Asa would say, planting a kiss on her ebony satin brow, "west is where it is. And when did Auntie Mattie ever get left out of the action?" But the widow remained steadfast in her refusal to move. She grew roses, daisies, and marigolds around the border of the front lawn next to the fence. In the back yard she cultivated a garden of mustards and collards, sweet corn, and tomatoes. She knitted afghans for her grandchildren's beds, crocheted table cloths and doilies for her daughter-in-law, and filled Asa's freezer with home made yeast rolls and chicken pies. But she would not leave the house on Mott Avenue.

Asa's patience finally diminished. One night when he and his wife Callie were splitting a tall can of Schlitz and looking at "Duel in the Sun" on the late show, Asa's nasal voice suddenly filled the room. "Hell, if the old lady don't want to move, let her stay there. She owns the place. She gets income from that other roach-infested four-unit rental

even it is the county that foots the bill. I help her with the taxes. If she wants to stay around all those Mississippi coons, let her!"

Callie tore her hazel eyes away from the color splash on the television screen. A pink blush rose in her octaroon cheeks. She stared momentarily at Asa then returned to the handsome face of Gregory Peck.

The next day Asa closed the office early. At four-fifteen he sent home the pretty tawny high school senior who worked as his part-time secretary. The office usually stayed open until six-thirty, late enough to catch the hod carriers and longshoremen who might be working overtime. The more overtime the more green to invest in Asa Brown Investments: the service station over on Central and Clymar; the all-"blood"-run supermarket at Sample and Germaine Streets; the proposed Afro-American Progressive Credit Union. Asa watched the student close the door and start down the street toward the mid-city bus stop. He walked to the paned wall at the front of the two-desked office and followed her until the gold lambs wool sweater, the maiden hips swaying in the beige knit skirt, and the jaunty white kid boots were out of sight. Asa smiled. He didn't mind paying the girl a full salary when she didn't make all her hours. She was one of his people. He sure as hell wouldn't have one of the new breed of nappy-haired, bangle-wearing nationalists sitting behind his mahogany stenographer's desk.

It took five minutes to walk from the office to JoJo's. JoJo's was a landmark. As much a part of Sample Street as the empty cigarette packs and squashed beer cans gutter. As enduring as the Jewish pawn shops, the Jewish clothing stores, and the syndicate-owned, "blood"-run liquor stores. Asa slipped onto a stool at the far end of the bar beneath the Hamm's bear smiling out of his winter wonderland. Dock waved from the cash register, picked up the familiar bottle of bourbon and a shot glass, and started towards his friend. Dock and Asa went way back together. They had gotten drunk for the first time out of the same fifth of wine that Asa slipped out of the Chink grocery down the street from the high school. They were fifteen years old and as Eastside as the youngsters who now sported tee shirts with Malcolm X's spectacled image.

Asa shook his head and pointed to the Hamm's bear.

Dock shrugged. "What's the matter? You sick? You ain't had nothing but beer all week!"

Asa waited until he had taken the foam from the glass before he spoke. "You know this Lottie Stevens you been investing for? Well, things are looking bad. She's got over $5,000 in and . . . hell, I've gotta fold."

Asa put the beer glass down and thought of his mother. An ancient lioness at sixty. Bent from scrubbing Mister Charlie's floor; numb after taking the crap his father dished out. Asa had been glad when the old man died and the gray face and snow hair were out of view forever. Lottie Stevens was probably like his mother. She had to be to think like Asa Brown people. Migrated to Los Angeles, from Louisiana or Texas, before the Japs ever thought of dropping a bomb at Pearl Harbor. Before niggers began writing home about the bread they made at North Amer-

ican and Lockheed. Asa emptied the glass of lager. He couldn't waste sympathy on Lottie Stevens. He had never even seen her. She didn't even have sense enough to tend to her own business. Dock of all people took care of it for her. Dock was still chitlins and black-eyed peas; twenty-five-dollar suits and turned-up-toed shoes; Saturday night broads, black as tar and ugly as sin. Coons, all of them. The Jews were getting fat off of them every day. It seemed fitting that one of their own ought to be in on the take too. Yes, Asa Brown had done all right. Callie had an account at Robinson's and Haggarty's. Her shoes came from Wetherby Kyser. Callie drove the Barracuda to her job at city hall. He drove the old Chevy when he made the east of Main Street trip to the office each morning, and he sported the new thick cut hair. Just another one of the brothers. The last time he and Callie went to Vegas they had stayed at The Flamingo and gambled at the plush casinos of The Dunes, The Alladin, and Caesar's Palace. The sidewalks of the strip belonged to Mister and Missus Asa Brown. The pale-skinned whiteys with their fat paunches and starved-for-glamour's-sake women could no longer hoard it for themselves. Yes, they left gaudy downtown Vegas to the poor whiteys and the niggers whose Vegas experience was a once in a lifetime dream.

Asa shook his head. He didn't have time to cry about Lottie Stevens or any of the others. The supermarket hadn't brought in the loot he anticipated. Coons still did their shopping at the rebuilt Safeways and Ralphs that they had leveled during the riots. It just went to show how dumb a nigger was. When Texaco showered credit cards onto the populace, the service station's business dropped. The niggers went for that sucker bait. He had hired two Muslims as mechanics in the station's body shop. But on a windy rain-promising-sky day late in November, the grease monkeys forgot Elijah Mohammed's teaching and got loaded on a half gallon of Gallo port. They messed up two cars. The word got around. The body and fender shop had been closed ever since. There was never any sweat though. When Callie had the miscarriage she wouldn't go to any other hospital but Cedars, and it had cost him nearly $1,000. Adjusting the books was simple; a new advertising expenditure for the market was created. The week of their fourth wedding anniversary party he gave Callie a blue mutation mink stole. So the study of economic conditions in the community was initiated, and business analyst A. Bowen was added to the payroll and placed on the books under miscellaneous necessities. That was the breaks.

Asa looked down the bar at Dock. Dock was making a couple of gin and tonics for two business-suited whiteys. Fuzz. Asa could spot them a mile away. Why had Dock put up for this Lottie Stevens? She probably belonged to the scrubbing floors, little-Charlie's-nanny sect like his mother. Put in over $5,000. Every penny her black hands had ever saved.

"Hey, Dock," Asa called, "I'll take that bourbon after all."

Seeing Mattie Brown's whitewashed frame house always did something to Asa. He sure as hell couldn't live in the house on Mott Avenue again.

Not if his life depended on it and there wasn't a fat chance of that. Still it was home. Mattie was bent over the ironing board, listening to Mahalia Jackson's thunder voice singing "Precious Lord Take My Hand." Mattie glided the iron back and forth over a starched white dress for Asa's baby girl.

"Mama, I told you, you don't have to iron perma-prest. That's what it means, permanently-pressed."

Mattie looked up. "Them little dresses don' look right, with no starch in 'em. Sit down. I jus' took a sweet potato pie out of the stove. I'll make you a cup of hot tea."

Asa shook his head. But it didn't do any good. He'd have to eat even if he wasn't hungry. When he got home he was going to raise Cain with Callie. He'd told her when the kids slept over, not to leave their clothes. His Mama's back was too painful to be bending over an ironing board. His Mama ironed everything, even the dish towels.

While Asa sat at the oak table near the kitchen, half of the sweet potato pie in his stomach, the remaining portion tempting from the pyrex dish in front of him, he could hear Mattie putting gumbo into an empty half-gallon pickle jar. The smell of crab paws and large pink shrimp, okra, and tomatoes filled his nostrils. Callie at least appreciated receiving Mattie's New Orleans dishes.

"Mama, the business is finished," Asa muttered, his mouth full of pie. Mattie dropped the ladle with which she had been spooning the gumbo. Asa looked into the kitchen. The pupils of Mattie's eyes moved crazily back and forth and her thin dark lips trembled in her equally dark face. Asa thought of Lottie Stevens.

"Don't worry, Mama. I got something else going. Your son is going to be one of the new mayor's field deputies. It's a seventeen-thousand-a-year job. Think about it, Auntie Mattie. Seventeen thousands bucks a year."

"You mean you done sold us out for a white man at City Hall?"

Asa washed down the last of the pie with the cold tea. He brushed the flakey crumbs from the side of his mouth and thought about Lottie Stevens.

"Where's your savings, Mama?"

"Asa, I ask you somethin' first. I don' care how old you git. Answer me first."

"I ain't sold out to nobody. I'll be right down here in this area. Sample Street always been my home, ain't it? I'll be right here serving my people jest like always. Making Los Angeles a better place to live, like always." Conning Mattie was hard. She knew a lie a yard away. Asa's head was starting to hurt. "Where's your savings, Mama? You raised that bastard-in-Beverly-Hills' children for that money."

Mattie picked up the ladle and finished filling the pickle jar. "It's in the bank. Where you think it was?"

Mattie Brown died the day after the investors were notified of the demise of Asa Brown Investments. Asa had never been able to cry even as a kid when his father beat him with a cat-o-nine tails. The passing of his mother was the deepest wound he had ever sustained. But there were no tears. He grieved and thought about Lottie Stevens. The night of the rosary for Mattie, family and friends gathered in the flower-

wreathed living room on Mott Avenue after returning from the mortuary. Neighbor women moved in and out of the kitchen, pouring coffee, refilling the dishes on the table with fried chicken, potato salad, cake, and cookies. Black hands pumped Asa's arm until it ached. Teary-eyed brown women hugged his neck, planting kisses on his moist cheeks, community expressions of sympathy for the loss of his mother, the failure of his business, and nigger bewilderment at another example of God's inexplicable will. Dock stood by the front door never taking his angry eyes from Asa's face, and Asa thought of Lottie Stevens.

Asa went through Mattie's papers ten times. He found the deed for the house on Mott Avenue; the rent subsidy agreement with the county for the units north of Sample and Germaine Streets; a yellow scotch-tape-mended paper attesting to the fact that one Mattie Marie St. Amant had married Asa Joseph Brown in the city of New Orleans at the Church of the Blessed Sacrament on January 30th, 1932. The birth certificate of Asa Joseph Anthony Brown, Junior, born July 10th, 1930, was attached to the marriage certificate with a rusted paper clip. Asa wanted Mattie's bank book and he couldn't find it. Asa could feel Callie's eyes on his back again as he went through the papers for the last time. That's what he had told her when he had opened the box again after dinner the night before. "Listen, Callie, if Mama was Lottie Stevens, I took her life as sure as if I had smothered her in her bed." Callie was holding the evening paper in her hands. She threw it into the den next to Asa's feet. But Asa was already opening a paper-filled manila envelope.

"So what if Auntie Mattie was Lottie Stevens?" Callie said, anger reddening her yellow skin.

Asa looked up. "Well, I guess I'll buy that stained glass window she always wanted for St. Gregory's."

"Have you seen Dock? Have you asked him about Lottie Stevens?" Callie wiped the tears from her eyes. "That window would cost $1500. All the savings we've got left."

Asa looked blankly at Callie. He'd heard that Dock was dabbling in politics. Someone else was taking his place at JoJo's. Asa picked up the marriage certificate and read it over again for the eleventh time.

Callie really flipped when Asa told her that the mayor had given the field deputy job to an Eastside nigger who owned a cleaning shop on Sample Street. "Some unfavorable community sentiment just now, with your business holdings going sour," the mayor said. Asa tried to explain to Callie that being the mayor's economic board representative paid less but in time he'd be back on top again. Callie pursed her pink lips and spat in his face. Asa slapped her good. Later he apologized. Sure, if he had known the field deputy job would fall through he wouldn't have donated the window to St. Gregory's. But the Mattie Brown-Lottie Stevens-filled days and nights had ceased. Asa knew that if he had to do it over again, St. Gregory's would get Mattie Brown's window.

Asa walked up Sample Street. Past the loan shops, the small dirty-floored groceries, the beauty shops, and the second-hand stores. Past the old Asa Brown Investment office. Large white letters outlined in red paint decorated the windows now: LOOK TO HIM WHO SAVES. THE MACEDONIA BAPTIST CHURCH. REVEREND B. B. BLAKELY, PASTOR. Asa laughed. When he opened his next office it would be further down on Sample Street. Maybe near the furniture store or the new Woolworth's.

JoJo's was empty except for Dock and a woman who was sitting at the bar drinking a glass of sherry. Dock nodded as Asa pointed to the Hamm's bear.

"Come on down here," Dock called.

Asa took a seat next to the woman. She was wrapped in a heavy black coat and he noticed that her hands shook when she reached for the sherry. Gray wool escaped the confines of a faded blue bandana tied around her head. Dock put the bottle of beer on the bar.

"This is my Auntie, Asa," Dock said.

Asa turned and looked into a pair of faded brown eyes. The pupils changed to blue-black and he was looking at his Mama again. Mattie Brown smiled.

"Auntie Lottie, this is Asa Brown," Dock said.

A chill seized Asa and tears slid down his cheeks. Mattie Brown's glazed eyes bore into him.

"Lottie Stevens?" Asa muttered.

"Lottie Stevens," Dock said.

Asa put a half dollar on the bar and hurried out to Sample Street. He had to get home. Back to Asa Brown people. It just went to show you what happened when you started thinking like a coon. Damn it! You acted like 'em too.

Old Man Minick

Edna Ferber

His wife had always spoiled him outrageously. No doubt of that. Take, for example, the matter of the pillows merely. Old man Minick slept high. That is, he thought he slept high. He liked two plump pillows on his side of the great, wide, old-fashioned cherry bed. He would sink into them with a vast grunting and sighing and puffing expressive of nerves and muscles relaxed and gratified. But in the morning there was always one pillow on the floor. He had thrown it there. Always, in the morning, there it lay, its plump white cheek turned reproachfully up at him from the side of the bed. Ma Minick knew this, naturally, after forty years of the cherry bed. But she never begrudged him that extra pillow. Each morning, when she arose, she picked it up on her way to shut the window. Each morning the bed was made up with two pillows on his side of it, as usual.

Then there was the window. Ma Minick liked it open wide. Old man Minick, who rather prided himself on his modernism (he called it being up to date) was distrustful of the night air. In the folds of its sable mantle lurked a swarm of dread things—cold, clammy miasmas, fevers.

"Night air's just like any other air," Ma Minick would say, with some asperity. Ma Minick was no worm; and as modern as he. So when they went to bed the window would be open wide. They would lie there, the two old ones, talking comfortably about commonplace things. The kind of talk that goes on between a man and a woman who have lived together in wholesome peace (spiced with occasional wholesome bickerings) for more than forty years.

"Remind me to see Gerson to-morrow about that lock on the basement door. The paper's full of burglars."

"If I think of it." She never failed to.

"George and Nettie haven't been over in a week now."

"Oh, well, young folks . . . Did you stop in and pay that Koritz the fifty cents for pressing your suit?"

"By golly, I forgot again! First thing in the morning."

A sniff. "Just smell the Yards." It was Chicago.

"Wind must be from the west."

Sleep came with reluctant feet, but they wooed her patiently. And presently she settled down between them and they slept lightly. Usually, some time during the night, he awoke, slid cautiously and with infinite stealth from beneath the covers and closed the wide-flung window to within a bare two inches of the sill. Almost invariably she heard him; but she was a wise old woman; a philosopher of parts. She knew better than to allow a window to shatter the peace of their marital felicity. As she lay there, smiling a little grimly in the dark and giving no sign of being awake, she thought, "Oh, well, I guess a closed window won't kill me either."

Still, sometimes, just to punish him a little, and to prove that she was nobody's fool, she would wait until he had dropped off to sleep again and then she, too, would achieve a stealthy trip to the window and would raise it slowly, carefully, inch by inch.

"How did that window come to be open?" he would say in the morning, being a poor dissembler.

"Window? Why, it's just the way it was when we went to bed." And she would stoop to pick up the pillow that lay on the floor.

There was little or no talk of death between this comfortable, active, sound-appearing man of almost seventy and this plump capable woman of sixty-five. But as always, between husband and wife, it was understood wordlessly (and without reason) that old man Minick would go first. Not that either of them had the slightest intention of going. In fact, when it happened they were planning to spend the winter in California and perhaps live there indefinitely if they liked it and didn't get too lonesome for George and Nettie, and the Chicago smoke, and Chicago noise, and Chicago smells and rush and dirt. Still, the solid sum paid yearly in insurance premiums showed clearly that he meant to leave her in comfort and security. Besides, the world is full of widows. Everyone sees that. But how many widowers? Few. Widows there are by the thousands; living alone; living in hotels; living with married daughters and sons-in-law or married sons and daughters-in-law. But of widowers in a like situation there are bewilderingly few. And why this should be no one knows.

So, then. The California trip never materialized. And the year that followed never was quite clear in old man Minick's dazed mind. In the first place, it was the year in which stocks tumbled and broke their backs. Gilt-edged securities showed themselves to be tinsel. Old man Minick had retired from active business just one year before, meaning to live comfortably on the fruit of a half-century's toil. He now saw that fruit rotting all about him. There was in it hardly enough nourishment to sustain them. Then came the day when Ma Minick went downtown to see Matthews about that pain right

here and came home looking shrivelled, talking shrilly about nothing,
and evading Pa's eyes. Followed
months that were just a jumble of
agony, X-rays, hope, despair, mor-
phia, nothingness.

After it was all over: "But I was
going first," old man Minick said,
dazedly.

The old house on Ellis near
Thirty-ninth was sold for what it
would bring. George, who knew Chicago real-estate if any one did, said
they might as well get what they
could. Things would only go lower.
You'll see. And nobody's going to
have money for years. Besides, look
at the neighborhood!

Old man Minick said George was
right. He said everybody was right.
You would hardly have recognized in
this shrunken figure and wattled face
the spruce and dressy old man whom
Ma Minick used to spoil so delight-
fully. "You know best, George. You
know best." He who used to stand up
to George until Ma Minick was
moved to say, "Now, Pa, you don't
know everything."

After Matthews' bills, and the hospital, and the nurses and the medicines and the thousand and one
things were paid there was left exactly five hundred dollars a year.

"You're going to make your home
with us, Father," George and Nettie
said. Alma, too, said this would be
best. Alma, the married daughter,
lived in Seattle. "Though you know
Ferd and I would be only too glad to
have you."

Seattle! The ends of the earth. Oh,
no. No! he protested, every fibre of
his old frame clinging to the accustomed. Seattle, at seventy! He turned
piteous eyes on his son George and
his daughter-in-law Nettie. "You're

going to make your home with us,
Father," they reassured him. He
clung to them gratefully. After it was
over Alma went home to her husband and their children.

So now he lived with George and
Nettie in the five-room flat on South
Park Avenue, just across from Washington Park. And there was no extra
pillow on the floor.

Nettie hadn't said he couldn't have
the extra pillow. He had told her he
used two and she had given him two
the first week. But every morning she
had found a pillow cast on the floor.

"I thought you used two pillows,
Father."

"I do."

"But there's always one on the
floor when I make the bed in the
morning. You always throw one on
the floor. You only sleep on one pillow, really."

"I use two pillows."

But the second week there was one
pillow. He tossed and turned a good
deal there in his bedroom off the
kitchen. But he got used to it in
time. Not used to it, exactly, but—
well——

The bedroom off the kitchen
wasn't as menial as it sounds. It was
really rather cosy. The five-room flat
held living room, front bedroom,
dining room, kitchen, and maid's
room but Nettie had no maid.
George's business had suffered with
the rest. George and Nettie had said,
"I wish there was a front room for
you, Father. You could have ours
and we'd move back here, only this
room's too small for twin beds and
the dressing table and the chif-
fonier." They had meant it—or
meant to mean it.

"This is fine," old man Minick had
said. "This is good enough for any-

body." There was a narrow white enamel bed and a tiny dresser and a table. Nettie had made gay cretonne covers and spreads and put a little reading lamp on the table and arranged his things. Ma Minick's picture on the dresser with her mouth sort of pursed to make it look small. It wasn't a recent picture. Nettie and George had had it framed for him as a surprise. They had often urged her to have her picture taken, but she had dreaded it. Old man Minick didn't think much of that photograph, though he never said so. He needed no photograph of Ma Minick. He had a dozen of them; a gallery of them; thousands of them. Lying on his one pillow he could take them out and look at them one by one as they passed in review, smiling, serious, chiding, praising, there in the dark. He needed no picture on his dresser.

A handsome girl, Nettie, and a good girl. He thought of her as a girl, though she was well past thirty. George and Nettie had married late. This was only the third year of their marriage. Alma, the daughter, had married young, but George had stayed on, unwed, in the old house on Ellis until he was thirty-six and all Ma Minick's friends' daughters had had a try at him in vain. The old people had urged him to marry, but it had been wonderful to have him around the house, just the same. Somebody young around the house. Not that George had stayed around very much. But when he was there you knew he was there. He whistled while dressing. He sang in the bath. He roared down the stairway, "Ma, where's my clean shirts?" The telephone rang for him. Ma Minick prepared special dishes for him. The servant girl said, "Oh, now, Mr. George, look what you've done! Gone and spilled the grease all over my clean kitchen floor!" and wiped it up adoringly while George laughed and gobbled his bit of food filched from pot or frying pan.

They had been a little surprised about Nettie. George was in the bond business and she worked for the same firm. A plump, handsome, eyeglassed woman with fine fresh coloring, a clear skin that old man Minick called appetizing, and a great coil of smooth dark hair. She wore plain tailored things and understood the bond business in a way that might have led you to think hers a masculine mind if she hadn't been so feminine, too, in her manner. Old man Minick had liked her better than Ma Minick had.

Nettie had called him Pop and joked with him and almost flirted with him in a daughterly sort of way. He liked to squeeze her plump arm and pinch her soft cheek between thumb and forefinger. She would laugh up at him and pat his shoulder and that shoulder would straighten spryly and he would waggle his head doggishly.

"Look out there, George!" the others in the room would say. "Your dad'll cut you out. First thing you know you'll lose your girl, that's all."

Nettie would smile. Her teeth were white and strong and even. Old man Minick would laugh and wink, immensely pleased and flattered. "We understand each other, don't we, Pop?" Nettie would say.

During the first years of their married life Nettie stayed home. She fussed happily about her little flat, gave parties, went to parties, played bridge. She seemed to love the ease, the relaxation, the small luxuries. She and George were very much in love. Before her marriage she had lived in a boarding house on Michigan Avenue. At mention of it now she puckered up her face. She did not attempt to conceal her fondness for these five rooms of hers, so neat, so quiet, so bright, so cosy. Over-stuffed velvet in the living room, with silk lampshades, and small tables holding books and magazines and little boxes containing cigarettes or hard handies. Very modern. A gate-legged table in the dining room. Caramel-colored walnut in the bedroom, rich and dark and smooth. She loved it. An orderly woman. Everything in its place. Before eleven o'clock the little apartment was shining, spotless; cushions plumped, crumbs brushed, vegetables in cold water. The telephone. "Hello! . . . Oh, hello, Bess! Oh, hours ago . . . Not a thing . . . Well, if George is willing . . . I'll call him up and ask him. We haven't seen a show in two weeks. I'll call you back within the next half hour . . . No, I haven't done my marketing yet . . . Yes, and have dinner downtown. Meet at seven."

Into this orderly smooth-running mechanism was catapulted a bewildered old man. She no longer called him Pop. He never dreamed of squeezing the plump arm or pinching the smooth cheek. She called him Father. Sometimes George's Father. Sometimes, when she was telephoning, there came to him—"George's father's living with us now, you know. I can't."

They were very kind to him, Nettie and George. "Now just you sit right down here, Father. What do you want to go poking off into your own room for?"

He remembered that in the last year Nettie had said something about going back to work. There wasn't enough to do around the house to keep her busy. She was sick of afternoon parties. Sew and eat, that's all, and gossip, or play bridge. Besides, look at the money. Business was awful. The two old people had resented this idea as much as George had—more, in fact. They were scandalized.

"Young folks nowdays!" shaking their heads. "Young folks nowdays. What are they thinking of! In my day when you got married you had babies."

George and Nettie had had no babies. At first Nettie had said, "I'm so happy. I just want a chance to rest. I've been working since I was seventeen. I just want to rest, first." One year. Two years. Three. And now Pa Minick.

Ma Minick, in the old house on Ellis Avenue, had kept a loose sort of larder; not lavish, but plentiful. They both ate a great deal as old people are likely to do. Old man Minick, especially, had liked to nibble. A handful of raisins from the box on the shelf. A couple of nuts from the dish on the sideboard. A bit of candy rolled beneath the tongue. At dinner (sometimes, towards the last, even at noon-time) a plate of steaming soup, hot, revivifying, stimulating. Plenty of this and plenty of that. "What's the matter, Jo? You're not eating." But he was, amply. Ma Minick had liked to see him eat too much. She was wrong, of course.

But at Nettie's things were different. Hers was a sufficient but stern ménage. So many mouths to feed; just so many lamb chops. Nettie knew about calories and vitamins and mysterious things like that, and talked about them. So many calories in this. So many calories in that. He never was quite clear in his mind about these things said to be lurking in his food. He had always thought of spinach as spinach, chops as chops. But to Nettie they were calories. They lunched together, these two. George was, of course, downtown. For herself Nettie would have one of those feminine pick-up lunches; a dab of apple sauce, a cup of tea, and a slice of cold toast left from breakfast. This she would eat while old man Minick guiltily supped up his cup of warmed-over broth, or his coddled egg. She always pressed upon him any bit of cold meat that was left from the night before, or any remnants of vegetable or spaghetti. Often there was quite a little fleet of saucers and sauce plates grouped about his main plate. Into these he dippel and swooped uncomfortably, and yet with a relish. Sometimes, when he had finished, he would look about, furtively.

"What'll you have, Father? Can I get you something?"

"Nothing, Nettie, nothing. I'm doing fine." She had finished the last of her wooden toast and was waiting for him, kindly.

Still, this balanced and scientific fare seemed to agree with him. As the winter went on he seemed actually to have regained most of his former hardiness and vigor. A handsome old boy he was, ruddy, hale, with the zest of a juicy old apple, slightly withered but still sappy. It should be mentioned that he had a dimple in his cheek which flashed unexpectedly when he smiled. It gave him a roguish—almost boyish—effect most appealing to the beholder. Especially the feminine beholder. Much of his spoiling at the hands of Ma Minick had doubtless been due to this mere depression of the skin.

Spring was to bring a new and welcome source of enrichment into his life. But these first six months of his residence with George and Nettie were hard. No spoiling there. He missed being made much of. He got kindness, but he needed love. Then, too, he was rather a gabby old man. He liked to hold forth. In the old house on Ellis there had been visiting back and forth between men and women of his own age, and Ma's. At these gatherings he had waxed oratorical or argumentative, and they had heard him, some in agreement, some in disagreement, but always respectfully, whether he prated of real estate or social depravity; prohibition or European exchange.

"Let me tell you, here and now, something's got to be done before you can get a country back on a sound financial basis. Why, take Russia alone, why . . ." Or: "Young people nowadays! They don't know what respect means. I tell you there's got to be a change and there will be, and it's the older generation that's got to bring it about. What do they know of hardship! What do they know about work—real work. Most of 'em's never done a real day's work in their life. All they think of is dancing and gambling and drinking. Look at the way they dress! Look at . . ."

Ad lib.

"That's so," the others would agree. "I was saying only yesterday . . ."

Then, too, until a year or two before, he had taken active part in business. He had retired only at the urging of Ma and the children. They said he ought to rest and play and enjoy himself.

Now, as his strength and good spirits gradually returned he began to go downtown, mornings. He would dress, carefully, though a little shakily. He had always shaved himself and he kept this up. All in all, during the day, he occupied the bathroom literally for hours, and this annoyed Nettie to the point of frenzy, though she said nothing. He liked the white cheerfulness of the little tiled room. He puddled about in the water endlessly. Snorted and splashed and puffed and snuffled and blew. He was one of those audible washers who emerge dripping and whose ablutions are distributed impartially over ceiling, walls, and floor.

Nettie, at the closed door: "Father, are you all right?"

Splash! Prrrf! "Yes. Sure. I'm all right!"

"Well, I didn't know. You've been in there so long."

He was a neat old man, but there was likely to be a spot or so on his vest or his coat lapel, or his tie. Ma used to remove these, on or off him, as the occasion demanded, rubbing carefully and scolding a little, making a chiding sound between tongue and teeth indicative of great impatience of his carelessness. He had rather enjoyed these sounds, and this rubbing and scratching on the cloth with the fingernail and a moistened rag. They indicated that someone cared. Cared about the way he looked. Had pride in him. Loved him. Nettie never removed spots. Though infrequently she said, "Father, just leave that suit out, will you? I'll send it to the cleaner's with George's. The man's coming tomorrow morning." He would look down at himself, hastily, and attack a spot here and there with a futile fingernail.

His morning toilette completed, he would make for the Fifty-first Street L. Seated in the train he would assume an air of importance and testy haste; glance out of the window; look at his watch. You got the impression of a handsome and well-preserved old gentleman on his way downtown to consummate a shrewd business deal. He had been familiar with Chicago's downtown for fifty years and he could remember when State Street was a tree-shaded cottage district. The noise and rush and clangor of the Loop had long been familiar to him. But now he seemed to find the downtown trip arduous, even hazardous. The roar of the elevated trains, the hoarse hoots of the motor horns, the clang of the street cars, the bedlam that is Chicago's downtown district bewildered him, frightened him almost. He would skip across the street like a harried hare, just missing a motor truck's nose and all unconscious of the stream of invective directed at him by his charioteer. "Heh! Watcha! . . . Look!"—Sometimes a policeman came to his aid, or attempted to, but he resented this proffered help.

"Say, look here, my lad," he would say to the tall, tired, and not at all burly (standing on one's feet directing traffic at Wabash and Madison for eight hours a day does not make for burliness) policeman, "I've been coming downtown since long before you were born. You don't need to help me. I'm no jay from the country."

He visited the Stock Exchange. This depressed him. Stocks were lower than ever and still going down. His five hundred a year was safe, but the rest seemed doomed for his lifetime, at least. He would drop in at George's office. George's office was pleasantly filled with dapper, neat young men and (surprisingly enough) dapper, slim young women, seated at desks in the big light-flooded room. At one corner of each desk stood a polished metal placard on a little standard, and bearing the name of the desk's occupant. Mr. Owens. Mr. Satterlee. Mr. James. Miss Rauch. Mr. Minick.

"Hello, Father," Mr. Minick would say, looking annoyed. "What's bringing you down?"

"Oh, nothing. Nothing. Just had a little business to tend to over at the Exchange. Thought I'd drop in. How's business?"

"Rotten."

"I should think it was!" Old man Minick would agree. "I—should—think—it—was! Hm."

George wished he wouldn't. He couldn't have it, that's all. Old man Minick would stroll over to the desk marked Satterlee, or Owens, or James. These brisk young men would toss an upward glance at him and concentrate again on the sheets and files before them. Old man Minick would stand, balancing from heel to toe and blowing out his breath a little. He looked a bit yellow and granulated and wavering, there in the cruel morning light of the big plate glass windows. Or per-

haps it was the contrast he presented with these slim, slick young salesmen.

"Well, h'are you to-day, Mr.—uh —Satterlee? What's the good word?"

Mr. Satterlee would not glance up this time. "I'm pretty well. Can't complain."

"Good. Good."

"Anything I can do for you?"

"No-o-o. No. Not a thing. Just dropped in to see my son a minute."

"I see." Not unkindly. Then, as old man Minick still stood there, balancing, Mr. Satterlee would glance up again, frowning a little. "Your son's desk is over there, I believe. Yes."

George and Nettie had a bedroom conference about these visits and Nettie told him, gently, that the bond house head objected to friends and relatives dropping in. It was against office rules. It had been so when she was employed there. Strictly business. She herself had gone there only once since her marriage.

Well, that was all right. Business was like that nowadays. Rush and grab and no time for anything.

The winter was a hard one, with a record snowfall and intense cold. He stayed indoors for days together. A woman of his own age in like position could have occupied herself usefully and happily. She could have hemmed a sash-curtain; knitted or crocheted; tidied a room; taken a hand in the cooking or preparing of food; ripped an old gown; made over a new one; indulged in an occasional afternoon festivity with women of her own years. But for old man Minick there were no small tasks. There was nothing he could do to make his place in the household justifiable. He wasn't even particularly good at those small jobs of hammering, or painting, or general "fixing." Nettie could drive a nail more swiftly, more surely than he. "Now, Father, don't you bother. I'll do it. Just you go and sit down. Isn't it time for your afternoon nap?"

He waxed a little surly. "Nap! I just got up. I don't want to sleep my life away."

George and Nettie frequently had guests in the evening. They played bridge, or poker, or talked.

"Come in, Father," George would say. "Come in. You all know Dad, don't you, folks?" He would sit down, uncertainly. At first he had attempted to expound, as had been his wont in the old house on Ellis. "I want to say, here and now, that this country's got to . . ." But they went on, heedless of him. They interrupted or refused, politely, to listen. So he sat in the room, yet no part of it. The young people's talk swirled and eddied all about him. He was utterly lost in it. Now and then Nettie or George would turn to him and with raised voice (he was not at all deaf and prided himself on it) would shout, "It's about this or that, Father. He was saying . . ."

When the group roared with laughter at a sally from one of them he would smile uncertainly but amiably, glancing from one to the other in complete ignorance of what had passed, but not resenting it. He took to sitting more and more in his kitchen bedroom, smoking a comforting pipe and reading and rereading the evening paper. During that winter he and Canary, the Negro washwoman, became quite good friends. She washed down in the basement once a week but came up to the kitchen for her massive lunch. A walrus-waisted black woman, with a rich throaty voice, a rolling eye, and a kindly heart. He actually waited for her appearance above the laundry stairs.

"Weh, how's Mist' Minick to-day! Ah nev' did see a gemun spry's you ah fo' yo' age. No, suh! nev' did."

At this rare praise he would straighten his shoulders and waggle his head. "I'm worth any ten of these young sprats to-day." Canary would throw back her head in a loud and companionable guffaw.

Nettie would appear at the kitchen swinging door. "Canary's having her lunch, Father. Don't you want to come into the front room with me? We'll have our lunch in another half-hour."

He followed her obediently enough. Nettie thought of him as a troublesome and rather pathetic child—a child who would never grow up. If she attributed any thoughts to that fine old head they were ambling thoughts, bordering, perhaps, on senility. Little did she know how expertly this old one surveyed her and how ruthlessly he passed judgment. She never suspected the thoughts that formed in the active brain.

He knew about women. He had married a woman. He had had children by her. He looked at this woman—his son's wife—moving about her little five-room flat. She had theories about children. He had heard her expound them. You didn't have them except under such and such circumstances. It wasn't fair otherwise. Plenty of money for their education. Well. He and his wife had had three children. Paul, the second, had died at thirteen. A blow, that had been. They had not always planned for the coming of the three but they always had found a way, afterward. You managed, somehow, once the little wrinkled red ball had fought its way into the world. You managed. You managed. Look at George! Yet when he was born, thirty-nine years ago, Pa and Ma Minick had been hard put to it.

Sitting there, while Nettie dismissed him as negligible, he saw her clearly, grimly. He looked at her. She was plump, but not too short, with a generous width between the hips; a broad full bosom, but firm; round arms and quick slim legs; a fine sturdy throat. The curve between arm and breast made a graceful gracious line . . . Working in a bond office . . . Working in a bond office . . . There was nothing in the Bible about working in a bond office. Here was a woman built for child-bearing.

She thought him senile, negligible.

In March Nettie had in a sewing woman for a week. She had her two or three times a year. A hawk-faced woman of about forty-nine, with a blue-bottle figure and a rapacious eye. She sewed in the dining room and there was a pleasant hum of machine and snip of scissors and murmur of conversation and rustle of silky stuff; and hot savory dishes for lunch. She and old man Minick became great friends. She even let him take out bastings. This when Nettie had gone out from two to four, between fittings.

He chuckled and waggled his head. "I expect to be paid regular assistant's wages for this," he said.

"I guess you don't need any wages, Mr. Minick," the woman said. "I guess you're pretty well fixed."

"Oh, well, I can't complain." (Five hundred a year.)

"Complain! I should say not! If I was to complain it'd be different. Work all day to keep myself; and nobody to come home to at night."

"Widow, ma'am?"

"Since I was twenty. Work, work, that's all I've had. And lonesome! I suppose you don't know what lonesome is."

"Oh, don't I!" slipped from him. He had dropped the bastings.

The sewing woman flashed him a look from the cold hard eye. "Well, maybe you do. I suppose living here like this, with sons and daughters, ain't so grand, for all your money. Now me, I've always managed to keep my own little place that I could call home, to come back to. It's only two rooms, and nothing to rave about, but it's home. Evenings I just cook and fuss around. Nobody to fuss for, but I fuss, anyway. Cooking, that's what I love to do. Plenty of good food, that's what folks need to keep their strength up." Nettie's lunch that day had been rather scant.

She was there a week. In Nettie's absence she talked against her. He protested, but weakly. Did she give him egg-nogs? Milk? Hot toddy? Soup? Plenty of good rich gravy and meat and puddings? Well! That's what folks needed when they weren't so young any more. Not that he looked old. My, no. Sprier than many young boys, and handsomer than his own son if she did say so.

He fed on it, hungrily. The third day she was flashing meaning glances at him across the luncheon table. The fourth she pressed his foot beneath the table. The fifth, during

Nettie's afternoon absence, she got up, ostensibly to look for a bit of cloth which she needed for sewing, and, passing him, laid a caressing hand on his shoulder. Laid it there and pressed his shoulder ever so little. He looked up, startled. The glances across the luncheon had largely passed over his head; the foot beneath the table might have been an accident. But this—this was unmistakable. He stood up, a little shakily. She caught his hand. The hawk-like face was close to his.

"You need somebody to love you," she said. "Somebody to do for you, and love you." The hawk face came nearer. He leaned a little toward it. But between it and his face was Ma Minick's face, plump, patient, quizzical, kindly. His head came back sharply. He threw the woman's hot hand from him.

"Woman!" he cried. "Jezebel!"

The front door slammed. Nettie. The woman flew to her sewing. Old man Minick, shaking, went into his kitchen bedroom.

"Well," said Nettie, depositing her bundles on the dining room table, "did you finish that fagotting? Why, you haven't done so very much, have you!"

"I ain't feeling so good," said the woman. "That lunch didn't agree with me."

"Why, it was a good plain lunch. I don't see——"

"Oh, it was plain enough, all right."

Next day she did not come to finish her work. Sick, she telephoned. Nettie called it an outrage. She finished the sewing herself, though she hated sewing. Pa Minick said nothing, but there was a light in his eye.

Now and then he chuckled, to Nettie's infinite annoyance, though she said nothing.

"Wanted to marry me!" he said to himself, chuckling. "Wanted to marry me! The old rip!"

At the end of April, Pa Minick discovered Washington Park, and the Club, and his whole life was from that day transformed.

He had taken advantage of the early spring sunshine to take a walk, at Nettie's suggestion.

"Why don't you go into the Park, Father? It's really warm out. And the sun's lovely. Do you good."

He had put on his heaviest shirt, and a muffler, and George's old red sweater with the great white "C" on its front, emblem of George's athletic prowess at the University of Chicago; and over all, his greatcoat. He had taken warm mittens and his cane with the greyhound's-head handle, carved. So equipped he had ambled uninterestedly over to the Park across the way. And there he had found new life.

New life in old life. For the Park was full of old men. Old men like himself, with greyhound's-head canes, and mufflers and somebody's sweater worn beneath their greatcoats. They wore arctics, though the weather was fine. The skin of their hands and cheek-bones was glazed and had a tight look though it lay in fine little folds. There were splotches of brown on the backs of their hands, and on the temples and forehead. Their heavy grey or brown socks made comfortable folds above their ankles. From that April morning until winter drew on the Park saw old man Minick daily. Not only daily but by the day. Except for his meals, and a brief hour for his after-luncheon nap, he spent all his time there.

For in the Park old man Minick and all the old men gathered there found a Forum—a safety valve—a means of expression. It did not take him long to discover that the Park was divided into two distinct sets of old men. There were the old men who lived with their married sons and daughters-in-law or married daughters and sons-in-law. Then there were the old men who lived in the Grant Home for Aged Gentlemen. You saw its fine red-brick façade through the trees at the edge of the Park

And the slogan of these first was: "My son and my da'ter they wouldn't want me to live in any public Home. No, siree! They want me right there with them. In their own home. That's the kind of son and daughter I've got!"

The slogan of the second was: "I wouldn't live with any son or daughter. Independent. That's me. My own boss. Nobody to tell me what I can do and what I can't. Treat you like a child. I'm my own boss! Pay my own good money and get my keep for it."

The first group, strangely enough, was likely to be spotted of vest and a little frayed as to collar. You saw them going on errands for their daughters-in-law. A loaf of bread. Spool of white No. 100. They took their small grandchildren to the duck pond and between the two toddlers hand in hand—the old and infirm and the infantile and infirm —was hard to tell which led which.

The second group was shiny as to shoes, spotless as to linen, dapper as to clothes. They had no small errands. Theirs was a magnificent leisure. And theirs was magnificent conversation. The questions they discussed and settled there in the Park—these old men—were not international merely. They were cosmic in scope.

The War? Peace? Disarmament? China? Free love? Mere conversational bubbles to be tossed in the air and disposed of in a burst of foam. Strong meat for old man Minick who had so long been fed on pap. But he soon got used to it. Between four and five in the afternoon, in a spot known as Under The Willows, the meeting took the form of a club—an open forum. A certain group made up of Socialists, Free thinkers, parlor anarchists, bolshevists, had for years drifted there for talk. Old man Minick learned high-sounding phrases. "The Masters . . . democracy . . . toil of the many for the good of the few . . . the ruling class . . . free speech . . . the People. . . .

The strong-minded ones held forth. The weaker ones drifted about on the outskirts, sometimes clinging to the moist and sticky paw of a round-eyed grandchild. Earlier in the day—at eleven o'clock, say—the talk was not so general nor so inclusive. The old men were likely to drift into groups of two or three or four. They sat on sun-bathed benches and their conversation was likely to be rather smutty at times, for all they looked so mild and patriarchal and dessicated. They paid scant heed to the white-haired old women who, like themselves, were sunning in the park. They watched the young women switch by, with appreciative glances at their trim figures and slim ankles. The day of the short skirt was a grand time for them. They chuckled among themselves and made wicked comment. One saw only white-haired, placid, tremulous old men, but their minds still worked with be-lated masculinity like naughty small boys talking behind the barn.

Old man Minick early achieved a certain leadership in the common talk. He had always liked to hold forth. This last year had been one of almost unendurable bottling up. At first he had timidly sought the less assertive ones of his kind. Mild old men who sat in rockers in the pavil-ion waiting for lunch time. Their conversation irritated him. They re-marked everything that passed be-fore their eyes.

"There's a boat. Fella with a boat."

A silence. Then, heavily: "Yeh." Five minutes.

"Look at those people laying on the grass. Shouldn't think it was warm enough for that . . . Now they're getting up."

A group of equestrians passed along the bridle path on the oppo-site side of the lagoon. They made a frieze against the delicate spring greenery. The coats of the women were scarlet, vivid green, arresting, stimulating.

"Riders."

"Yes."

"Good weather for riding."

A man was fishing near by. "Good weather for fishing."

"Yes."

"Wonder what time it is, anyway." From a pocket, deep-buried, came forth a great gold blob of a watch. "I've got one minute to eleven."

Old man Minick dragged forth a heavy globe. "Mm. I've got eleven."

"Little fast, I guess."

Old man Minick shook off this con-versation. This was oral death, though he did not put it thus. He joined the other men. They were discussing Spiritualism. He listened, ventured an opinion, was heard re-spectfully and then combated merci-lessly. He rose to the verbal fight, and won it.

"Let's see," said one of the old men. "You're not living at the Grant Home, are you?"

"No," old man Minick made re-ply, proudly. "I live with my son and his wife. They wouldn't have it any other way."

"Hm. Like to be independent my-self."

"Lonesome, ain't it? Over there?"

"Lonesome! Say, Mr.—what'd you say your name was? Minick? Mine's Hughes—I never was lonesome in my life 'cept for six months when I lived with my daughter and her husband and their five children. Yes, sir. That's what I call lonesome, in an eight-room flat."

George and Nettie said, "It's doing you good, Father, being out in the air so much." His eyes were brighter, his figure straighter, his color better. It was that day he had held forth so eloquently on the emigration question. He had read a lot—papers and magazines and one thing and another—to keep up. He devoured all the books and pamphlets about bond issues and national finances brought home by George. In the Park he was considered an authority on bonds and banking. He and a re-tired real-estate man named Mowry sometimes debated a single question for weeks. George and Nettie, believed, thought he ambled to the Park and spent senile hours with his drooling old friends discussing nothing amiably and witlessly. This while he was eating strong meat, drinking strong drink.

Summer sped. Was past. Autumn held a new dread for old man Minick. When winter came where should he go? Where should he go? Not back to the five-room flat all day, and the little back bedroom, and nothingness. In his mind there rang a childish old song they used to sing at school. A silly song:

Where do all the birdies go?
I know. *I* know.

But he didn't know. He was terror-stricken. October came and went. With the first of November the Park became impossible, even at noon, and with two overcoats and the sweater. The first frost was a black frost for him. He scanned the heavens daily for rain or snow. There was a cigar store and billiard room on the corner across the boulevard and there he sometimes went, with a few of his Park cronies, to stand behind the players' chairs and watch them at pinochle or rum. But this was a dull business. Besides, the Grant men never came there. They had card rooms of their own.

He turned away from this smoky little den on a drab November day, sick at heart. The winter. He tried to face it, and at what he saw he shrank and was afraid.

He reached the apartment and went around to the rear, dutifully. His rubbers were wet and muddy and Nettie's living room carpet was a fashionable grey. The back door was unlocked. It was Canary's day

downstairs, he remembered. He took off his rubbers in the kitchen and passed into the dining room. Voices. Nettie had company. Some friends, probably, for tea. He turned to go to his room, but stopped at hearing his own name. Father Minick. Father Minick. Nettie's voice.

"Of course, if it weren't for Father Minick I would have. But how can we as long as he lives with us? There isn't room. And we can't afford a bigger place now, with the rents what they are. This way it wouldn't be fair to the child. We've talked it over, George and I. Don't you suppose? But not as long as Father Minick is with us. I don't mean we'd use the maid's room for a—for the— if we had a baby. But I'd have to have someone in to help, then, and we'd have to have that extra room."

He stood there in the dining room, quiet. Quiet. His body felt queerly remote and numb, but his mind was working frenziedly. Clearly, too, in spite of the frenzy. Death. That was the first thought. Death. It would be easy. But he didn't want to die. Strange, but he didn't want to die. He liked Life. The Park, the trees, the Club, the talk, the whole show. . . . Nettie was a good girl . . . The old must make way for the young. They had the right to be born . . . Maybe it was just another excuse. Almost four years married. Why not three years ago? . . . The right to live. The right to live. . . .

He turned, stealthily, stealthily, and went back into the kitchen, put on his rubbers, stole out into the darkening November afternoon.

In an hour he was back. He entered at the front door this time, ringing the bell. He had never had a key. As if he were a child they would not trust him with one. Nettie's women friends were just leaving. In the air you smelled a mingling of perfume, and tea, and cakes, and powder. He sniffed it, sensitively.

"How do you do, Mr. Minick!" they said. "How are you! Well, you certainly look it. And how do you manage these gloomy days?"

He smiled genially, taking off his greatcoat and revealing the red sweater with the big white "C" on it. "I manage. I manage." He puffed out his cheeks. "I'm busy moving."

"Moving!" Nettie's startled eyes flew to his, held them. "Moving, Father?"

"Old folks must make way for the young," he said, gaily. "That's the law of life. Yes, sir! New ones. New ones."

Nettie's face was scarlet. "Father, what in the world——"

"I signed over at the Grant Home to-day. Move in next week." The women looked at her, smiling. Old man Minick came over to her and patted her plump arm. Then he pinched her smooth cheek with a quizzical thumb and forefinger. Pinched it and shook it ever so little.

"I don't know what you mean," said Nettie, out of breath.

"Yes, you do," said old man Minick, and while his tone was light and jesting there was in his old face something stern, something menacing. "Yes, you do."

When he entered the Grant Home a group of them was seated about the fireplace in the main hall. A neat, ruddy, septuagenarian circle. They greeted him casually, with delicacy of feeling, as if he were merely approaching them at their bench in the Park.

"Say, Minick, look here. Mowry here says China ought to have been included in the four-power treaty. He says——"

Old man Minick cleared his throat. "You take China, now," he said, "with her vast and practically, you might say, virgin country, why——"

An apple-cheeked maid in a black dress and a white apron stopped before him. He paused.

"Housekeeper says for me to tell you your room's all ready, if you'd like to look at it now."

"Minute. Minute, my child." He waved her aside with the air of one who pays five hundred a year for independence and freedom. The girl turned to go. "Uh—young lady! Young lady!" She looked at him. "Tell the housekeeper two pillows, please. Two pillows on my bed. Be sure."

"Yes, sir. Two pillows. Yes, sir. I'll be sure."

from

A Raisin in the Sun

Lorraine Hansberry

CHARACTERS

Ruth Younger
Travis Younger
Walter Lee Younger (Brother)
Beneatha Younger
Lena Younger (Mama)
Joseph Asagai
George Murchison
Karl Lindner
Bobo
Moving Men

ACT I

Scene One

The Younger living room would be a comfortable and well-ordered room if it were not for a number of indestructible contradictions to this state of being. Its furnishings are typical and undistinguished and their primary feature now is that they have clearly had to accommodate the living of too many people for too many years—and they are tired. Still, we can see that at some time, a time probably no longer remembered by the family (except perhaps for MAMA) the furnishings of this room were actually selected with care and love and even hope—and brought to this apartment and arranged with taste and pride.

That was a long time ago. Now the once loved pattern of the couch upholstery has to fight to show itself from under acres of crocheted doilies and couch covers which have themselves finally come to be more important than the upholstery. And here a table or a chair has been moved to disguise the worn places in the carpet; but the carpet has fought back by showing its weariness, with depressing uniformity, elsewhere on its surface.

Weariness has, in fact, won in this

room. *Everything has been polished, washed, sat on, used, scrubbed too often. All pretenses but living itself have long since vanished from the very atmosphere of this room.*

Moreover, a section of this room, for it is not really a room unto itself, though the landlord's lease would make it seem so, slopes backward to provide a small kitchen area, where the family prepares the meals that are eaten in the living room proper, which must also serve as dining room. The single window that has been provided for these "two" rooms is located in this kitchen area. The sole natural light the family may enjoy in the course of a day is only that which fights its way through this little window.

At left, a door leads to a bedroom which is shared by MAMA *and her daughter,* BENEATHA. *At right, opposite, is a second room (which in the beginning of the life of this apartment was probably a breakfast room) which serves as a bedroom for* WALTER *and his wife,* RUTH.

Time: Sometime between World War II and the present.

Place: Chicago's Southside.

At Rise: It is morning dark in the living room. TRAVIS *is asleep on the make-down bed at center. An alarm clock sounds from within the bedroom at right, and presently* RUTH *enters from that room and closes the door behind her. She crosses sleepily toward the window. As she passes her sleeping son she reaches down and shakes him a little. At the window she raises the shade and a dusky Southside morning light comes in feebly. She fills a pot with water and puts it on to boil. She calls to the boy, between yawns, in a slightly muffled voice.*

RUTH *is about thirty. We can see that she was a pretty girl, even exceptionally so, but now it is apparent that life has been little that she expected, and disappointment has already begun to hang in her face. In a few years, before thirty-five even, she will be known among her people as a "settled woman."*

She crosses to her son and gives him a good, final, rousing shake.

RUTH: Come on now, boy, it's seven thirty! *(Her son sits up at last, in a stupor of sleepiness)* I say hurry up, Travis! You ain't the only person in the world got to use a bathroom! *(The child, a sturdy, handsome little boy of ten or eleven, drags himself out of the bed and almost blindly takes his towels and "today's clothes" from drawers and a closet and goes out to the bathroom, which is in an outside hall and which is shared by another family or families on the same floor.* RUTH *crosses to the bedroom door at right and opens it and calls in to her husband)* Walter Lee! . . . It's after seven thirty! Lemme see you do some waking up in there now! *(She waits)* You better get up from there, man! It's after seven thirty I tell you. *(She waits again)* All right, you just go ahead and lay there and next thing you know Travis be finished and Mr. Johnson'll be in there and you'll be fussing and cussing round here like a mad man! And be late too! *(She waits, at the end of patience)* Walter Lee—it's time for you to get up! *(She waits another second and then starts to go into the bedroom, but is apparently satisfied that her husband has begun to get up. She stops, pulls the door to, and returns to the kitchen area. She wipes her face with a moist cloth and runs her fingers through her sleep-disheveled hair in a vain effort and ties an apron around her housecoat. The bedroom door at right opens and her husband stands in the doorway in his pajamas, which are rumpled and mismated. He is a lean, intense young man in his middle thirties, inclined to quick nervous movements and erratic speech habits—and always in his voice there is a quality of indictment)*

WALTER: Is he out yet?

RUTH: What you mean *out?* He ain't hardly got in there good yet.

WALTER: *(Wandering in, still more oriented to sleep than to a new day)* Well, what was you doing all that yelling for if I can't get in there yet? *(Stopping and thinking)* Check coming today?

RUTH: They *said* Saturday and this is just Friday and I hopes to God you ain't going to get up here first thing this morning and start talking to me 'bout no money—'cause I 'bout don't want to hear it.

WALTER: Something the matter with you this morning?

RUTH: No—I'm just sleepy as the devil. What kind of eggs you want?

WALTER: Not scrambled. *(RUTH starts to scramble eggs)* Paper come? (RUTH *points impatiently to the rolled up* Tribune *on the table, and he gets it and spreads it out and vaguely reads the front page)* Set off another bomb yesterday.

RUTH: *(Maximum indifference)* Did they?

WALTER: *(Looking up)* What's the matter with you?

RUTH: Ain't nothing the matter with me. And don't keep asking me that this morning.

WALTER: Ain't nobody bothering

you. *(Reading the news of the day absently again)* Say Colonel McCormick is sick.

RUTH: *(Affecting tea-party interest)* Is he now? Poor thing.

WALTER: *(Sighing and looking at his watch)* Oh, me. *(He waits)* Now what is that boy doing in that bathroom all this time? He just going to have to start getting up earlier. I can't be being late to work on account of him fooling around in there.

RUTH: *(Turning on him)* Oh, no he ain't going to be getting up no earlier no such thing! It ain't his fault that he can't get to bed no earlier nights 'cause he got a bunch of crazy good-for-nothing clowns sitting up running their mouths in what is supposed to be his bedroom after ten o'clock at night . . .

WALTER: That's what you mad about, ain't it? The things I want to talk about with my friends just couldn't be important in your mind, could they? *(He rises and finds a cigarette in her handbag on the table and crosses to the little window and looks out, smoking and deeply enjoying this first one)*

RUTH: *(Almost matter of factly, a complaint too automatic to deserve emphasis)* Why you always got to smoke before you eat in the morning?

WALTER: *(At the window)* Just look at 'em down there . . . Running and racing to work . . *(He turns and faces his wife and watches her a moment at the stove, and then, suddenly)* You look young this morning, baby.

RUTH: *(Indifferently)* Yeah?

WALTER: Just for a second—stirring them eggs. It's gone now—just for a second it was—you looked real young again. *(Then, drily)* It's gone

pride) That's *my* boy. *(She looks at him in disgust and turns back to her work)* You know what I was thinking 'bout in the bathroom this morning?

RUTH: No.

WALTER: How come you always try to be so pleasant!

RUTH: What is there to be pleasant 'bout!

WALTER: You want to know what I was thinking 'bout in the bathroom or not!

RUTH: I know what you thinking 'bout.

WALTER: *(Ignoring her)* 'Bout what me and Willy Harris was talking about last night.

RUTH: *(Immediately—a refrain)* Willy Harris is a good-for-nothing loud mouth.

WALTER: Anybody who talks to me has got to be a good-for-nothing loud mouth, ain't he? And what you know about who is just a good-for-nothing loud mouth? Charlie Atkins was just a "good-for-nothing loud mouth" too, wasn't he? When he wanted me to go in the dry-cleaning business with him. And now—he's grossing a hundred thousand a year. A hundred thousand dollars a year! You still call *him* a loud mouth!

RUTH: *(Bitterly)* Oh, Walter Lee . . . *(She folds her head on her arms over the table)*

WALTER: *(Rising and coming to her and standing over her)* You tired, ain't you? Tired of everything. Me, the boy, the way we live—this beat-up hole—everything. Ain't you? *(She doesn't look up, doesn't answer)* So tired—moaning and groaning all the time, but you wouldn't do nothing to help, would you? You couldn't be on my side that long for nothing, could you?

RUTH: Walter, please leave me alone.

WALTER: A man needs for a woman to back him up . . .

RUTH: Walter—

WALTER: Mama would listen to you. You know she listen to you more than she do me and Bennie. She think more of you. All you have to do is just sit down with her when you drinking your coffee one morning and talking 'bout things like you do and—*(He sits down beside her and demonstrates graphically what he thinks her methods and tone should be)*—you just sip your coffee, see, and say easy like that you been thinking 'bout that deal Walter Lee is so interested in, 'bout the store and all, and sip some more coffee, like what you saying ain't really that important to you— And the next thing you know, she be listening good and asking you questions and when I come home—I can tell her the details. This ain't no fly-by-night proposition, baby. I mean we figured it out, me and Willy and Bobo.

RUTH: *(With a frown)* Bobo?

WALTER: Yeah. You see, this little liquor store we got in mind cost seventy-five thousand and we figured the initial investment on the place be 'bout thirty thousand, see. That be ten thousand each. Course, there's a couple of hundred you got to pay so's you don't spend your life just waiting for them clowns to let your license get approved—

RUTH: You mean graft?

WALTER: *(Frowning impatiently)* Don't call it that. See there, that just goes to show you what women understand about the world. Baby, don't *nothing* happen for you in this world 'less you pay *somebody* off!

RUTH: Walter, leave me alone! *(She*

raises her head and stares at him vigorously—then says, more quietly) Eat your eggs, they gonna be cold.

WALTER: *(Straightening up from her and looking off)* That's it. There you are. Man say to his woman: I got me a dream. His woman say: Eat your eggs. *(Sadly, but gaining in power)* Man say: I got to take hold of this here world, baby! And a woman will say: Eat your eggs and go to work. *(Passionately now)* Man say: I got to change my life, I'm choking to death, baby! And his woman say—*(In utter anguish as he brings his fists down on his thighs)*—Your eggs is getting cold!

RUTH: *(Softly)* Walter, that ain't none of our money.

WALTER: *(Not listening at all or even looking at her)* This morning, I was lookin' in the mirror and thinking about it . . . I'm thirty-five years old; I been married eleven years and I got a boy who sleeps in the living room—*(Very, very quietly)*—and all I got to give him is stories about how rich white people live . . .

RUTH: Eat your eggs, Walter.

WALTER: *Damn my eggs . . . damn all the eggs that ever was!*

RUTH: Then go to work.

WALTER: *(Looking up at her)* See— I'm trying to talk to you 'bout myself—*(Shaking his head with the repetition)*—and all you can say is eat them eggs and go to work.

RUTH: *(Wearily)* Honey, you never say nothing new. I listen to you every day, every night and every morning, and you never say nothing new. *(Shrugging)* So you would rather *be* Mr. Arnold than be his chauffeur. So—I would *rather* be living in Buckingham Palace.

WALTER: That is just what is wrong with the colored woman in this world . . . Don't understand about building their men up and making 'em feel like they somebody. Like they can do something.

RUTH: *(Drily, but to hurt)* There *are* colored men who do things.

WALTER: No thanks to the colored woman.

RUTH: Well, being a colored woman, I guess I can't help myself none. *(She rises and gets the ironing board and sets it up and attacks a huge pile of rough-dried clothes, sprinkling them in preparation for the ironing and then rolling them into tight fat balls)*

WALTER: *(Mumbling)* We one group of men tied to a race of women with small minds.

His sister BENEATHA *enters. She is about twenty, as slim and intense as her brother. She is not as pretty as her sister-in-law, but her lean, almost intellectual face has a handsomeness of its own. She wears a bright-red flannel nightie, and her thick hair stands wildly about her head. Her speech is a mixture of many things; it is different from the rest of the family's insofar as education has permeated her sense of English—and perhaps the Midwest rather than the South has finally—at last—won out in her inflection; but not altogether, because over all of it is a soft slurring and transformed use of vowels which is the decided influence of the Southside. She passes through the room without looking at either* RUTH *or* WALTER *and goes to the outside door and looks, a little blindly, out to the bathroom. She sees that it has been lost to the Johnsons. She closes the door with a sleepy vengeance and crosses to the table and sits down a little defeated.*

BENEATHA: I am going to start timing those people.

WALTER: You should get up earlier.

BENEATHA: *(Her face in her hands. She is still fighting the urge to go back to bed)* Really—would you suggest dawn? Where's the paper?

WALTER: *(Pushing the paper across the table to her as he studies her almost clinically, as though he has never seen her before)* You a horrible-looking chick at this hour.

BENEATHA: *(Drily)* Good morning, everybody.

WALTER: *(Senselessly)* How is school coming?

BENEATHA: *(In the same spirit)* Lovely. Lovely. And you know, biology is the greatest. *(Looking up at him)* I dissected something that looked just like you yesterday.

WALTER: I just wondered if you've made up your mind and everything.

BENEATHA: *(Gaining in sharpness and impatience)* And what did I answer yesterday morning—and the day before that?

RUTH: *(From the ironing board, like someone disinterested and old)* Don't be so nasty, Bennie.

BENEATHA: *(Still to her brother)* And the day before that and the day before that!

WALTER: *(Defensively)* I'm interested in you. Something wrong with that? Ain't many girls who decide—

WALTER *and* BENEATHA: *(In unison)* —"to be a doctor." *(Silence)*

WALTER: Have we figured out yet just exactly how much medical school is going to cost?

RUTH: Walter Lee, why don't you leave that girl alone and get out of here to work?

BENEATHA: *(Exits to the bathroom and bangs on the door)* Come on out of there, please! *(She comes back into the room)*

WALTER: *(Looking at his sister intently)* You know the check is coming tomorrow.

BENEATHA: *(Turning on him with a sharpness all her own)* That money belongs to Mama, Walter, and it's for her to decide how she wants to use it. I don't care if she wants to buy a house or a rocket ship or just nail it up somewhere and look at it. It's hers. Not ours—*hers*.

WALTER: *(Bitterly)* Now ain't that fine! You just got your mother's interest at heart, ain't you, girl? You such a nice girl—but if Mama got that money she can always take a few thousand and help you through school too—can't she?

BENEATHA: I have never asked anyone around here to do anything for me!

WALTER: No! And the line between asking and just accepting when the time comes is big and wide—ain't it!

BENEATHA: *(With fury)* What do you want from me, Brother—that I quit school or just drop dead, which!

WALTER: I don't want nothing but for you to stop acting holy 'round here. Me and Ruth done made some sacrifices for you—why can't you do something for the family?

RUTH: Walter, don't be dragging me in it.

WALTER: You are in it— Don't you get up and go work in somebody's kitchen for the last three years to help put clothes on her back?

RUTH: Oh, Walter—that's not fair . . .

WALTER: It ain't that nobody expects you to get on your knees and say thank you, Brother; thank you, Ruth; thank you, Mama—and thank you, Travis, for wearing the same pair of shoes for two semesters—

BENEATHA: *(Dropping to her knees)* Well—I *do*—all right?—thank everybody . . . and forgive me for ever wanting to be anything at all . . . forgive me, forgive me!

RUTH: Please stop it! Your mama'll hear you.

WALTER: Who the hell told you you had to be a doctor? If you so crazy 'bout messing 'round with sick people—then go be a nurse like other women—or just get married and be quiet . . .

BENEATHA: Well—you finally got it said . . . It took you three years but you finally got it said. Walter, give up; leave me alone—it's Mama's money.

WALTER: *He was my father, too!*

BENEATHA: So what? He was mine, too—and Travis' grandfather—but the insurance money belongs to Mama. Picking on me is not going to make her give it to you to invest in any liquor stores—*(Underbreath, dropping into a chair)*—and I for one say, God bless Mama for that!

WALTER: *(To RUTH) See*—did you hear? Did you hear!

RUTH: Honey, please go to work.

WALTER: Nobody in this house is ever going to understand me.

BENEATHA: Because you're a nut.

WALTER: Who's a nut?

BENEATHA: You—you are a nut. Thee is mad, boy.

WALTER: *(Looking at his wife and his sister from the door, very sadly)* The world's most backward race of people, and that's a fact.

BENEATHA: *(Turning slowly in her chair)* And then there are all those prophets who would lead us out of the wilderness— (WALTER *slams out of the house)*—into the swamps!

RUTH: Bennie, why you always gotta be pickin' on your brother? Can't you be a little sweeter sometimes? *(Door opens.* WALTER *walks in)*

WALTER: *(To Ruth)* I need some money for carfare.

RUTH: *(Looks at him, then warms; teasing, but tenderly)* Fifty cents? *(She goes to her bag and gets money)* Here, take a taxi.

WALTER *exits.* MAMA *enters. She is a woman in her early sixties, full-bodied and strong. She is one of those women of a certain grace and beauty who wear it so unobtrusively that it takes a while to notice. Her dark-brown face is surrounded by the total whiteness of her hair, and, being a woman who had adjusted to many things in life and overcome many more, her face is full of strength. She has, we can see, wit and faith of a kind that keep her eyes lit and full of interest and expectancy. She is, in a word, a beautiful woman. Her bearing is perhaps most like the noble bearing of the women of the Hereros of Southwest Africa—rather as if she imagines that as she walks she still bears a basket or a vessel upon her head. Her speech, on the other hand, is as careless as her carriage is precise—she is inclined to slur everything—but her voice is perhaps not so much quiet as simply soft.*

MAMA: Who that 'round here slamming doors at this hour? *(She crosses through the room, goes to the window, opens it, and brings in a feeble little plant growing doggedly in a small pot on the window sill. She feels the dirt and puts it back out)*

RUTH: That was Walter Lee. He and Bennie was at it again.

MAMA: My children and they tempers. Lord, if this little old plant don't get more sun than it's been getting it ain't never going to see spring again. *(She turns from the window)* What's the matter with you this morning, Ruth? You looks right peaked. You aiming to iron all them things? Leave some for me. I'll get to 'em this afternoon. Bennie honey, it's too drafty for you to be sitting 'round half dressed. Where's your robe?

BENEATHA: In the cleaners.

MAMA: Well, go get mine and put it on.

BENEATHA: I'm not cold, Mama, honest.

MAMA: I know—but you so thin . . .

BENEATHA: *(Irritably)* Mama, I'm not cold.

MAMA: *(Seeing the make-down bed as* TRAVIS *has left it)* Lord have mercy, look at that poor bed. Bless his heart —he tries, don't he? *(She moves to the bed* TRAVIS *has sloppily made up)*

RUTH: No—he don't half try at all 'cause he knows you going to come along behind him and fix everything. That's just how come he don't know how to do nothing right now—you done spoiled that boy so.

MAMA: Well—he's a little boy. Ain't supposed to know 'bout housekeeping. My baby, that's what he is. What you fix for his breakfast this morning?

RUTH: *(Angrily)* I feed my son, Lena!

MAMA: I ain't meddling—*(Under breath; busy-bodyish)* I just noticed all last week he had cold cereal, and when it starts getting this chilly in the fall a child ought to have some hot grits or something when he goes out in the cold—

RUTH: *(Furious)* I gave him hot oats—is that all right!

MAMA: I ain't meddling. *(Pause)* Put a lot of nice butter on it? (RUTH *shoots her an angry look and does not reply)* He likes lots of butter.

RUTH: *(Exasperated)* Lena—

MAMA: *(To* BENEATHA. MAMA *is inclined to wander conversationally sometimes)* What was you and your brother fussing 'bout this morning?

BENEATHA: It's not important, Mama. *(She gets up and goes to look out at the bathroom, which is apparently free, and she picks up her towels and rushes out)*

MAMA: What was they fighting about?

RUTH: Now you know as well as I do.

MAMA: *(Shaking her head)* Brother still worrying hisself sick about that money?

RUTH: You know he is.

MAMA: You had breakfast?

RUTH: Some coffee.

MAMA: Girl, you better start eating and looking after yourself better. You almost thin as Travis.

RUTH: Lena—

MAMA: Un-hunh?

RUTH: What are you going to do with it?

MAMA: Now don't you start, child. It's too early in the morning to be talking about money. It ain't Christian.

RUTH: It's just that he got his heart set on that store—

MAMA: You mean that liquor store

that Willy Harris want him to invest in?

RUTH: Yes—

MAMA: We ain't no business people, Ruth. We just plain working folks.

RUTH: Ain't nobody business people till they go into business. Walter Lee say colored people ain't never going to start getting ahead till they start gambling on some different kinds of things in the world—investments and things.

MAMA: What done got into you, girl? Walter Lee done finally sold you on investing.

RUTH: No. Mama, something is happening between Walter and me. I don't know what it is—but he needs something—something I can't give him any more. He needs this chance, Lena.

MAMA: *(Frowning deeply)* But liquor, honey—

RUTH: Well—like Walter say—I spec people going to always be drinking themselves some liquor.

MAMA: Well—whether they drinks it or not ain't none of my business. But whether I go into business selling it to 'em *is,* and I don't want that on my ledger this late in life. *(Stopping suddenly and studying her daughter-in-law)* Ruth Younger, what's the matter with you today? You look like you could fall over right there.

RUTH: I'm tired.

MAMA: Then you better stay home from work today.

RUTH: I can't stay home. She'd be calling up the agency and screaming at them, "My girl didn't come in today—send me somebody! My girl didn't come in!" Oh, she just have a fit . . .

MAMA: Well, let her have it. I'll just call her up and say you got the flu—

RUTH: *(Laughing)* Why the flu?

MAMA: 'Cause it sounds respectable to 'em. Something white people get, too. They know 'bout the flu. Otherwise they think you been cut up or something when you tell 'em you sick.

RUTH: I got to go in. We need the money.

MAMA: Somebody would of thought my children done all but starved to death the way they talk about money here late. Child, we got a great big old check coming tomorrow.

RUTH: *(Sincerely, but also self-righteously)* Now that's your money. It ain't got nothing to do with me. We all feel like that—Walter and Bennie and me—even Travis.

MAMA: *(Thoughtfully, and suddenly very far away)* Ten thousand dollars—

RUTH: Sure is wonderful.

MAMA: Ten thousand dollars.

RUTH: You know what you should do, Miss Lena? You should take yourself a trip somewhere. To Europe or South America or someplace—

MAMA: *(Throwing up her hands at the thought)* Oh, child!

RUTH: I'm serious. Just pack up and leave! Go on away and enjoy yourself some. Forget about the family and have yourself a ball for once in your life—

MAMA: *(Drily)* You sound like I'm just about ready to die. Who'd go with me? What I look like wandering 'round Europe by myself?

RUTH: Shoot—these here rich white women do it all the time. They don't think nothing of packing up they suitcases and piling on one of them big steamships and—swoosh!—they gone, child.

MAMA: Something always told me I wasn't no rich white woman.

RUTH: Well—what are you going to do with it then?

MAMA: I ain't rightly decided. *(Thinking. She speaks now with emphasis)* Some of it got to be put away for Beneatha and her schoolin'—and ain't nothing going to touch that part of it. Nothing. *(She waits several seconds, trying to make up her mind about something, and looks at* RUTH *a little tentatively before going on)* Been thinking that we maybe could meet the notes on a little old two-story somewhere, with a yard where Travis could play in the summertime, if we use part of the insurance for a down payment and everybody kind of pitch in. I could maybe take on a little day work again, few days a week—

RUTH: *(Studying her mother-in-law furtively and concentrating on her ironing, anxious to encourage without seeming to)* Well, Lord knows, we've put enough rent into this here rat trap to pay for four houses by now . . .

MAMA: *(Looking up at the words "rat trap" and then looking around and leaning back and sighing—in a suddenly reflective mood—)* "Rat trap"—yes, that's all it is. *(Smiling)* I remember just as well the day me and Big Walter moved in here. Hadn't been married but two weeks and wasn't planning on living here no more than a year. *(She shakes her head at the dissolved dream)* We was going to set away, little by little, don't you know, and buy a little place out in Morgan Park. We had

even picked out the house. *(Chuckling a little)* Looks right dumpy today. But Lord, child, you should know all the dreams I had 'bout buying that house and fixing it up and making me a little garden in the back—*(She waits and stops smiling)* And didn't none of it happen. *(Dropping her hands in a futile gesture)*

RUTH: *(Keeps her head down, ironing)* Yes, life can be a barrel of disappointments, sometimes.

MAMA: Honey, Big Walter would come in here some nights back then and slump down on that couch there and just look at the rug, and look at me and look at the rug and then back at me—and I'd know he was down then . . . really down. *(After a second very long and thoughtful pause; she is seeing back to times that only she can see)* And then, Lord, when I lost that baby—little Claude—I almost thought I was going to lose Big Walter too. Oh, that man grieved hisself! He was one man to love his children.

RUTH: Ain't nothin' can tear at you like losin' your baby.

MAMA: I guess that's how come that man finally worked hisself to death like he done. Like he was fighting his own war with this here world that took his baby from him.

RUTH: He sure was a fine man, all right. I always liked Mr. Younger.

MAMA: Crazy 'bout his children! God knows there was plenty wrong with Walter Younger—hard-headed, mean, kind of wild with women—plenty wrong with him. But he sure loved his children. Always wanted them to have something—be something. That's where Brother gets all these notions, I reckon. Big Walter used to say, he'd get right wet in the eyes sometimes, lean his head back with the water standing in his eyes and say, "Seem like God didn't see fit to give the black man nothing but dreams—but He did give us children to make them dreams seem worth while." *(She smiles)* He could talk like that, don't you know.

RUTH: Yes, he sure could. He was a good man, Mr. Younger.

MAMA: Yes, a fine man—just couldn't never catch up with his dreams, that's all.

BENEATHA *comes in, brushing her hair and looking up to the ceiling, where the sound of a vacuum cleaner has started up.*

BENEATHA: What could be so dirty on that woman's rugs that she has to vacuum them every single day?

RUTH: I wish certain young women 'round here who I could name would take inspiration about certain rugs in a certain apartment I could also mention.

BENEATHA: *(Shrugging)* How much cleaning can a house need, for Christ's sakes.

MAMA: *(Not liking the Lord's name used thus)* Bennie!

RUTH: Just listen to her—just listen!

BENEATHA: Oh, God!

MAMA: If you use the Lord's name just one more time—

BENEATHA: *(A bit of a whine)* Oh, Mama—

RUTH: Fresh—just fresh as salt, this girl!

BENEATHA: *(Drily)* Well—if the salt loses its savor—

MAMA: Now that will do. I just ain't going to have you 'round here reciting the scriptures in vain—you hear me?

BENEATHA: How did I manage to get on everybody's wrong side by just walking into a room?

RUTH: If you weren't so fresh—

BENEATHA: Ruth, I'm twenty years old.

MAMA: What time you be home from school today?

BENEATHA: Kind of late. *(With enthusiasm)* Madeline is going to start my guitar lessons today. (MAMA *and* RUTH *look up with the same expression)*

MAMA: Your *what* kind of lessons?

BENEATHA: Guitar.

RUTH: Oh, Father!

MAMA: How come you done taken it in your mind to learn to play the guitar?

BENEATHA: I just want to, that's all.

MAMA: *(Smiling)* Lord, child, don't you know what to do with yourself? How long it going to be before you get tired of this now—like you got tired of that little play-acting group you joined last year? *(Looking at Ruth)* And what was it the year before that?

RUTH: The horseback-riding club for which she bought that fifty-five-dollar riding habit that's been hanging in the closet ever since!

MAMA: *(To* BENEATHA) Why you got to flit so from one thing to another, baby?

BENEATHA: *(Sharply)* I just want to learn to play the guitar. Is there anything wrong with that?

MAMA: Ain't nobody trying to stop you. I just wonders sometimes why you has to flit so from one thing to another all the time. You ain't never done nothing with all that camera equipment you brought home—

BENEATHA: I don't flit! I—I experiment with different forms of expression—

RUTH: Like riding a horse?

BENEATHA: —People have to express themselves one way or another.

MAMA: What is it you want to express?

BENEATHA: *(Angrily)* Me! (MAMA *and* RUTH *look at each other and burst into raucous laughter)* Don't worry—I don't expect you to understand.

MAMA: *(To change the subject)* Who you going out with tomorrow night?

BENEATHA: *(With displeasure)* George Murchison again.

MAMA: *(Pleased)* Oh—you getting a little sweet on him?

RUTH: You ask me, this child ain't sweet on nobody but herself—*(Underbreath)* Express herself! *(They laugh)*

BENEATHA: Oh—I like George all right, Mama. I mean I like him enough to go out with him and stuff, but—

RUTH: *(For devilment)* What does *and stuff* mean?

BENEATHA: Mind your own business.

MAMA: Stop picking at her now, Ruth. *(A thoughtful pause, and then a suspicious sudden look at her daughter as she turns in her chair for emphasis)* What *does* it mean?

BENEATHA: *(Wearily)* Oh, I just mean I couldn't ever really be serious about George. He's—he's so shallow.

RUTH: Shallow—what do you mean he's shallow? He's *Rich!*

MAMA: Hush, Ruth.

BENEATHA: I know he's rich. He knows he's rich, too.

RUTH: Well—what other qualities a man got to have to satisfy you, little girl?

BENEATHA: You wouldn't even begin to understand. Anybody who married Walter could not possibly understand.

MAMA: *(Outraged)* What kind of way is that to talk about your brother?

BENEATHA: Brother is a flip—let's face it.

MAMA: *(To* RUTH, *helplessly)* What's a flip?

RUTH: *(Glad to add kindling)* She's saying he's crazy.

BENEATHA: Not crazy. Brother isn't really crazy yet—he—he's an elaborate neurotic.

MAMA: Hush your mouth!

BENEATHA: As for George. Well. George looks good—he's got a beautiful car and he takes me to nice places and, as my sister-in-law says, he is probably the richest boy I will ever get to know and I even like him sometimes—but if the Youngers are sitting around waiting to see if their little Bennie is going to tie up the family with the Murchisons, they are wasting their time.

RUTH: You mean you wouldn't marry George Murchison if he asked you someday? That pretty, rich thing? Honey, I knew you was odd—

BENEATHA: No I would not marry him if all I felt for him was what I feel now. Besides, George's family wouldn't really like it.

MAMA: Why not?

BENEATHA: Oh, Mama—The Murchisons are honest-to-God-real-*live*-rich colored people, and the only people in the world who are more snobbish than rich white people are rich colored people. I thought everybody knew that. I've met Mrs. Murchison. She's a scene!

MAMA: You must not dislike people 'cause they well off, honey.

BENEATHA: Why not? It makes just as much sense as disliking people 'cause they are poor, and lots of people do that.

RUTH: *(A wisdom-of-the-ages manner. To* MAMA*)* Well, she'll get over some of this—

BENEATHA: Get over it? What are you talking about, Ruth? Listen, I'm going to be a doctor. I'm not worried about who I'm going to marry yet—if I ever get married.

MAMA *and* RUTH: *If!*

MAMA: Now, Bennie—

BENEATHA: Oh, I probably will . . . but first I'm going to be a doctor, and George, for one, still thinks that's pretty funny. I couldn't be bothered with that. I am going to be a doctor and everybody around here better understand that!

MAMA: *(Kindly)* 'Course you going to be a doctor, honey, God willing.

BENEATHA: *(Drily)* God hasn't got a thing to do with it.

MAMA: Beneatha—that just wasn't necessary.

BENEATHA: Well—neither is God. I get sick of hearing about God.

MAMA: Beneatha!

BENEATHA: I mean it! I'm just tired of hearing about God all the time. What has He got to do with anything? Does he pay tuition?

MAMA: You 'bout to get your fresh little jaw slapped!

RUTH: That's just what she needs, all right!

BENEATHA: Why? Why can't I say what I want to around here, like everybody else?

MAMA: It don't sound nice for a young girl to say things like that—you wasn't brought up that way. Me and your father went to trouble to get you and Brother to church every Sunday.

BENEATHA: Mama, you don't understand. It's all a matter of ideas, and God is just one idea I don't accept. It's not important. I am not going out and be immoral or commit crimes because I don't believe in God. I don't even think about it. It's just that I get tired of Him getting credit for all the things the human race achieves through its own stubborn effort. There simply is no blasted God—there is only man and it is he who makes miracles! *(*MAMA *absorbs this speech, studies her daughter and rises slowly and crosses to* BENEATHA *and slaps her powerfully across the face. After, there is only silence and the daughter drops her eyes from her mother's face, and* MAMA *is very tall before her)*

MAMA: Now—you say after me, in my mother's house there is still God. *(There is a long pause and* BENEATHA *stares at the floor wordlessly.* MAMA *repeats the phrase with precision and cool emotion)* In my mother's house there is still God.

BENEATHA: In my mother's house there is still God. *(A long pause)*

MAMA: *(Walking away from* BENEATHA, *too disturbed for triumphant posture. Stopping and turning back to her daughter)* There are some ideas we ain't going to have in this house. Not long as I am at the head of this family.

BENEATHA: Yes, ma'am. *(*MAMA *walks out of the room)*

RUTH: *(Almost gently, with profound understanding)* You think you a woman, Bennie—but you still a little girl. What you did was childish —so you got treated like a child.

BENEATHA: I see. *(Quietly)* I also

see that everybody thinks it's all right for Mama to be a tyrant. But all the tyranny in the world will never put a God in the heavens! *(She picks up her books and goes out)*

RUTH: *(Goes to* MAMA's *door)* She said she was sorry.

MAMA: *(Coming out, going to her plant)* They frightens me, Ruth. My children.

RUTH: You got good children, Lena. They just a little off sometimes—but they're good.

MAMA: No—there's something come down between me and them that don't let us understand each other and I don't know what it is. One done almost lost his mind thinking 'bout money all the time and the other done commence to talk about things I can't seem to understand in no form or fashion. What is it that's changing, Ruth?

RUTH: *(Soothingly, older than her years)* Now . . . you taking it all too seriously. You just got strong-willed children and it takes a strong woman like you to keep 'em in hand.

MAMA: *(Looking at her plant and sprinkling a little water on it)* They spirited all right, my children. Got to admit they got spirit—Bennie and Walter. Like this little old plant that ain't never had enough sunshine or nothing—and look at it . . . *(She has her back to* RUTH, *who has had to stop ironing and lean against something and put the back of her hand to her forehead)*

RUTH: *(Trying to keep* MAMA *from noticing)* You . . . sure . . . loves that little old thing, don't you? . . .

MAMA: Well, I always wanted me a garden like I used to see sometimes at the back of the houses down home. This plant is close as I ever got to having one. *(She looks out of the window as she replaces the plant)* Lord, ain't nothing as dreary as the view from this window on a dreary day, is there? Why ain't you singing this morning, Ruth? Sing that "No Ways Tired." That song always lifts me up so—*(She turns at last to see that* RUTH *has slipped quietly into a chair, in a state of semiconsciousness)* Ruth! Ruth honey—what's the matter with you . . . Ruth!

Curtain

part 4

Losing...

Frida Kahlo de Rivera. *Child with Skeleton's Mask.*

from

The Death of Ivan Ilyich

LEO N. TOLSTOY

I

"Gentlemen," he said, "Ivan Ilyich has died!"

"You don't say so!"

"Here, read it yourself," replied Peter Ivanovich, handing Fedor Vassilievich the paper still damp from the press. Surrounded by a black border were the words: "Praskovya Fedorovna Golovina, with profound sorrow, informs relatives and friends of the demise of her beloved husband Ivan Ilyich Golovin, Member of the Court of Justice, which occurred on February 4 of this year 1882. The funeral will take place on Friday at one o'clock in the afternoon."

Ivan Ilyich had been a colleague of the gentlemen present and was liked by them all. He had been ill for some weeks with an illness said to be incurable. His post had been

kept open for him. but there had been conjectures that in case of his death Alexeiev might receive his appointment, and that either Vinnicov or Shtabel would succeed Alexeiev. So on receiving the news of Ivan Ilyich's death the first thought of each of the gentlemen in those chambers was of the changes and promotions it might occasion among themselves or their acquaintances.

"I'll be sure to get Shtabel's place or Vinnicov's," thought Fedor Vassilievich. "I was promised that long ago, and the promotion means an extra eight hundred rubles a year for me besides the allowance."

"Now I must apply for my brother-in-law's transfer from Kaluga," thought Peter Ivanovich. "My wife will be very glad, and then she won't be able to say that I never do anything for her relatives."

"I thought he'd never leave his

bed again," Peter Ivanovich said aloud. "It's very sad."

"But what really was the matter with him?"

"The doctors couldn't say—at least they could, but each of them said something different. When last I saw him I thought he was getting better."

"And I haven't been to see him since the holidays. I always meant to go."

"Had he any property?"

"I think his wife had a little—but something quite trifling."

"We'll have to go to see her, but they live so terribly far away."

"Far away from you, you mean. Everything's far away from your place."

"You see, he never can forgive my living on the other side of the river," said Peter Ivanovich, smiling at Shebek. Then, still talking of the distances between different parts of the city, they returned to the Court.

Besides considerations as to the possible transfers and promotions likely to result from Ivan Ilyich's death, the mere fact of the death of a near acquaintance aroused, as usual, in all who heard of it the complacent feeling that "It's he who's dead and not I."

Each one thought or felt: "Well, he's dead, but I'm alive!" But the more intimate of Ivan Ilyich's acquaintances, his so-called friends, could not help thinking also that they would now have to fulfill the very tiresome demands of propriety by attending the funeral service and paying a visit of condolence to the widow.

Fedor Vassilievich and Peter Ivanovich had been his nearest acquaintances. Peter Ivanovich had studied law with Ivan Ilyich and had considered himself under obligations to him.

Having told his wife at dinnertime of Ivan Ilyich's death, and of his conjecture that it might be possible to get her brother transferred to their circuit, Peter Ivanovich sacrificed his usual nap, put on his evening clothes, and drove to Ivan Ilyich's house.

At the entrance stood a carriage and two cabs. Leaning against the wall in the hall downstairs near the coat rack was a coffin lid covered with cloth of gold, ornamented with gold cord and tassels that had been polished up with metal powder. Two ladies in black were taking off their fur cloaks. Peter Ivanovich recognized one of them as Ivan Ilyich's sister, but the other was a stranger to him. His colleague Schwartz was just coming downstairs, but on seeing Peter Ivanovich enter he stopped and winked at him, as if to say: "Ivan Ilyich has made a mess of things—not like you and me."

Schwartz's face with his fashionable whiskers, and his slim figure in evening dress, had as usual an air of elegant solemnity which contrasted with the playfulness of his character and had a special piquancy here, or so it seemed to Peter Ivanovich. He allowed the ladies to precede him and slowly followed them upstairs. Schwartz did not come down but remained where he was, and Peter Ivanovich understood that he wanted to arrange where they should play bridge that evening. The ladies went upstairs to the widow's room, and Schwartz, with seriously compressed lips but a playful look in his eyes, indicated by a twitch of his eyebrows the room to the right where the body lay.

Peter Ivanovich, like everyone else on such occasions, entered feeling uncertain what he would have to do. All he knew was that at such times it is always safe to cross oneself. But he was not quite sure whether one should make obeisances while doing so. He therefore adopted a middle course. On entering the room he began crossing himself and made a slight movement resembling a bow. At the same time, as far as the motion of his head and arm allowed, he surveyed the room. Two young men —apparently nephews, one of whom was a high-school pupil—were leaving the room, crossing themselves as they did so. An old woman was standing motionless, and a lady with strangely arched eyebrows was saying something to her in a whisper. A vigorous, resolute clerical person in a frock coat was reading something in a loud voice with an expression that precluded any contradiction. The butler's assistant, Gerasim, stepping lightly in front of Peter Ivanovich, was strewing something on the floor. Noticing this, Peter Ivanovich was immediately aware of the faint odor of a decomposing body.

The last time he had called on Ivan Ilyich, Peter Ivanovich had seen Gerasim in the study. Ivan Ilyich had been particularly fond of him and he had been performing the duty of a sick nurse.

Peter Ivanovich continued to make the sign of the cross, slightly inclining his head in an intermediate direction between the coffin, the reader, and the icons on the table in a corner of the room. Afterward, when it seemed to him that this movement of his arm in crossing himself had gone on too long, he stopped and began to look at the corpse.

The dead man lay, as dead men always lie, in a specially cumbrous way, his rigid limbs sunk in the soft cushions of the coffin, with the head forever bowed on the pillow. His yellow waxen brow with bald patches over his sunken temples was thrust up in the way peculiar to the dead, the protruding nose seeming to press on the upper lip. He was much changed and had grown even thinner since Peter Ivanovich had last seen him, but, as is always the case with the dead, his face was handsomer and above all more dignified than when he was alive. The expression on the face said that what was necessary had been accomplished, and accomplished rightly. Besides this, there was in that expression a reproach and a warning to the living. This warning seemed to Peter Ivanovich out of place, or at least not applicable to him. He felt a certain discomfort and so he hurriedly crossed himself once more and turned and went out of the door—too hurriedly and too regardless of propriety, as he himself was aware.

Schwartz was waiting for him in the adjoining room with legs spread wide apart and both hands toying with his top hat behind his back. The mere sight of that playful, well-groomed, and elegant figure refreshed Peter Ivanovich. He felt that Schwartz was above all these happenings and would not surrender to any depressing influences. His very look said that this incident of a church service for Ivan Ilyich could not be a sufficient reason for infringing the order of the session—in other words, that it would certainly not prevent his unwrapping a new pack of cards and shuffling them that evening while a footman placed four fresh candles on the table: in fact, that there was no reason for supposing that this incident would hinder their spending the evening agreeably. Indeed, he said this in a whisper as Peter Ivanovich passed him, proposing that they should meet for a game at Fedor Vassilievich's. But apparently Peter Ivanovich was not destined to play bridge that evening. Praskovya Fedorovna (a short, fat woman who, despite all efforts to the contrary, had continued to broaden steadily from her shoulders downward and who had the same extraordinarily arched eyebrows as the lady who had been standing by the coffin), dressed all in black, her head covered with lace, came out of her own room with some other ladies, conducted them to the room where the dead body lay, and said: "The service will begin immediately. Please go in."

Schwartz, making an indeterminate bow, stood still, evidently neither accepting nor declining this invitation. Praskovya Fedorovna, recognizing Peter Ivanovich, sighed, went close up to him, took his hand, and said: "I know you were a true friend to Ivan Ilyich—" and looked at him, awaiting some suitable response. And Peter Ivanovich knew that, just as it had been the right thing to cross himself in that room, what he had to do here was to press her hand, sigh, and say: "Believe me—" So he did all this and as he did it felt that the desired result had been achieved: that both he and she were touched.

"Come with me. I want to speak to you before it begins," said the widow. "Give me your arm."

Peter Ivanovich gave her his arm, and they went to the inner rooms, passing Schwartz, who winked at Peter Ivanovich commiseratingly.

"That kills our bridge game. Don't object if we find another player. Perhaps you can cut in when you do escape," said his playful look.

Peter Ivanovich sighed still more deeply and despondently, and Praskovya Fedorovna pressed his arm gratefully. When they reached the drawing room, upholstered in pink cretonne and lighted by a dim lamp, they sat down at the table—she on a sofa and Peter Ivanovich on a low, soft ottoman, the springs of which yielded spasmodically under his weight. Praskovya Fedorovna had been on the point of warning him to take another seat, but felt that such a warning was out of keeping with her present condition and so changed her mind. As he sat down on the ottoman Peter Ivanovich recalled how Ivan Ilyich had arranged this room and had consulted him regarding this pink cretonne with green leaves. The whole room was full of furniture and knickknacks, and on her way to the sofa the lace of the widow's black shawl caught on the carved edge of the table. Peter Ivanovich rose to detach it, and the springs of the ottoman, relieved of his weight, rose also and gave him a bounce. The widow began detaching her shawl herself, and Peter Ivanovich again sat down, suppressing the rebellious springs of the ottoman under him. But the widow had not quite freed herself, and Peter Ivanovich got up again, and again the ottoman rebelled and even creaked. When this was all over she took out a clean cambric handkerchief and began to weep. The episode with the shawl and the struggle with the ottoman had cooled Peter Ivanovich's emotions, and he sat there with a sullen look on his face. This awkward situation was interrupted by Sokolov, Ivan Ilyich's butler, who came to report that the plot in the cemetery that Praskovya Fedorovna had chosen would cost two hundred rubles. She stopped weeping and, looking at Peter Ivanovich with the air of a victim, remarked in French that it was very hard for her. Peter Ivanovich made a silent gesture signifying his full conviction that it must indeed be so.

"You may smoke," she said in a magnanimous yet crushed voice, and turned to discuss with Sokolov the price of the plot for the grave.

Peter Ivanovich, while lighting his cigarette, heard her inquiring very circumstantially into the prices of different plots in the cemetery and finally decide which she would take. When that was done she gave instructions about engaging the choir. Sokolov then left the room.

"I look after everything myself," she told Peter Ivanovich, shifting the albums that lay on the table; and noticing that the table was endangered by his cigarette ash, she immediately passed him an ash tray, saying as she did so: "I consider it an affectation to say that my grief prevents my attending to practical affairs. On the contrary, if anything can—I won't say console me, but rather distract me—it is seeing to everything concerning him." She again took out her handkerchief as if preparing to cry, but suddenly, as if mastering her feeling, she shook herself and began to speak calmly. "But there's something I want to talk to you about."

Peter Ivanovich bowed, keeping under control the springs of the ottoman, which immediately began quivering under him.

"He suffered terribly the last few days."

"Did he?" asked Peter Ivanovich.

"Oh, terribly! He screamed unceasingly, not for minutes but for hours. For the last three days he screamed incessantly. It was unendurable. I cannot understand how I bore it; you could hear him three rooms off. Oh, what I have suffered!"

"Is it possible that he was conscious all that time?" asked Peter Ivanovich.

"Yes," she whispered. "To the last moment. He took leave of us a quarter of an hour before he died, and asked us to take Volodya away."

The thought of the sufferings of this man he had known so intimately, first as a laughing little boy, then as a schoolmate, and later as a grown-up colleague, suddenly struck Peter Ivanovich with horror, despite an unpleasant consciousness of his own and this woman's dissimulation. He again saw that brow, and that nose pressing down on the upper lip, and felt personal fear.

"Three days of frightful suffering and then death! Why, that might suddenly, at any time, happen to me," he thought, and for a moment felt terrified. But—he did not himself know how—the customary reflection at once occurred to him that this had happened to Ivan Ilyich and not to him, and that it should not and could not happen to him, and that to think that it could would be yielding to depression, which he ought not to do, as Schwartz's expression plainly showed. After which reflection Peter Ivanovich felt reassured, and began to ask with interest about the details of Ivan Ilyich's death, as though death were an accident natural to Ivan Ilyich but certainly not to himself.

After many details of the really

dreadful physical sufferings Ivan Ilyich had endured (which details he learned only from the effect those sufferings had produced on Praskovya Federovna's nerves), the widow apparently found it necessary to get down to business.

"Oh, Peter Ivanovich, how hard it is! How terribly, terribly hard!" and she again began to weep.

Peter Ivanovich sighed and waited for her to finish blowing her nose. When she had done so he said: "Believe me—" and she again began talking and brought out what was evidently her chief concern with him —namely, to question him as to how she could obtain a grant of money from the Government on the occasion of her husband's death. She made it appear that she was asking Peter Ivanovich's advice about her pension, but he soon saw that she already knew about that to the minutest detail, even better than he did himself. She knew how much could be got out of the Government in consequence of her husband's death, but wanted to find out whether she could not possibly extract something more. Peter Ivanovich tried to think of some means of doing so, but after reflecting for a while and, out of propriety, condemning the Government for its niggardliness, he said he thought that nothing more could be got. Then she sighed and evidently began to devise means of getting rid of her visitor. Noticing this, he put out his cigarette, rose, pressed her hand, and went out into the anteroom.

In the dining room, where the clock stood that Ivan Ilyich had liked so much and had bought at an antique shop, Peter Ivanovich met a priest and a few acquaintances who had come to attend the service, and he recognized Ivan Ilyich's daughter, a handsome young woman. She was in black, and her slim figure appeared slimmer than ever. She had a gloomy, determined, almost angry expression, and bowed to Peter Ivanovich as though he were in some way to blame. Behind her, with the same offended look, stood a wealthy young man, an Examining Magistrate, whom Peter Ivanovich also knew and who was her fiancé, as he had heard. He bowed mournfully to them and was about to pass into the death chamber, when from under the stairs appeared the figure of Ivan Ilyich's schoolboy son, who was extremely like his father. He seemed a young Ivan Ilyich, such as Peter Ivanovich remembered him when they had studied law together. His tear-stained eyes had in them the look one sees in the eyes of boys of thirteen or fourteen who are not pure-minded. When he saw Peter Ivanovich he scowled morosely and shamefacedly. Peter Ivanovich nodded to him and entered the death chamber. The service began: candles, groans, incense, tears and sobs. Peter Ivanovich stood looking gloomily down at his feet. He did not look once at the dead man, did not yield to any depressing influence, and was one of the first to leave the room. There was no one in the anteroom, but Gerasim darted out of the dead man's room, rummaged with his strong hands among the fur coats to find Peter Ivanovich's, and helped him on with it.

"Well, friend Gerasim," said Peter Ivanovich, so as to say something. "It's a sad affair, isn't it?"

"It's God's will. We'll all come to it some day," answered Gerasim, displaying his teeth—the even, white teeth of a healthy peasant—and, like a man in the thick of urgent work, he briskly opened the front door, called the coachman, helped Peter Ivanovich into the sleigh, and sprang back to the front steps, as if in readiness for what he had to do next.

Peter Ivanovich found the fresh air particularly pleasant after the smells of incense, the dead body, and carbolic acid.

"Where to, Sir?" asked the driver.

"It's not too late late even now. I'll call on Fedor Vassilievich," thought Peter Ivanovich.

He accordingly drove there and found the players just finishing the first rubber, so that it was quite convenient for him to cut in.

II

Ivan Ilyich's life had been most simple and most ordinary, and therefore most terrible.

He had been a member of the Court of Justice, and died at the age of forty-five. His father had been an official who, after serving in various ministries and departments in Petersburg, had made the sort of career which brings men to positions from which by reason of their long service they cannot be dismissed, though they are obviously unfit to hold any responsible post, and for whom, therefore, posts are specially created which, though fictitious, carry salaries of from six to ten thousand rubles that are not fictitious, and in receipt of which they live on to a great age.

Such had been the Privy Councilor and superfluous member of various superfluous institutions, Ilya Ephimovich Golovin.

He had three sons, of which Ivan Ilyich was the second. The oldest son was following in his father's footsteps, only in another department, and was already approaching that stage in the service at which a similar sinecure would be reached. The third son was a failure. He had ruined his prospects in a number of posts and was now serving in the Department of Railroads. His father and brothers, and still more their wives, not merely disliked meeting him, but avoided remembering his existence unless compelled to do so. His sister had married Baron Greff, a Petersburg official of her father's type. Ivan Ilyich was *le phénix de la famille,* as people said. He had been neither as cold and formal as his elder brother nor as wild as the younger, but was a happy mean between them—an intelligent, polished, lively, and agreeable man. He had studied with his younger brother at the School of Law, but the latter had failed to complete the course and was expelled when he was in the fifth class. Ivan Ilyich finished the course well. Even when he was at the School of Law he was just what he remained for the rest of his life: a capable, cheerful, good-natured, and sociable man, though strict in the fulfillment of what he considered to be his duty: and he considered his duty to be what was so considered by those in authority.

* * *

In the province he had an affair with a lady who made advances to the elegant young lawyer, and there had also been a milliner; and there were carousals with aides-de-camp who visted the district, and after-supper visits to a certain outlying street of doubtful reputation; and there was, too, some obsequiousness to his chief and even to his chief's wife, but all this was done with such a tone of good breeding that no hard names could be applied to it. It all came under the heading of the French saying: *"Il faut que jeunesse se passe."* It was all done with clean hands, in clean linen, with French phrases, and, above all, among people of the best society and consequently with the approval of people of rank.

Thus did Ivan Ilyich serve for five years, and then there came a change in his official life. The new and reformed judicial institutions were introduced, and new men were needed. Ivan Ilyich became such a new man. He was offered the post of Examining Magistrate, and he accepted it, though the post was in another province and obliged him to give up the connections he had formed and to make new ones. His friends met to give him a send-off; they had a group photograph taken and presented him with a silver cigarette case, and he set off to his new post.

* * *

After living there for two years he met his future wife, Praskovya Fedorovna Mihel, who was the most attractive, clever, and brilliant girl of the set in which he moved, and among other amusements and relaxations from his labors as Examining Magistrate, Ivan Ilyich established light and playful relations with her.

While he had been an official on special service he had been accustomed to dance, but now as an Examining Magistrate it was exceptional for him to do so. If he danced now, he did it as if to show that though he served under the reformed order of things, and had reached the fifth official rank, yet when it came to dancing he could do it better than most people. So at the end of an evening he sometimes danced with Praskovya Fedorovna, and it was chiefly during these dances that he captivated her. She fell in love with him. He had had at first no definite intention of marrying, but when the girl fell in love with him he said to himself: "Really, why shouldn't I marry?"

Praskovya Fedorovna came of a good family, was not bad-looking, and had some little property. Ivan Ilyich might have aspired to a more brilliant match, but even this was good. He had his salary, and she, he hoped, would have an equal income. She was well connected and was a sweet, pretty, and thoroughly correct young woman. To say that he had married because he had fallen in love with her and found that she sympathized with his views of life would be as incorrect as to say that he married because his social circle approved of the match. He was swayed by both these considerations: the marriage gave him personal satisfaction, and at the same time it was considered the right thing by the most highly placed of his associates.

So Ivan Ilyich got married.

The preparations for marriage and the beginning of married life, with its conjugal caresses, the new furniture, new crockery, and new linen, were very pleasant until his wife became pregnant—so that he had begun to think that marriage would not impair the easy, agreeable, gay, and always decorous character of his life, approved of by society and regarded by himself as natural, but would even improve it. But from the first months of his wife's pregnancy, something new, unpleasant, depressing, and unseemly, and from which there was no way of escape, unexpectedly showed itself.

His wife, without any reason—*de gaieté de cœur,* as Ivan Ilyich expressed it to himself—began to disturb the pleasure and propriety of their life. She began to be jealous without any cause, expected him to devote all his attention to her, found fault with everything, and made coarse and ill-mannered scenes.

At first he hoped to escape from the unpleasantness of this state of affairs by the same easy and decorous relation to life that had served him heretofore: he tried to ignore his wife's disagreeable moods, continued to live in his usual easy and pleasant way, invited friends to his house for a game of cards, and also tried going out to his club or spending his evenings with friends. But one day his wife began upbraiding him so vigorously, using such coarse words, and continued to abuse him every time he did not fulfill her demands so resolutely and with such evident determination not to give way till he submitted—that is, till he stayed at home and was bored just as she was— that he became alarmed. He now realized that matrimony—at any rate with Praskovya Fedorovna—was not always conducive to the pleasures and amenities of life, but on the contrary often infringed both upon comfort and propriety, and that he must therefore entrench himself against such infringement. And he began to seek for means of doing so. His official duties were the one thing that impressed Praskovya Federovna, and by means of his official work and the duties attached to it he began struggling with his wife to secure his own independence.

With the birth of their child, the attempts to feed it and the various failures in doing so, and with the real and imaginary illnesses of mother and child, in which Ivan Ilyich's sympathy was demanded but about which he understood nothing, the need of securing for himself an existence outside his family life became still more imperative.

As his wife grew more irritable and exacting and Ivan Ilyich transferred the center of gravity of his life more and more to his official work, so did he grow to like his work better and become more ambitious than before.

Very soon, within a year of his wedding, he had realized that marriage, though it may add some comforts to life, is in fact a very intricate and difficult affair toward which, in order to perform one's duty—that is, to lead a decorous life approved of by society—one must adopt a definite attitude, just as toward one's official duties.

And he evolved such an attitude toward married life. He only required of it those conveniences— dinner at home, housewife, and bed —which it could give him, and, above all, that propriety of external forms required by public opinion. For the rest, he looked for lighthearted pleasure and propriety, and was very thankful when he found them, but if he met with antagonism and quer- ulousness he at once retired into his separate fenced-off world of official duties, where he found satisfaction.

Ivan Ilyich was esteemed a good official, and after three years was made Assistant Public Prosecutor. His new duties, their importance, the possibility of indicting and im- prisoning anyone he chose, the pub- licity his speeches received, and the success he had in all these things, made his work still more attractive.

More children came. His wife be- came more and more querulous and ill-tempered, but the attitude Ivan Ilyich had adopted toward his home life rendered him almost impervious to her grumbling.

After seven years' service in that town he was transferred to another province as Public Prosecutor. They moved, but were short of money, and his wife did not like the place they moved to. Though the salary was higher, the cost of living was greater, besides which, two of their children died, and family life became still more unpleasant for him.

Praskovya Fedorovna blamed her husband for every inconvenience they encountered in their new home. Most of the conversations between husband and wife, especially as to the children's education, led to top- ics which recalled former disputes, and those disputes were apt to flare up again at any moment. There re- mained only those rare periods of amorousness which still came to them at times but did not last long. These were islets at which they an- chored for a while and then again set out upon that ocean of veiled hostility which showed itself in their aloofness from each other. This aloofness might have grieved Ivan

Ilyich had he considered that it ought not to exist, but he now re- garded the position as normal, and even made it the goal at which he aimed in family life. His aim was to free himself more and more from those unpleasantnesses and to give them a semblance of harmlessness and propriety. He attained this by spending less and less time with his family, and when obliged to be at home he tried to safeguard his posi- tion by the presence of outsiders.

The chief thing, however, was that he had his official duties. The whole interest of his life now centered in the official world, and that interest absorbed him. The consciousness of his power, being able to ruin any- body he wished to ruin, the impor- tance, even the external dignity of his entry into court, or meetings with his subordinates, his success with su- periors and inferiors, and, above all, his masterly handling of cases, of which he was conscious—all this gave him pleasure and filled his life, to- gether with chats with his colleagues, dinners, and bridge. So that on the whole his life continued to flow as he considered it should do—pleasantly and properly.

Thus things continued for another seven years. His eldest daughter was already sixteen, another child had died, and only one son was left, a schoolboy and a subject of dissen- sion. Ivan Ilyich wanted to enter him in the School of Law, but to spite him Praskovya Federovna entered him at the High School. The daugh- ter had been educated at home and had turned out well: the boy did not learn badly either.

Thus had Ivan Ilyich lived for seventeen years after his marriage. He was already a public prosecutor of long standing, and had declined several proposed transfers while awaiting a more desirable post, when an unanticipated and unpleasant occurrence quite upset the peaceful course of his life. He was expecting to be offered the post of Presiding Judge in a university town, but Happe somehow came to the front and obtained the appointment instead. Ivan Ilyich became irritable, reproached Happe, and quarreled both with him and with his immediate superiors—who became colder to him and again passed him over when other appointments were made.

This was in 1880, the hardest year of Ivan Ilyich's life. It was then that it became evident, on the one hand, that his salary was insufficient for his family to live on, and, on the other, that he had been forgotten, and not only this, but that what was for him the greatest and most cruel injustice appeared to others a quite ordinary occurrence. Even his father did not consider it his duty to help him. Ivan Ilyich felt himself abandoned by everyone, and also felt that all regarded his position with a salary of thirty-five hundred rubles [about $1,750] as quite normal and even fortunate. He alone knew that with the consciousness of the injustices done him, with his wife's incessant nagging, and with the debts he had contracted by living beyond his means, his position was far from normal.

In order to save money that summer he obtained leave of absence and went with his wife to live in the country at her brother's place.

In the country, without his work, he experienced ennui for the first time in his life, and not only ennui but intolerable depression, and he decided that it was impossible to go on living like that, and that it was necessary to take energetic measures.

Having passed a sleepless night pacing up and down the veranda, he decided to go to Petersburg and bestir himself, in order to punish those who had failed to appreciate him and to get transferred to some other ministry.

Next day, despite many protests from his wife and her brother, he started for Petersburg with the sole object of obtaining a post with a salary of five thousand rubles a year. He was no longer bent on any particular department, or tendency, or kind of activity. All he now wanted was an appointment to another post with a salary of five thousand rubles, either in the administration, in the banks, with the Department of Railroads, in one of the Empress Maria's institutions, or even in the Customs —but it had to carry with it a salary of five thousand rubles and be in a ministry other than that in which they had failed to appreciate him.

And this quest of his was crowned with remarkable and unexpected success.

*   *   *

Thanks to [a] change of personnel, Ivan Ilyich had unexpectedly obtained an appointment in his former ministry which placed him two stages above his former colleagues, besides giving him five thousand rubles salary and three thousand five hundred rubles for expenses connected with his removal. All his ill humor toward his former enemies and the whole department vanished, and Ivan Ilyich was completely happy.

He returned to the country more cheerful and contented than he had been for a long time. Praskovya Fedorovna also cheered up, and a truce was arranged between them. Ivan Ilyich told her how he had been feted by everybody in Petersburg, how all those who had been his enemies were put to shame and now fawned on him, how envious they were of his appointment, and how much everybody in Petersburg had liked him.

Praskovya Federovna listened to all this and appeared to believe it. She did not contradict anything, but merely went ahead with plans for their life in the town into which they were going. Ivan Ilyich saw with delight that these plans were his plans, that he and his wife agreed, and that, after a stumble, his life was regaining its due and natural character of pleasant lightheartedness and decorum.

He had come back for a short time only, for he had to take up his new duties on the tenth of September. Moreover, he needed time to settle into the new place, to move all his belongings from the province, and to buy and order many additional things: in a word, to make such arrangements as he had resolved on, which were almost exactly what Praskovya Fedorovna, too, had decided on.

Now that everything had happened so fortunately, and that he and his wife were at one in their aims, and moreover saw so little of one another, they got on together better than they had done since the first years of marriage. Ivan Ilyich had thought of taking his family away with him at once, but the insistence of his wife's brother and her sister-in-law, who had suddenly become particularly amiable and friendly to him and his family, induced him to depart alone.

So he departed, and the cheerful state of mind induced by his success and by the harmony between his wife and himself, the one intensifying the other, did not leave him. He found a delightful house, just the thing both he and his wife had dreamed of. Spacious, lofty reception rooms in the old style, a convenient and dignified study, rooms for his wife and daughter, a study for his son—it might have been specially built for them. Ivan Ilyich himself superintended the arrangements, chose the wallpapers, supplemented the furniture (preferably with antiques, which he considered particularly *comme il faut*), and supervised the upholstering.

* * *

All this so absorbed him that his new duties—though he liked his official work—interested him less than he had expected. Sometimes he even had moments of absent-mindedness during the Court sessions, and would consider whether he should have straight or curved cornices for his curtains. He was so interested in all this that he often did things himself, rearranging the furniture or rehanging the curtains. Once when mounting a stepladder to show the upholsterer, who did not understand, how he wanted the hangings draped, he made a false step and slipped, but being a strong and agile man he had clung on and merely bruised his side against the window frame. The bruised place was painful, but the pain soon passed, and he felt particularly bright and well just then. He wrote: "I feel fifteen years younger." He thought he would have everything ready by September, but things dragged on till mid-October. But the result was charming not only in his eyes but in those of everyone who saw it.

* * *

He was very happy when he met his family at the station and brought them to the newly furnished house, all lit up, with a footman in a white tie opening the door into a hall decorated with plants, and when they went on into the drawing room and his study, uttering exclamations of delight. He conducted them everywhere, drank in their praises eagerly, and beamed with pleasure. At tea that evening, when Praskovya Fedorovna, among other things, asked him about his fall, he laughed and showed them how he had gone flying and had frightened the upholsterer.

"It's a good thing I'm something of an athlete. Another man might have been killed, but I merely bruised myself, right here; it hurts when you touch it, but it's passing off already—it's only a bruise."

So they began living in their new home—which, as always happens, when they got thoroughly settled, they found to be just one room short; the increased income, too, was bound to be, as is always the case, just a trifle (some five hundred rubles) too small. But it was all very fine.

Things went particularly well at first, before everything was finally arranged and while something had still to be done: this thing bought, that thing ordered, another thing moved, and something else adjusted. Though there were some disputes between husband and wife, they were both so well satisfied and had so much to do that it all passed off without any serious quarrels. When nothing was left to arrange, things became rather dull and something seemed to be lacking, but by that time they were making acquaintances, forming habits, and life was growing fuller.

Ivan Ilyich spent his mornings at the law court and came home to dinner, and at first he was generally in a good humor, though he occasionally became irritable, precisely on account of his house. (Every spot on the tablecloth or the upholstery, and every broken window-blind string, irritated him. He had devoted so much trouble to arranging it all that every disturbance of it distressed him.) But on the whole his life ran its course as he believed life should do: easily, pleasantly, and decorously.

* * *

After dinner, if they had no visitors, Ivan Ilyich sometimes read a book that was being much discussed at the time, and in the evening settled down to work—that is, he read official papers, compared the depositions of witnesses, and noted the paragraphs of the Code applying to them. This was neither dull nor amusing. It was dull when he might have been playing bridge, but if no bridge were available it was at any rate better than doing nothing or sitting with his wife. His chief pleasure was giving little dinners to which he invited men and women of good social position, and just as his drawing room resembled all other drawing rooms, so did his enjoyable little parties resemble all other such parties.

Once they even gave a dance. He enjoyed it, and everything went off well, except that it led to a violent quarrel with his wife about the cakes and sweets. Praskovya Fedorovna had made her own plans, but Ivan Ilyich had insisted on getting everything from an expensive confectioner and ordered too many cakes, and the quarrel occurred because some of those cakes were left over and the confectioner's bill had come to forty-five rubles. It was a great and disagreeable quarrel. Praskovya Fedorovna called him "a fool and an imbecile," and he had clutched at his head and made angry allusions to divorce.

But the dance itself had been enjoyable. The best people attended, and Ivan Ilyich had danced with Princess Trufonova, a sister of the distinguished founder of the "Bear my Burden" Society.

The pleasures connected with his work were pleasures of ambition; his social pleasures were those of vanity; but his greatest pleasure was playing bridge. He acknowledged that whatever disagreeable incident happened in his life, the pleasure that beamed like a ray of light above everything else was to sit down to bridge with good players, not noisy partners, and of course to four-handed bridge (with five players it was annoying to be the dummy, though one pretended not to mind), to play a clever and serious game (when the cards allowed it) and then to sup and have a glass of wine. After a game of bridge, especially if he had won a little (to win a large sum was unpleasant), Ivan Ilyich went to bed in specially good humor.

Thus did they live. They had formed a circle of acquaintances among the best people and were visited by people of importance and by young folk. In their views as to their acquaintances, husband, wife and daughter were in entire agreement, and tacitly and unanimously kept at arm's length and shook off the various shabby friends and relations who, with much show of affection, gushed into the drawing room with Japanese plates hung on the walls. Soon these shabby friends ceased to obtrude themselves and only the best people remained in the Golovins' set.

Young men made up to Lisa, and Petrischev, an Examining Magistrate and Dmitri Ivanovich Petrischev's son and sole heir, began to be so attentive to her that Ivan Ilyich had already spoken to Praskovya Fedorovna about it, and considered whether they should not arrange a party for them, or get up some private theatricals.

Thus did they live, and all went well, without change, and life flowed pleasantly.

IV

They were all in good health. It could hardly be called ill health if Ivan Ilyich sometimes complained that he had a queer taste in his mouth and felt some discomfort in his left side.

But this discomfort increased and, though not exactly painful, grew into a sense of pressure in his side, accompanied by ill humor. And his irritability became worse and worse and began to mar the agreeable, easy, and correct life that had become established in the Golovin family. Quarrels between husband and wife became more and more frequent, and soon the ease and amenity disappeared and even the decorum was barely maintained. Scenes again became frequent, and very few of those islets remained on which husband and wife could meet without an explosion. Praskovya Fedorovna now had good reason to say that her husband's temper was trying. With characteristic exaggeration she said he had always had a dreadful temper, and that it had needed all her good nature to put up with it for twenty years. It was true that now the quarrels were started by him. His bursts of temper always came just before dinner, often just as he began on his soup. Sometimes he noticed that a plate or dish was chipped, or that the food was not right, or his son put his elbow on the table, or his daughter's hair was not done as he liked it, and for all this he blamed Praskovya Fedorovna.

At first she used to answer and say disagreeable things to him, but once or twice he fell into such a rage at the beginning of dinner that she realized it was due to some physical derangement brought on by taking food, and so she restrained herself and did not answer, but merely hurried to get the dinner over with. She regarded this self-restraint as highly praiseworthy. Having come to the conclusion that her husband had a dreadful temper and made her life miserable, she began to feel sorry for herself, and the more she pitied herself the more she hated her husband. She began to wish he would die; yet she did not want him to die, because then his salary would cease. And this irritated her against him still more. She considered herself dreadfully unhappy just because not even his death could save her, and though she concealed her exasperation, that hidden exasperation of hers increased his irritaton also.

After one scene in which he had been particularly unfair and after which he had said in explanation that he certainly was irritable but that it was due to his not being well, she said that if he was ill it should be attended to, and insisted on his going to see a celebrated doctor.

He went. Everything took place as he had expected and as it always does. There was the usual waiting and the important air assumed by the doctor, with which he was so familiar (resembling that which he himself assumed in court), and the sounding and listening, and the questions which called for answers that were foregone conclusions and were evidently unnecessary, and the look of importance which implied that "if only you put yourself in our hands we will arrange everything— we know indubitably how it has to be done, always in the same way for everybody alike." It was all just as it was in the law court. The doctor put on just the same air toward him as he himself put on toward an accused person.

The doctor said that so-and-so indicated that this and that was going on inside the patient, but if the investigation of so-and-so did not confirm this, then he must assume this-that-and-the-other. If he assumed this-that-and-the-other, then—and so on. To Ivan Ilyich only one question was important: Was his case serious or not? But the doctor ignored that inappropriate question. From his point of view it was not the one under consideration; the real question was to decide between a floating kidney, chronic catarrh, or appendicitis. It was not a question of Ivan Ilyich's life or death, but one between a floating kidney and an inflamed appendix. And that question the doctor solved brilliantly, as it seemed to Ivan Ilyich, in favor of the appendix, with the reservation that should an examination of the urine give fresh indications the matter would be reconsidered. All this was just what Ivan Ilyich had himself brilliantly accomplished a thousand times in dealing with men on trial. The doctor summed up just as brilliantly, looking over his spectacles triumphantly and even gaily at the accused. From the doctor's summing up Ivan Ilyich concluded that things were bad, but that for the doctor, and perhaps for everybody else, it was a matter of indifference, though for him it was bad. And this conclusion struck him painfully, arousing in him a great feeling of pity for himself and of bitterness toward the doctor's indifference to a matter of such importance.

He said nothing of this, but arose, placed the doctor's fee on the table, and remarked with a sigh: "We sick people probably often put inappropriate questions. But tell me, in general, is this complaint dangerous or not?"

The doctor looked at him sternly over his spectacles with one eye, as if

to say: "Prisoner, if you will not keep to the questions put to you, I shall be obliged to have you removed from the court."

"I've already told you what I consider necessary and proper. The analysis may show something more." And the doctor bowed.

Ivan Ilyich went out slowly, seated himself disconsolately in his sleigh, and drove home. All the way home he was going over what the doctor had said, trying to translate those complicated, obscure, scientific phrases into plain language and find in them an answer to the question: "Is my condition bad? Is it very bad? Or is there as yet nothing much wrong?" And it seemed to him that the meaning of what the doctor had said was that it was very bad. Everything in the streets seemed depressing. The cabmen, the houses, the passers-by, and the shops were all dismal. His ache, this dull gnawing ache that never ceased for a moment, seemed to have acquired a new and more serious significance from the doctor's dubious remarks. Ivan Ilyich now watched that ache with a new and oppressive feeling.

He reached home and began to tell his wife about it. She listened, but in the middle of his account his daughter came in with her hat on, ready to go out with her mother. She sat down reluctantly to listen to this tedious recital but could not stand it long, and her mother, too, did not hear him to the end.

"Well, I'm very glad you went," she said. "Mind, now, and take your medicine regularly. Give me the prescription and I'll send Gerasim to the druggist's." And she went to get ready for going out.

While she was in the room he had hardly taken time to breathe, but he sighed deeply when she left it.

"Well," he thought, "perhaps it isn't so bad after all."

*   *   *

The pain in his side oppressed him and seemed to grow worse and more incessant, while the taste in his mouth grew more and more peculiar. It seemed to him that his breath had a disgusting smell, and he was conscious of a loss of appetite and strength. There was no deceiving himself: something terrible, new, and more important than anything before in his life, was taking place within him, of which he alone was aware. Those about him did not understand or would not understand it, but thought everything in the world was going on as usual. This tormented Ivan Ilyich above all. He saw that his household, especially his wife and daughter, who were in a perfect whirl of visiting, did not understand anything of all this and were annoyed that he was so depressed and so exacting, as if he were to blame for it.

Though they tried to disguise it, he saw that he was an obstacle in their path, and that his wife had adopted a definite line in regard to his illness and kept to it regardless of anything he said or did. Her attitude was this: "You know," she would say to her friends, "Ivan Ilyich can't do as other people do, and keep to the treatment prescribed for him. One day he'll take his drops and keep strictly to his diet and go to bed in good time, but the next day, unless I watch him, he'll suddenly forget his medicine, eat sturgeon—which is forbidden—and sit up playing cards till one o'clock in the morning."

"Oh, come, when was that?" Ivan Ilyich would ask in vexation. "Only once, at Peter Ivanovich's."

"And yesterday, with Shebek."

"Well, even if I hadn't stayed up, this pain would have kept me awake."

"Be that as it may, you'll never get well like that, but will always make us wretched."

Praskovya Fedorovna's attitude to Ivan Ilyich's illness, as she expressed it both to others and to him, was that it was his own fault and was another of the annoyances he caused her. Ivan Ilyich felt that this opinion escaped her involuntarily—but that did not make it any easier for him.

At the law court, too, he noticed, or thought he noticed, a strange attitude toward himself. It sometimes seemed to him that people were watching him inquisitively, as a man whose place might soon be vacant. Then again, his friends would suddenly begin to chaff him in a friendly way about his low spirits, as if the awful, horrible, and unheard-of-thing that was going on within him, incessantly gnawing at him and irresistibly drawing him away, was a very agreeable subject for jests.

With this consciousness, and with physical pain besides the terror, he must go to bed, often to lie awake the greater part of the night. Next morning he had to get up again, dress, go to the law court, speak, and write; or, if he did not go out, spend at home those twenty-four hours a day each of which was a torture. And he had to live thus all alone on the brink of an abyss, with no one who understood or pitied him.

V

Thus one month passed, and then another. Just before the New Year his brother-in-law came to town and stayed at their house. Ivan Ilyich was at the law court, and Praskovya Fedorovna had gone shopping. When Ivan Ilyich came home and entered his study he found his brother-in-law there—a healthy, florid man—unpacking his portmanteau. He raised his head on hearing Ivan Ilyich's footsteps and looked up at him for a moment without a word. That stare told Ivan Ilyich everything. His brother-in-law opened his mouth to utter an exclamation of surprise, but checked himself, and that action confirmed it all.

"I have changed, eh?"

"Yes, there is a change."

And after that, try as he would to get his brother-in-law to return to the subject of his looks, the latter would say nothing about it.

Praskovya Fedorovna came home, and her brother went out to her. Ivan Ilyich locked the door and began to examine himself in the glass, first full face, then in profile. He took up a portrait of himself taken with his wife and compared it with what he saw in the glass. The change in him was immense. Then he bared his arms to the elbow, looked at them, drew the sleeves down again, sat down on an ottoman, and grew blacker than night.

"No, no, this won't do!" he said to himself, and jumped up, went to the desk, took up some law papers and began to read them, but could not

continue. He unlocked the door and went into the reception room. The door leading to the drawing room was shut. He approached it on tiptoe and eavesdropped.

"No, you're exaggerating!" Praskovya Fedorovna was saying.

"Exaggerating! Don't you see it? Why, he's a dead man! Look at his eyes—there's no light in them. But what's wrong with him?"

"No one knows. Nicolaevich [that was another doctor] said something, but I don't know what. And Leshchetitski [this was the celebrated specialist] said quite the contrary—"

* * *

"The vermiform appendix! The kidney!" he said to himself. "It's not a question of appendix or kidney, but of life and—death. Yes, life was there and now it's going, going, and I can't stop it. Yes. Why deceive myself? Isn't it obvious to everyone but me that I'm dying, and that it's only a question of weeks, days—it may happen this moment. There was light, and now there is darkness. I was here and now I'm going there! Where?" A chill came over him, his breathing ceased, and he felt only the throbbing of his heart.

"When I am no more, what will there be? There'll be nothing. Then where shall I be when I am no more? Can this be dying? No, I don't want to die!" He jumped up and tried to light the candle, felt for it with trembling hands, dropped candle and candlestick on the floor, and fell back on his pillow.

"What's the use? It makes no difference," he said to himself, staring with wide-open eyes into the darkness. "Death. Yes, death. And none of them knows or wishes to know it, and they have no pity for me. Now they're playing." (He heard through the door the distant sound of a song and its accompaniment.) "It's all the same to them, but they will die, too! Fools! I first, and they later, but it'll be the same for them. And now they are having a fine time—the beasts!"

Anger choked him, and he was agonizingly, unbearably miserable. "It's impossible that all men have been doomed to suffer this awful horror!" He raised himself.

"Something must be wrong. I must calm myself—must think it all over from the beginning." And he again began thinking. "Yes, the beginning of my illness: I hit my side, but I was still quite well that day and the next. It hurt a little, then rather more. I saw the doctors; this was followed by despondency and anguish, and more doctors, and I drew nearer to the abyss. My strength grew less and I kept coming nearer and nearer, and now I have wasted away and there's no light in my eyes. I think of the appendix—but this is death! I think of my appendix getting better, and all the while death is right here! Can it really be death?"

Again terror seized him, and he gasped for breath. He leaned down and began feeling for the matches, pressing with his elbow against the stand beside the bed. It was in his way and hurt him, he grew furious with it, pressed on it still harder, and upset it. Breathless and in despair, he fell on his back, expecting death to come immediately.

Meanwhile the visitors were leaving. Praskovya Fedorovna was seeing them off. She heard something fall and came in.

"What's happened?"

"Nothing. I knocked the stand over accidentally."

She went out and returned with a candle. He lay there panting heavily, like a man who has run a thousand yards, and stared upward at her with a fixed look.

"What is it, Jean?"

"No-o-thing. I upset it." ("Why say anything? She won't understand," he thought.)

And in truth she did not understand. She picked up the stand, lit his candle, and hurried away to see another visitor off. When she came back he still lay on his back, looking upward.

"What is it? Do you feel worse?"

"Yes."

She shook her head and sat down.

"Do you know, Jean, I think we must ask Leshchetitski to come and see you here."

This meant calling in the famous specialist, regardless of expense. He smiled malevolently and said "No." She remained a little longer and then went up to him and kissed his forehead.

While she was kissing him he hated her from the bottom of his soul and with difficulty refrained from pushing her away.

"Good night. God send you sound sleep."

"Yes."

VI

Ivan Ilyich saw that he was dying, and he was in continual despair.

In the depth of his heart he knew he was dying, but not only was he not accustomed to the thought, he simply did not and could not grasp it.

The syllogism he had learnt from Kiezewetter's *Logic:* "Caius is a man, men are mortal, therefore Caius is mortal," had always seemed to him correct as applied to Caius, but certainly not as applied to Ivan Ilyich. That Caius—man in the abstract—was mortal was perfectly correct, but he wasn't Caius, not an abstract man, but a creature quite, quite apart from all others. He had been little Vanya, with a mamma and a papa, with Mitya and Volodya, with toys, a coachman and a nurse, afterward with Katenka and with all the joys, griefs, and delights of childhood, boyhood, and youth. What did Caius know of the smell of that striped leather ball Vanya had been so fond of? Had Caius kissed his mother's hand like that, and did the silk of her dress rustle so for Caius? Had he rioted like that at school when the pastry was bad? Had Caius been in love like that? Could Caius preside at a session the way he did? "Caius really was mortal, and it was right for him to die; but for me, little Vanya, Ivan Ilyich, with all my thoughts and emotions, it's altogether a different matter. It cannot be that I ought to die. That would be too terrible."

That was how he felt.

"If I had to die like Caius I'd have known it was so. An inner voice would have told me so, but there was nothing of the sort in me and I and all my friends felt that our case was quite different from that of Caius. And now here it is!" he said to himself. "It can't be. It's impossible! Yet here it is. How is that? How is one to understand it?"

He could not understand it, and tried to drive this false, incorrect, morbid thought away and to replace it by other proper and healthy thoughts. But that thought, and not the thought only but the reality itself, seemed to come and confront him.

And to replace that thought he called up a succession of others, hoping to find in them some support. He tried to get back into the former current of thoughts that had once screened the thought of death from him. But strange to say, all that had formerly shut off, hidden, and destroyed his consciousness of death no longer had that effect.

He now spent most of his time in attempting to re-establish that old current. He would say to himself: "I will take up my duties again—after all, I used to live by them." And banishing all doubts he would go to the law court, enter into conversation with his colleagues, and sit at ease as was his wont, scanning the crowd with a thoughtful look and leaning both his emaciated arms on the arms of his oak chair; bending over as usual to a colleague and drawing his papers nearer, he would interchange whispers with him, and then suddenly raising his eyes and sitting erect would pronounce certain words and open the proceedings.

But suddenly in the midst of those proceedings the pain in his side, regardless of the stage the proceedings had reached, would begin its own gnawing work. He would turn his attention to it and try to drive the thought of it away, but without success. *It* would come and stand before him and look at him, and he would be petrified and the light would die out of his eyes, and he would again begin asking himself whether *It* alone was true. And his colleagues and subordinates would see with surprise and distress that he, the brilliant and subtle judge, was becoming confused and making mistakes. He would shake himself, try to pull himself together, manage somehow to bring the sitting to a close, and return home with the sorrowful consciousness that his judicial labors could not as formerly hide from him what he wanted them to hide, and could not deliver him from *It*. And what was worst of all was that *It* drew his attention to *Itself*, not in order to make him take some action, but only that he should look at *It*, look *It* straight in the face: look at *It* and, without doing anything, suffer inexpressibly.

And to save himself from this condition Ivan Ilyich looked for consolations—new screens—and new screens were found and for a while seemed to save him, but then they immediately fell to pieces or rather became transparent, as if *It* penetrated them and nothing could veil *It*.

In these latter days he would go into the drawing room he had arranged—that drawing room where he had fallen and for the sake of which (how bitterly ridiculous it seemed) he had sacrificed his life—for he knew that his illness had originated with that injury. He would enter and see that something had scratched the polished table. He would look for the cause of this and find that it was an album, whose bronze ornamentation had got bent. He would take up the expensive album which he had lovingly arranged, and feel vexed with his daughter and her friends for their untidiness—for the album was torn here and there and some of the photographs had been reinserted upside down. He would

put it carefully in order and bend the ornamentation back into position. Then it would occur to him to place all those thing in another corner of the room, near the plants. He would call the footman, but his daughter or wife would come to help him. They would not agree, and his wife would contradict him, and he would dispute and grow angry. But that was all right, for then he did not think about *It. It* was invisible then.

But then, when he was moving something himself, his wife would say: "Let the servants do it. You'll hurt yourself again." And suddenly *It* would flash through the screen and he would see *It*. Just a flash of *It,* and he hoped *It* would disappear, but he would involuntarily pay attention to his side. *"It* sits there as before, gnawing just the same!" And he could no longer forget *It,* but could distinctly see *It* looking at him from behind the flowers. "What's all this for?"

"It really is so! I lost my life over that curtain as I might have done when storming a fort. Is that possible? How terrible and how stupid. It can't be true! It can't—but it is."

He would go to his study, lie down, and again be alone with *It:* face to face with *It*. And nothing could be done with *It* except to look at *It* and shudder.

VII

How it happened is impossible to say, because it came about step by step, unnoticed, but in the third month of Ivan Ilyich's illness his wife, his daughter, his son, his acquaintances, the doctors, the servants, and above all he himself, were aware that the whole interest he had for other people was whether he would soon vacate his place, and at last release the living from the discomfort caused by his presence, and he himself released from his sufferings.

He slept less and less. He was given opium and hyopdermic injections of morphine, but this did not relieve him. The dull depression he experienced in a somnolent condition at first gave him a little relief, but only as something new; afterward it became as distressing as the pain itself or even more so.

Special foods were prepared for him by the doctors' orders, but all those foods became increasingly distasteful and disgusting to him.

For his excretions also special arrangements had to be made, and this was a torment to him every time—a torment from the uncleanliness, the unseemliness, and the smell, and from knowing that another person had to take part in it.

But it was just through this most unpleasant matter that Ivan Ilyich obtained comfort. Gerasim, the butler's young assistant, always came in to carry away the bedpan. Gerasim was a clean, fresh peasant lad, grown stout on citified fare, and always cheerful and bright. At first the sight of him, in his clean Russian peasant costume, engaged in that disgusting task, embarrassed Ivan Ilyich.

Once, when he got up from the pan, too weak to draw up his trousers, he dropped into a soft armchair and looked with horror at his bare, flaccid thighs with the muscles so sharply marked on them.

Gerasim, with a firm light tread, his heavy boots emitting a pleasant smell of tar and fresh winter air, came in wearing a clean Hessian apron, the sleeves of his print shirt tucked up over his strong bare young arms, and refraining from looking at his sick master out of consideration for his feelings, restraining the joy of life that beamed from his face, he went up to the pan.

"Gerasim!" Ivan Ilyich called him in a weak voice.

Gerasim started, evidently afraid he might have committed some blunder, and with a rapid movement turned his fresh, kind, simple young face, which was just showing the first downy signs of a beard.

"Yes, Sir?"

"That must be very unpleasant for you. You must forgive me. I'm helpless."

"Oh, why, Sir?" and Gerasim's eyes beamed and he showed his glistening white teeth. "What's a little trouble? It's a case of illness with you, Sir."

And his deft strong hands did their accustomed task, and he went out of the room stepping lightly. Five minutes later he as lightly returned.

Ivan Ilyich was still sitting in the same position in the armchair.

"Gerasim," he said when the latter had replaced the freshly washed utensil, "please come here and help me." Gerasim went up to him. "Lift me up. It is hard for me to get up, and I have sent Dmitrii away."

Gerasim went up to him, grasped his master with his strong arms deftly but gently, in the same way that he walked—lifted him, supported him with one hand, and with the other drew up his trousers and would have set him down again, but Ivan Ilyich asked to be led to the sofa. Gerasim, without an effort and without apparent pressure, led him, almost lifting him, to the sofa and placed him on it.

"Thank you. How easily and well you do it all!"

Gerasim smiled again and turned to leave the room. But Ivan Ilyich felt such a comfort in his presence that he did not want to let him go.

"One thing more, please move up that chair. No, the other one—under my feet. It is easier for me when my feet are raised."

Gerasim brought the chair, set it down gently in place, and raised Ivan Ilyich's legs on to it. It seemed to Ivan Ilyich that he felt better while Gerasim was holding up his legs.

"It's better when my legs are higher," he said. "Place that cushion under them."

Gerasim did so. He again lifted the legs and placed them back, and again Ivan Ilyich felt better while Gerasim held his legs. When he set them down Ivan Ilyich fancied he felt worse.

"Gerasim," he said, "are you busy now?"

"Not at all, Sir," said Gerasim, who had learned from the townspeople how to speak to gentlefolk.

"What have you still to do?"

"What have I to do? I've done everything except chopping the logs for tomorrow."

"Then hold my legs up a bit higher, can you?"

"Of course I can. Why not?" And Gerasim raised his master's legs higher, and Ivan Ilyich thought that in that position he did not feel any pain at all.

"And how about the logs?"

"Don't trouble about that, Sir. There's plenty of time."

Ivan Ilyich told Gerasim to sit down and hold his legs, and began to talk to him. And strange to say it seemed to him that he felt better while Gerasim held his legs up.

After that Ivan Ilyich would sometimes call Gerasim and get him to hold his legs on his shoulders, and he liked talking to him. Gerasim did it all easily, willingly, simply, and with a good nature that touched Ivan Ilyich. Health, strength, and vitality in other people were offensive to him, but Gerasim's strength and vitality did not mortify but soothed him.

What tormented Ivan Ilyich most was the deception, the lie, which for some reason they all accepted, that he was not dying but was simply ill, and that he only need keep quiet and undergo treatment and then the results would be very good. He, however, knew that, do what they would, nothing would come of it, only still more agonizing suffering and death. This deception tortured him—their not wishing to admit what they all knew and what he knew, but wanting to lie to him concerning his terrible condition, and wishing and forcing him to participate in that lie. Those lies—lies enacted over him on the eve of his death and destined to degrade this awful, solemn act to the level of their visitings, their curtains, their sturgeon for dinner—were a terrible agony for Ivan Ilyich. And, strangely enough, many times when they were going through their antics over him he had been within a hair's-breadth of calling out to them: "Stop lying! You know and I know that I'm dying. Then at least stop lying about it!" But he had never had the spirit to do it.

The awful, terrible act of his dying was, he could see, reduced by those about him to the level of a casual, unpleasant, and almost indecorous incident (as if someone entered a drawing room diffusing an unpleasant odor), and this was done by that very decorum which he had served all his life long. He saw that no one felt for him, because no one even wished to grasp his position. Only Gerasim recognized it and pitied him, and so Ivan Ilyich felt at ease only with him. He felt comforted when Gerasim supported his legs (sometimes all night long) and refused to go to bed, saying: "Don't you worry, Ivan Ilyich, I'll get sleep enough later on," or when he suddenly became familiar and exclaimed: "If you weren't sick it would be another matter, but as it is, why should I grudge a little trouble?" Gerasim alone did not lie; everything showed that he alone understood the facts of the case and did not consider it necessary to disguise them, but simply felt sorry for his emaciated and enfeebled master. Once when Ivan Ilyich was sending him away he even said straight out: "We shall all of us die, so why should I grudge a little trouble?"—expressing the fact that he did not think his work burdensome, because he was doing it for a dying man and hoped someone would do the same for him when his time came.

Apart from this lying, or because of it, what most tormented Ivan Ilyich was that no one pitied him as he wished to be pitied. At certain moments after prolonged suffering he wished most of all (though he would have been ashamed to confess it) for someone to pity him as a sick child is pitied. He longed to be petted and comforted. He knew he was an important functionary, that he

had a beard turning gray, and that therefore what he longed for was impossible, but still he longed for it. And in Gerasim's attitude toward him there was something akin to what he wished for, and so that attitude comforted him. Ivan Ilyich wanted to weep, wanted to be petted and cried over, and then his colleague Shebek would come, and instead of weeping and being petted Ivan Ilyich would assume a serious, severe, and profound air, and by force of habit would express his opinion on a decision of the Court of Appeals and would stubbornly insist on that view. This falsity around him and within him did more than anything else to poison his last days.

VIII

It was morning. He knew it was morning because Gerasim had gone, and Peter the footman had come and put out the candles, drawn back one of the curtains, and begun quietly to tidy up. Whether it was morning or evening, Friday or Sunday, made no difference, it was all just the same: the gnawing, unmitigated, agonizing pain, never ceasing for an instant, the consciousness of life inexorably waning but not yet extinguished, the approach of that ever dreaded and hateful Death which was the only reality, and always the same falsity. What were days, weeks, hours, in such a case?

* * *

Ivan Ilyich knows quite well and definitely that all this is nonsense and pure humbug, but when the doctor, getting down on one knee, leans over him, putting his ear first higher then lower, and performs various gymnastic movements over him with a significant expression on his face, Ivan Ilyich submits to it all as he used to submit to the speeches of the lawyers, though he knew very well that they were all lying and why they were lying.

The doctor, kneeling on the sofa, was still sounding him when Praskovya Fedorovna's silk dress swished at the door and she was heard scolding Peter for not having let her know of the doctor's arrival.

She came in, kissed her husband, and at once proceeded to prove that she had been up a long time already, and only owing to a misunderstanding failed to be there when the doctor arrived.

Ivan Ilyich looked at her, scrutinized her from head to toe, and set down against her the whiteness and plumpness and cleanness of her hands and neck, the gloss of her hair, and the sparkle of her vivacious eyes. He hated her with his whole soul. And the thrill of hatred he felt for her made him suffer from her touch.

Her attitude toward him and his disease was still the same. Just as the doctor had adopted a certain relation to his patient which he could not abandon, so had she formed one toward her husband—that he wasn't doing something he ought to do and was himself to blame, and that she reproached him lovingly for this— and she could not now change that attitude.

"You see, he doesn't listen to me and doesn't take his medicine at the proper time. And, above all, he lies in a position that is no doubt bad for him—with his legs up."

She described how he made Gerasim hold up his legs.

The doctor smiled with a disdainful affability that said: "What's to be done? These sick people do have foolish fancies of that kind, but we must forgive them."

When the examination was over the doctor looked at his watch, and then Praskovya Fedorovna announced to Ivan Ilyich that of course he could do as he pleased, but she had sent today for a celebrated specialist who would examine him and have a consultation with Michael Danilovich (their regular doctor).

"Please don't raise any objections. I'm doing this for my own sake," she said ironically, letting it be felt that she was doing it all for his sake and said this only not to leave him any right to refuse. He remained silent, knitting his brows. He felt that he was so surrounded and involved in a mesh of falsity that it was hard to unravel anything.

*   *   *

After dinner, at seven o'clock, Praskovya Fedorovna came into the room in evening dress, her full bosom pushed up by her corset, and with traces of powder on her face. She had reminded him in the morning that they were going to the theater. Sarah Bernhardt was visiting the town, and they had a box, which he had insisted on their taking. Now he had forgotten about it, and her evening gown offended him, but he concealed his vexation when he remembered that he had himself insisted on their securing a box and going because it would be an instructive and aesthetic pleasure for the children.

Praskovya Fedorovna had come in,

self-satisfied but yet with a rather guilty air. She sat down and asked how he was but, as he saw, only for the sake of asking and not in order to learn about it, knowing that there was nothing to learn—and then went on to what she really wanted to say: that she would not on any account have gone, but that the box had been taken, and Helen and their daughter were going, as well as Petrishchev (the Examining Magistrate, their daughter's fiancé), and that it was out of the question to let them go alone; but that she would have much preferred to sit with him for a while; and he must be sure to follow the doctor's orders while she was away.

"Oh, and Fedor Petrovich [the fiancé] would like to come in. May he? And Lisa?"

"All right."

Their daughter came in in full evening dress, her fresh young flesh exposed (making a show of that very flesh which in his own case caused so much suffering), strong, healthy, evidently in love, and having but little patience with illness, suffering, and death, because they interfered with her happiness.

Fedor Petrovich came in, too, in full dress, his hair curled à la Capoul, a tight stiff collar round his long sinewy neck, an enormous white shirt front and narrow black trousers tightly stretched over his strong thighs. He had one white glove tightly drawn on, and was holding his opera hat.

Following him the schoolboy crept in unnoticed, in a new uniform, poor little fellow, and wearing gloves. Terribly dark shadows showed under his eyes, the meaning of which Ivan Ilyich knew well.

His son had always seemed pathetic to him, and now it was dreadful to see the boy's frightened look of pity. It seemed to Ivan Ilyich that Vassya was the only one besides Gerasim who understood and pitied him.

They all sat down and again asked how he was. A silence ensued. Lisa asked her mother about the opera glasses, and there was an altercation between mother and daughter as to who had taken them and where they had been put. This occasioned some unpleasantness.

Fedor Petrovich inquired of Ivan Ilyich whether he had ever seen Sarah Bernhardt. Ivan Ilyich did not at first catch the question, but then replied: "No; have you seen her before?"

"Yes, in *Adrienne Lecouvreur*."

Praskovya Fedorovna mentioned some roles in which Sarah Bernhardt was particularly good. Her daughter disagreed. Conversation sprang up as to the splendor and realism of her acting—the sort of conversation that is always repeated and is always the same.

In the midst of the conversation Fedor Petrovich glanced at Ivan Ilyich and became silent. The others also looked at him and grew silent. Ivan Ilyich was staring with glittering eyes straight before him, evidently indignant with them. This had to be rectified, but it was impossible to do it. The silence had to be broken, but for a time no one dared to break it, and they all became afraid that the conventional deception would suddenly become obvious and the truth become plain to all. Lisa was the first to pluck up courage and break that silence, but by trying to hide what everybody was feeling she betrayed it:

"Well, if we're going, it's time to start," she said, looking at her watch, a present from her father, and with a faint and significant smile at Fedor Petrovich pertaining to something known only to them. She got up with a swish of her dress.

They all rose, said good night, and went away.

When they had gone it seemed to Ivan Ilyich that he felt better; the falsity had gone with them. But the pain remained—that same pain and that same fear that made everything monotonously alike, nothing harder and nothing easier. Everything was worse.

Again minute followed minute and hour followed hour. Everything remained the same and there was no cessation. And the inevitable end of it all became more and more terrible.

"Yes, send Gerasim here," he replied to a question Peter asked.

IX

His wife returned late at night. She came in on tiptoe, but he heard her, opened his eyes, and made haste to close them again. She wished to send Gerasim away and to sit with him herself, but he opened his eyes and said: "No, go away."

"Are you in great pain?"

"It's always the same."

"Take some opium."

He agreed and took some. She went away.

Till about three in the morning he was in a state of stupefied misery. It seemed to him that he and his pain were being thrust into a narrow, deep black sack, but though they were pushed further and further in they could not be pushed to the bottom. And this, terrible enough in itself, was accompanied by suffering. He was frightened, yet wanted to fall through the sack; he struggled yet co-operated. And suddenly he broke through, fell, and regained consciousness. Gerasim was sitting at the foot of the bed dozing quietly and patiently, while he himself lay with his emaciated stockinged legs resting on Gerasim's shoulders; the same shaded candle was there and the same unceasing pain.

"Go away, Gerasim," he whispered.

"It's all right, Sir. I'll stay a while."

"No. Go away."

He removed his legs from Gerasim's shoulders, turned sideways onto his arm, and felt sorry for himself. He only waited till Gerasim had gone into the next room and then restrained himself no longer but wept like a child. He wept on account of his helplessness, his terrible loneliness, the cruelty of man, the cruelty of God, and the absence of God.

"Why hast Thou done all this? Why hast Thou brought me here? Why, why dost Thou torture me so terribly?"

He did not expect any answer and yet wept because there was no answer and could be none. The pain again grew more acute, but he did not stir and did not call. He said to himself: "Go on! Strike me! But what is it for? What have I done to Thee? What is it for?"

Then he grew quiet and not only ceased weeping but even held his breath and became all attention. It was as though he were listening not to an audible voice but to the voice of his soul, to the current of thoughts arising within him.

"What is it you want?" was the first clear conception capable of expression in words that he heard.

"What do you want? What do you want?" he repeated to himself.

"What do I want? To live and not to suffer," he answered.

And again he listened with such concentrated attention that even his pain did not distract him.

"To live? How?" asked his inner voice.

"Why, to live as I used to—well and pleasantly."

"As you lived before, well and pleasantly?" the voice repeated.

And in imagination he began to recall the best moments of his pleasant life. But, strange to say, none of those best moments of his pleasant life now seemed at all what they had then seemed—none of them except the first recollections of childhood. There, in childhood, there had been something really pleasant with which it would be possible to live if it could return. But the child who had experienced that happiness existed no longer; it was like a reminiscence of someone else.

As soon as the period began which had produced the present Ivan Ilyich, all that had then seemed joys now melted before his sight and turned into something trivial and often nasty.

And the further he departed from childhood, and the nearer he came to the present, the more worthless and doubtful were the joys. This began with the School of Law. A little that was really good was still found there—there was lightheartedness, friendship, and hope. But in the upper classes there had already been fewer of such good moments.

Then during the first years of his official career, when he was in the service of the Governor, some pleasant moments again occurred: they were the memories of love for a woman. Then all became confused and there was still less of what was good; later on again there was still less that was good, and the further he went, the less there was. His marriage, a mere accident, then the disenchantment that followed it, his wife's bad breath and her sensuality and hypocrisy; then that deadly official life and those preoccupations about money, a year of it, and two, and ten, and twenty, and always the same thing. And the longer it lasted, the more deadly it became. "It's just as if I'd been going downhill while I imagined I was going up. And that's really what it was. I was going up in public opinion, but to the same extent life was ebbing away from me. And now it's all done, and there is only death.

"Then what does it mean? Why? It can't be that life is so senseless and horrible. But if it really has been so horrible and senseless, why must I die and die in agony? There's something wrong!

"Maybe I didn't live as I ought to have done," it suddenly occurred to him. "But how could that be, when I did everything properly?" he replied, and immediately dismissed from his mind this, the sole solution of all the riddles of life and death, as something quite impossible.

"Then what do you want now? To live? Live how? Live as you lived in the law court when the court attendant proclaimed: 'His Honor, His Honor, the Judge!' he repeated to himself. 'Here he is—His Honor! His

Honor, the Judge!' But I'm not guilty!" he exclaimed angrily. "What is it for?" And he ceased crying but, turning his face to the wall, continued to ponder on the same question: Why, and for what purpose, was there all this horror? But however much he pondered, he found no answer. And whenever the thought occurred to him, as it often did, that it all resulted from his not having lived as he ought to have lived, he at once recalled the correctness of his whole life and dismissed so strange an idea.

X

Another fortnight passed. Ivan Ilyich no longer left his sofa now. He would not lie in bed but lay on the sofa, facing the wall nearly all the time. He suffered ever the same unceasing agonies and in his loneliness pondered always on the same insoluble question: "What is this? Can it be that this is death?" And an inner voice answered: "Yes, it is death."

"Why these sufferings?" And the voice answered: "For no reason—they just are." Beyond and besides this there was nothing.

From the very beginning of his illness, ever since he had first been to see the doctor, Ivan Ilyich's life had been divided between two contrary and alternating moods: now it was despair and the expectation of this uncomprehended and terrible death, and now hope and an intently interested observation of the functioning of his organs. Now before his eyes there was only a kidney or an intestine that temporarily evaded its duty, and now only that incomprehensible and dreadful death from which it was impossible to escape.

These two states of mind had alternated from the very beginning of his illness, but the further it progressed, the more doubtful and fantastic became his conception of the kidney, and the more real his sense of impending death.

He had but to call to mind what he had been three months before and what he was now, to see with what regularity he had been going downhill, for every possibility of hope to be shattered.

Latterly, during that loneliness in which he found himself as he lay with his face to the back of the sofa, a loneliness in the midst of a populous town and surrounded by numerous acquaintances and relations but that yet could not have been more complete anywhere—either at the bottom of the sea or under the earth—during that terrible loneliness he had lived only in memories of the past. Pictures of his past rose before him one after the other. They always began with what was nearest in time and then went back to what was most remote—to his childhood—and rested there. If he thought of the stewed prunes that had been offered him that day, his mind went back to the raw shriveled French plums of his childhood, their peculiar flavor and the flow of saliva when he sucked their stones, and along with the memory of that taste came a whole series of memories of those days: his nurse, his brother, and their toys. "No, I mustn't think of that. It's too painful," he said to himself, and brought himself back to the present—to the button on the back of the sofa and the creases in its morocco. "Morocco is expensive, but it does not wear well: there had been a quarrel about it. It was a different kind of quarrel and a different kind of morocco that time when we tore father's portfolio and were punished, and mamma brought us some tarts." And again his thoughts dwelt on his childhood, and again it was painful, and he tried to banish them and fix his mind on something else.

Then again together with that chain of memories another series passed through his mind—of how his illness had progressed and grown worse. There also, the farther back he looked, the more life there had been. There had been more of what was good in life and more of life itself. The two merged together. "Just as the pain went on getting worse and worse, so my life grew worse and worse," he thought. "There's one bright spot there at the back, at the beginning of life, and afterward all becomes blacker and blacker and proceeds more and more rapidly—in inverse ratio to the square of the distance from death," thought Ivan Ilyich. And the example of a stone falling downward with increasing velocity entered his mind. Life, a series of increasing sufferings, flies farther and farther toward its end—the most terrible suffering. "I'm flying—" He shuddered, shifted his body, and tried to resist, but was already aware that resistance was impossible, and again with eyes weary of gazing but unable to cease seeing what was before them, he stared at the back of the sofa and waited—awaited that dreadful fall and shock and destruction.

"Resistance is impossible!" he said to himself. "If I could only understand what it's all for! But that, too, is impossible. An explanation would be possible if it could be said that I haven't lived as I ought to. But it's impossible to say that," and he remembered all the legality, rectitude and propriety of his life. "That, at any rate, certainly can't be admitted," he thought, and his lips smiled ironically as if someone could see that smile and be taken in by it. "There is no explanation! Agony, death. . . . What for?"

XI

Another two weeks went by in this way, and during that fortnight an event occurred that Ivan Ilyich and his wife had desired. Petrishchev formally proposed. It happened in the evening. The next day Praskovya Fedorovna came into her husband's room considering how best to inform him of it, but that very night there had been a fresh change for the worse in his condition. She found him still lying on the sofa but in a different position. He lay on his back, groaning and staring fixedly straight in front of him.

She began to remind him of his medicines, but he turned his eyes toward her with such a look that she did not finish what she was saying, so great an animosity, to her in particular, did that look express.

"For Christ's sake, let me die in peace!" he said.

She would have gone away, but just then their daughter came in and went up to say good morning. He looked at her as he had done at his wife, and in reply to her inquiry about his health said drily that he would soon free them all of himself. They were both silent, and after sitting with him for a while went away.

"Is it our fault?" Lisa said to her mother. "Just as if we were to blame! I'm sorry for papa, but why should we be tortured?"

The doctor came at his usual time. Ivan Ilyich answered "Yes" and "No," never taking his angry eyes from him, and at last said: "You know you can do nothing for me, so leave me alone."

"We can ease your sufferings."

"You can't even do that. Let me be."

The doctor went into the drawing room and told Praskovya Federovna that the case was very serious, and that the only recourse left was opium to allay her husband's sufferings, which must be terrible.

It was true, as the doctor said, that Ivan Ilyich's physical sufferings were terrible, but worse than the physical sufferings were his mental sufferings, which were his chief torture.

His mental sufferings were due to the fact that that night, as he looked at Gerasim's sleepy, good-natured face with its prominent cheek bones, the question suddenly occurred to him: "What if my whole life *has* really been wrong?"

It occurred to him that what had appeared perfectly impossible before—namely, that he had not spent his life as he should have done—might after all be true. It occurred to him that his scarcely perceptible attempts to struggle against what was considered good by the most highly placed people, those scarcely noticeable impulses which he had immediately suppressed, might have been the real thing, and all the rest false. And his professional duties and the whole arrangement of his life and of his family, and all his social and official interests, might all have been false. He tried to defend all those things to himself and suddenly felt the weakness of what he was defending. There was nothing to defend.

"But if that's so," he said to himself, "and I'm leaving this life with the consciousness that I have lost all that was given me, and it's impossible to rectify it—what then?"

He lay on his back and began to pass his life in review in quite a new way. In the morning when he saw first his footman, then his wife, then his daughter, and then the doctor, their every word and movement confirmed to him the awful truth that had been revealed to him during the night. In them he saw himself—all that for which he had lived—and saw clearly that it was not real at all, but a terrible and huge deception which had hidden both life and death. This consciousness intensified his physical suffering tenfold. He groaned and tossed about, and pulled at his clothing which choked and stifled him. And he hated them on that account.

He was given a large dose of opium and became unconscious, but at noon his sufferings began again. He drove everybody away and tossed from side to side.

His wife came to him and said:

"Jean, my dear, do this for me. It can't do any harm and often helps. Healthy people often do it—"

He opened his eyes wide.

"What? Take communion? Why? It's unnecessary! However—"

She began to cry.

"Yes, do, my dear. I'll send for our priest. He's such a lovely man."

"All right. Very well," he muttered.

When the priest came and heard

his confession, Ivan Ilyich was softened and seemed to feel a relief from his doubts, and consequently from his sufferings, and for a moment there came a ray of hope. He again began to think of his vermiform appendix and the possibility of correcting it. He received the sacrament with tears in his eyes.

When they laid him down again afterward he felt a moment's ease, and the hope that he might live awoke in him again. He began to think of the operation that had been suggested to him. "To live! I want to live!" he said to himself.

His wife came in to congratulate him after his communion, and when, uttering the usual conventional words, she added:

"You feel better, don't you?"—he, without looking at her, said: "Yes."

Her dress, her figure, the expression of her face, the tone of her voice, all revealed the same thing. "This is wrong, it's not as it should be. All you have lived for, and still live for, is falsehood and deception, hiding life and death from you." And as soon as he admitted that thought, his hatred and his agonizing physical suffering again sprang up, and with that suffering a consciousness of the ineluctable, approaching end. And to this was added a new sensation of grinding, shooting pain and a feeling of suffocation.

The expression of his face when he had uttered that "Yes" had been dreadful. Having uttered it, he looked her straight in the eyes, turned over on his face with a rapidity extraordinary in his weak state, and shouted:

"Go away! Go away and leave me alone!"

XII

From that moment the screaming began that continued for three days, and was so terrible that one could not hear it without horror even through two closed doors. At the moment he answered his wife he realized that he was lost, that there was no return, that the end had come, the very end, and his doubts were still unsolved and remained doubts.

"Oh! Oh! Oh!" he cried in various intonations. He had begun by screaming: "I won't have it! I won't!" and continued screaming on the letter *o*.

For three whole days, during which time did not exist for him, he struggled in that black sack into which he was being thrust by an invisible, resistless force. He struggled as a man condemned to death struggles in the hands of the executioner, though knowing that he cannot save himself. And every moment he felt that despite all his efforts he was drawing nearer and nearer to what terrified him. He felt that his agony was due to his being thrust into that black opening and still more to his not being able to get right into it. He was hindered from getting into it by his conviction that his life had been a good one. That very justification of his life held him fast and prevented his moving forward, and it caused him most torment of all.

Suddenly some force struck him in the chest and side, making it still harder to breathe, and he fell through the hole, and there at the bottom was a light. What had happened to him was like the sensation one sometimes experiences in a railway carriage when one thinks one is going backward, while one is really going forward, and suddenly becomes aware of the real direction.

"Yes, it was all not the right thing," he said to himself, "but that doesn't matter. It can be so. But what *is* the right thing?" he asked himself, and suddenly grew quiet.

This occurred at the end of the third day, two hours before his death. Just then his schoolboy son had crept softly in and gone up to the bedside. The dying man was still screaming desperately and waving his arms. His hand fell on the boy's head, and the boy caught it, pressed it to his lips, and began to cry.

At that very moment Ivan Ilyich fell through and caught sight of the light, and it was revealed to him that though his life had not been what it should have been, this could still be rectified. He asked himself: "What *is* the right thing?" and grew still, listening. Then he felt that someone was kissing his hand. He opened his eyes, looked at his son, and felt sorry for him. His wife came up to him, and he glanced at her. She was gazing at him open-mouthed, with undried tears on her nose and cheeks and a despairing look on her face. He felt sorry for her, too.

"Yes, I'm making them wretched," he thought. "They're sorry, but it will be better for them when I die." He wished to say this but had not the strength to utter it. "Besides, why speak? I must act," he thought. With a look at his wife, he indicated his son and said: "Take him away—sorry for him—sorry for you, too—" He tried to add: "Forgive me," but said

"Forego—" and waved his hand, knowing that He whose understanding mattered would understand.

And suddenly it grew clear to him that what had been oppressing him and would not leave him was all dropping away at once from two sides, from ten sides, and from all sides. He was sorry for them, he must act so as not to hurt them: release them and free himself from these sufferings. "How good and how simple!" he thought. "And the pain?" he asked himself. "What has become of it? Where are you, my pain?"

He turned his attention to it.

"Yes, here it is. Well, what of it? Let the pain be. And death—where is it?"

He sought his former accustomed fear of death and did not find it. "Where is it? What death?" There was no fear because he could not find death.

In place of death there was light.

"So that's what it is!" he suddenly exclaimed aloud. "What joy!"

To him all this happened in a single instant, and the meaning of that instant did not change. For those present his agony continued for another two hours. Something rattled in his throat, his emaciated body twitched, then the gasping and rattle became less and less frequent.

"It's all over!" said someone near him.

He heard these words and repeated them in his soul.

"Death is all over," he said to himself. "It's no more!"

He drew in a breath, stopped in the midst of a sigh, stretched out, and died.

William Hogarth. *The Unhappy Pair at Home.*

Summer in the Mountains

Roger Angell

Meg Porter had a look of determination on her face as she listened to her mother. Her eyes were fixed on her mother, and occasionally she raised her eyebrows or nodded slightly at a pause in the conversation, but her lips were clamped firmly together, as if to hold back any words she might later regret, as if no one could wring from her the mildest complaint or even a sigh of boredom or resignation. It was, nevertheless, a look of remarkable hostility, particularly from a woman of thirty whose face was normally pretty and animated. Both her husband and her mother had seen this look before; her mother, Mrs. Brockway, had, in fact, come to think it was her daughter's normal expression, and it worried her, but somehow she had never brought herself to speak of it. Larry Porter had once mentioned it to his wife. She ought to try, he had said, to appear happier and more interested when her mother was with them. She had no idea how bad it looked, he said, and Mrs. Brockway would think that that was the way her only daughter *felt* about her, which, of course, wasn't true. And besides, it made her look old. But Meg Porter somehow couldn't shake the habit. Now Larry, sitting across the room, tried to get her attention, to remind her again, but Meg wouldn't take her rigid stare away from her mother.

"And since Dr. Stout couldn't see me at any other time," Mrs. Brockway was saying, "naturally I had to call Mrs. Lincoln and tell her I wouldn't be able to come to the symphony after all. It was really a shame, because I do get out so little, you know. But the doctor had said to come back for a little checkup when I noticed that shortness of breath again, and Tuesday morning I just felt as if I couldn't take any air into my lungs. Of course, it wouldn't worry me if there were someone living with me in the hotel, but even with the elevator boys right outside that you told me I could call any time, it just isn't the same. I sometimes wonder what would happen if anything *did* happen to me and nobody there with me." Meg Porter opened her mouth suddenly and took a breath, but her mother held her hand up and smiled. "No," she said. "I know what both Dr. Stout and that other one said. I'm fine, perfectly fine, and there's really nothing to worry about. For my age. I'm not trying to scare you, Meg. I know you call up, and you know I love that. Nearly every day. And there's always the phone, in case anything . . . I really don't worry at all." She smiled bravely and shook her head. "It's just being alone, that's all. I'm afraid I'll never really be used to it. Of course, I don't have to tell you how much I appreciate—No, you told me not to use that word, didn't you?" She looked slyly at her daughter. "Well, how much I *love* that apartment and how glad I am knowing that it's mine, or almost mine. I always want you and Larry to know that I don't forget that or take it for granted. Ever. Well!" She smoothed her skirt carefully and then looked at Larry. "Where are you two going tonight?"

Larry Porter took his eyes off his wife. "It's just to cocktails with the Bradleys. You know—you met them here Christmas Eve. And then we thought we'd go to dinner at Luchow's. But we won't be late.

Shouldn't be later than ten or so. Of course, we can always come back before then if you want to get home earlier. We could just pick up a hamburger."

"Don't be silly, Larry," Mrs. Brockway said. "I wouldn't dream of it. Jane and I always have a wonderful time together. You know I love looking after her. It's so much like the old days, with Meg. After her father died and before she went away · to college. I'll bet she doesn't even remember how close we used to be."

"Of course I remember, Mother," Meg Porter said. She leaned forward and took a cigarette out of a box on the table beside her chair. "We used to play games. And we went to the theater. And—"

"And every summer to Chocorua," Mrs. Brockway said. "Larry, you have no idea how she used to adore it there. Sometimes I wonder how you all ever started going to the beach."

"Jane is crazy about the beach, Mother," Meg said. "She wouldn't go any place else. And neither would I."

"Of course, dear," Mrs. Brockway said gently. "It isn't the place that counts. It's just being with people that counts. People you love. I just hope, Meg, that you're *with* Jane enough. She's such a responsive little girl, and you have no idea how suddenly it all ends. Before you know it, she'll be grown up and off married to somebody, with a busy life of her own and lots of friends and parties. And then you'll start thinking back about your life and you'll remember every hour you had together."

"Mother, for heaven's sake! You don't honestly think I neglect—"

"Oh, no, Meg," Mrs. Brockway said, smiling again. "I just mean that Jane responds so. I hope you appreciate it. Why, I always look forward to these evenings we have together. Jane and I have such fine times it makes me feel like a real member of the family, just for a few hours, while you and Larry have a good time by yourselves. What I mean is, really, that at my age you know enough to appreciate a little child. They just *give* themselves, the way grownups can't. Nobody really understands that until they're as old as I am and their own children have gone."

"Oh, *Moth*-er," Meg said, helplessly dropping her hands.

Larry Porter stood up quickly, looking at his watch. "We'd better go, darling. It's past five-thirty. I'll tell Jane we're leaving. I think she's in her room watching that damned kid show again."

In the cab, a few minutes later, Meg was almost crying. "Oh, God, darling," she said to her husband, "why is she so awful, and why am I so awful to her? I've just got so that I think that every word is a hint or a slam or something, and I just can't stand it any more. There ought to be a compatibility test for parents and grown-up children to take together, and if they fail, there would be a law that said they couldn't see each other or couldn't live in the same city together. No, I don't mean that, either. I don't *really* feel that way—it's too hateful. It's just that Mother gets me all mixed up."

"I know," Larry Porter said. He lit a cigarette and handed it to her. "You had that look again. You never look that way except when she's there."

"I know I did. I can feel it coming over me and I can't stop it. Every time before she comes, I say to myself, This time will be different, this time we'll really get along, and she won't talk about herself and how lonely she is and how dreadful I am. But then the minute I see her I know it's going to be just the same."

"It's not your fault, Meg," Larry said gently. "It's just that she's old. That and the money."

"I know it. But we can't help the money. She needs it, and we want to give it to her. But why does she have to be so damned *grateful*? I've told her not to do that. It just ruins everything, and somehow it makes me feel as if we weren't doing enough for her, either. We all know she couldn't live with us, even if we had room. She doesn't really want that—she'd loathe it. And I'd go absolutely crazy and so would you."

"I wouldn't allow it," Larry said. "I'll never get us trapped with that."

"But what *does* she want?" Meg almost shouted. "I call her up. I have lunch with her. I go to the movies with her. We have her for dinner. We have her in to sit with Jane, because I honestly think she likes it. We get the best doctors for her and try to persuade her that she really is as well as she can expect. Why does she keep talking at me, then? Why does she make me feel so damned ungrateful and mean?"

Larry stared out at the passing lights for a moment before he answered. "I think," he said finally, "it's just that she's old and afraid, and she wants somebody to share that with her. And nobody can do it. You can't share being old with anybody."

Meg turned in the seat and quickly took hold of her husband's arm. "Don't let me get that way, Larry," she said urgently. She squeezed his arm hard. "Promise me you'll do that, Larry. Don't ever let me get like that. Don't let me do that to Jane, ever."

In the apartment, after she and the little girl had eaten the supper she had cooked, Mrs. Brockway went to the living-room windows to draw the curtains. She felt tired, and there was a vague pain in her side. Before dinner, she and her granddaughter had played a game on the floor, throwing dice that sent little racing cars around a numbered course. She hadn't understood the game well and the floor had made her knees ache, but Jane had seemed excited and happy, and they had played it twice. Now she stood for a moment by the window and looked out at the darkness that meant that soon the little girl would have to go to bed. After that, after an hour or two, the Porters would be back. She would hear them laughing and talking as they came out of the elevator, and they would come in looking excited and happy, and Larry would offer her a drink, which she would refuse, and then it would be time for her to leave and go back to the barren hotel apartment, with the uncomfortable bed, and the furniture that was not hers, and the steam heat that killed every plant she bought, and the silent telephone, and the strange voices in the hall late at night—all the ugly and frightening familiarities that had somehow become the place where she lived. There were a few of her possessions left there—the set of Thackeray of her husband's, and the Spode teacups, which she had to keep on the mantelpiece—but they no longer seemed like her own. Sitting there by herself in the evenings after she had turned off the loud, confusing television, she often studied these treasures of hers and tried to remember how they had looked back in her own apartment, the one she had shared with Meg—where in the bookcase the set of Thackeray had stood and how the cups had looked among the other familiar china on the shelves in the dining room. But she could no longer remember. When her furniture had gone (Meg had pointed out that it would be foolish and expensive to keep up her old place or to store all her books and pictures and furniture), her remaining treasures had somehow lost their identity. Out of place and without companions, they had become exactly the same as everything else in her bare rooms—somebody else's belongings, which she was now expected to live with for a time.

It was this she tried to explain to her daughter when she came to Meg's house and saw it full of the warmth of lived-with and familiar objects: an album of records that Jane had left open on the floor, the dachshund's worn cushion beside the fireplace, the clamshell ashtrays, which the three of them had collected on the beach the summer before—all the evidence of plans made long ago and happily accomplished. Seeing this after the cold emptiness of her own long afternoons, after the terrors of a sudden pain in the night with no one to call out to, Mrs.

Brockway always wanted to tell her daughter what she had here in her home, to cry desperately to her, "You don't know! You're so lucky here and you don't know it. You can't know what it is like to be without it, when it's all forgotten, when your own belongings are gone, and your house empty, and there are no more plans." But somehow it never came out that way. When she tried (without complaining) to explain all this to Meg, it ended in a recital of her days, of what she had eaten and what the doctor had said. Meg always became impatient and angry, and Mrs. Brockway went away feeling ashamed, because she said it all wrong.

"Grandma, look!"

The little girl had made a small tower out of the piled-up dice from the game, and now she sent one of the tiny racing cars crashing into it, knocking the dice onto the rug. "Bang!" she cried. "He hit the pylon on that turn and now he's out of the race. Dang! Dang! Dang! Here comes the ambulance!"

Mrs. Brockway sighed and closed the curtains. She would have to hurry. The little girl would have to be put to bed soon. There was less than an hour for the two of them to be together.

"Let's read now, Jane, shall we?" Mrs. Brockway said. "I'll sit on the couch and you can lie beside me, and we'll read anything you want. We'll have a nice, quiet time together, won't we, darling?"

"O.K.," Jane said. She dropped the racing car and ran over to the bookcase. "Here," she said, pulling out a book. "This one. Read me *The Tinder Box,* Grandma."

Smiling, Mrs. Brockway sat down on the sofa. Sometimes Jane was exactly the way her mother had been as a little girl. Like Meg, she was suggestible; at one moment she could appear frighteningly violent, but she could change her mood and her interest in a second. The similarity made the grandmother feel warm and happy. Now she took the book from the eight-year-old. "All right," she said comfortably, turning the pages. *"The Tinder Box* it is. I know that story. It has those wonderful dogs with the big eyes. Now you sit down beside me, and I'll start." The little girl quickly curled her legs on the sofa, and Mrs. Brockway put her arm around her, pulled her close, and began to read: "A soldier came marching along the highroad. One, two! One, two!"

As she read the familiar story, Mrs. Brockway kept glancing at the little girl's face, intent on the page. Looking at her, the grandmother felt calm. At these moments, she knew that the fears and pains of her hotel bedroom were not really part of her at all. She was the same as she had always been, capable of anything as long as there was someone near her whom she loved and could make happy. Meg could no longer accept that from her, but this little girl, her granddaughter, did. She knew Jane depended on her. All that was needed was for them to be together more often, not just for a few hours on an occasional evening.

When Mrs. Brockway finished the story, Jane didn't move but kept staring at the page with an intent and abstracted expression. Suddenly, before the child sat up, before it was time for her to go to bed, Mrs. Brockway wanted to do something to preserve the moment, plan something that she could take back with her to her room, like a possession of her own—a guarantee of love and happiness to come.

She tightened her arm about her granddaughter. "Jane," she said softly, "how would you like to come and live with me for a little while? Just we two together. Wouldn't that be fun? We do have such nice, happy times together it would be fun to do it for a longer time. I know a lovely place we could go, perhaps in the summer, when your school is over and Mum and your father like to take a little trip by themselves. Would you like that, Jane?"

"Where would Mum and Dad go?" Jane asked quietly.

"Oh, I don't know. Perhaps to Europe for a few weeks or on a motor trip somewhere in the summer. I haven't asked them, but perhaps, if you would like it, I could suggest it and they might think it was a good idea. They haven't been off by themselves for so long that they might say yes. And then you and I could go to the mountains together, to a place I know. I used to go there every single summer with your mother. We could get a lovely cottage, and there are mountains all around, all with long Indian names. And there's a lake, where you could learn to swim, right where your mother did. And you can climb mountains, and we could take little trips. Would you like that, Jane?"

"Could I have a bulldog?" Jane asked. "Jennifer French, at the beach, has a bulldog, and Dad said maybe I could have one sometime— a dog of my very own. Could I have one in the mountains?"

"Well, maybe, Jane. It all depends." Mrs. Brockway gave her another little hug. "We'd have to see, but we certainly could try. Wouldn't it be fun?"

"Yes, Grandma," Jane said. She sat up and looked at Mrs. Brockway. "And we can take the bulldog to climb the mountains and take him to the lake, and he'll learn to swim with me, won't he?"

Mrs. Brockway laughed aloud; actually, the plan didn't seem impossible at all. They might do it this very summer. It wasn't too late. She would be very practical and write and ask about cottages first and then speak to Meg after she had heard, so that there could be no difficulties or arguments. It was the simplest thing in the world. A whole summer with Jane, every moment of the long months with someone she loved. She pulled the child close. "We'll make a little secret of this, Jane, won't we? You won't tell Mum tomorrow, and then we'll talk to her about it together some day soon and surprise her, and she'll be so glad you can go to the mountains just the way she did."

"Not tell Mum?" Jane said doubtfully.

"Not right away, darling. It'll be a surprise, see? And then when we do tell her, you can explain how very much you want to go and spend the whole summer with Grandma. And after that, who knows? Perhaps we can do it again and we can stay even longer together. And we could go for a little trip next Christmas, just the two of us, on the train. Oh, you'll see! We'll have fine times together, and we'll have a lot of them!"

Mrs. Brockway had been hugging the child close to her for a long time, and now Jane began to struggle under her arm. She pulled herself

away and stood up, and Mrs. Brockway was astonished to see that there
were tears in her eyes.

"Go away for *Christmas?*" Jane
said in a frightened voice. "Where
will Mum and Dad be? Won't they
be here for Christmas?"

Mrs. Brockway was frightened
now, too. "Of course, darling," she
said quickly, putting her hands out
toward the child. "Of course they
will. You don't have to go away for
Christmas if you don't want to." She
hadn't meant to go so far with her
talk and plans. All that could come
later. She caught Jane's hand and
pulled her back to the couch. "You
don't have to go anywhere, Jane,"
she said, trying to make her voice
warm and soothing again. "Just to
the mountains. Just this summer in
the mountains, darling. You and I
together."

But Jane roughly jerked her arm
free and took three quick and de-
fiant backward steps away from her.
Mrs. Brockway again reached her
hands out toward her granddaughter, and as she searched hopelessly for
the words to recapture what she had
lost, she suddenly thought she saw
on the red, tear-streaked face of the
little girl a flicker of expression she
had noticed before in the eyes of
the young and the strong—a bright,
animal look of rejection and fear.

Judith Somerville. *Aged Woman.*

from

A Doll's House

Henrik Ibsen

CHARACTERS

Torvald Helmer
Nora, *his wife*
Doctor Rank
Mrs. Linde
Nils Krogstad
Helmer's Three Young Children
Anne, *their nurse*
A Housemaid
A Porter

The action takes place in Helmer's house.

ACT I

A bell rings in the hall; shortly afterwards the door is heard to open. Enter Nora, *humming a tune and in high spirits. She is in out-door dress and carries a number of parcels; these she lays on the table to the right. She leaves the outer door open after her, and through it is seen a* porter *who is carrying a Christmas Tree and a basket, which he gives to the* maid *who has opened the door.*

Nora: Hide the Christmas Tree carefully, Helen. Be sure the children do not see it till this evening, when it is dressed. (*To the* porter, *taking out her purse.*) How much?

porter: Sixpence.

Nora: There is a shilling. No, keep the change. (*The* porter *thanks her, and goes out.* nora *shuts the door. She is laughing to herself, as she takes off her hat and coat. She takes a packet of macaroons from her pocket and eats one or two; then goes cautiously to her husband's door and listens.*) Yes, he is in. (*Still humming, she goes to the table on the right.*)

HELMER: *(calls out from his room)* Is that my little lark twittering out there?

NORA: *(busy opening some of the parcels)* Yes, it is!

HELMER: Is it my little squirrel bustling about?

NORA: Yes!

HELMER: When did my squirrel come home?

NORA: Just now. *(Puts the bag of macaroons into her pocket and wipes her mouth.)* Come in here, Torvald, and see what I have bought.

HELMER: Don't disturb me. *(A little later, he opens the door and looks into the room, pen in hand.)* Bought, did you say? All these things? Has my little spendthrift been wasting money again?

NORA: Yes, but, Torvald, this year we really can let ourselves go a little. This is the first Christmas that we have not needed to economize.

HELMER: Still, you know, we can't spend money recklessly.

NORA: Yes, Torvald, we may be a bit more reckless now, mayn't we? Just a tiny wee bit! You are going to have a big salary and earn lots and lots of money.

HELMER: Yes, after the New Year; but then it will be a whole quarter before the salary is due.

NORA: Pooh! we can borrow till then.

HELMER: Nora! *(Goes up to her and takes her playfully by the ear.)* The same little featherhead! Suppose, now, that I borrowed fifty pounds today, and you spent it all in the Christmas week and then on New Year's Eve a slate fell on my head and killed me, and—

NORA: *(putting her hands over his mouth)* Oh! don't say such horrid things.

HELMER: Still, suppose that happened,—what then?

NORA: If that were to happen, I don't suppose I should care whether I owed money or not.

HELMER: Yes, but what about the people who had lent it?

NORA: They? Who would bother about them? I should not know who they were.

HELMER: That is like a woman! But seriously, Nora, you know what I think about that. No debt, no borrowing. There can be no freedom or beauty about a home life that depends on borrowing and debt. We two have kept bravely on the straight road so far and we will go on the same way for the short time longer than there need be any struggle.

NORA: *(moving towards the stove)* As you please, Torvald.

HELMER: *(following her)* Come, come, my little skylark must not droop her wings. What is this! Is my little squirrel out of temper? *(Taking out his purse.)* Nora, what do you think I have got here?

NORA: *(turning round quickly)* Money!

HELMER: There you are. *(Gives her some money.)* Do you think I don't know what a lot is wanted for housekeeping at Christmas-time?

NORA: *(counting)* Ten shillings—a pound—two pounds! Thank you, thank you, Torvald; that will keep me going for a long time.

HELMER: Indeed it must.

NORA: Yes, yes, it will. But come here and let me show you what I have bought. And all so cheap! Look, here is a new suit for Ivar, and a sword; and a horse and a trumpet for Bob; and a doll and dolly's bedstead for Emmy,—they are very plain, but anyway she will soon break them in pieces. And here are dress-lengths and handkerchiefs for the maids; old Anne ought really to have something better.

HELMER: And what is in this parcel?

NORA: *(crying out)* No, no! you mustn't see that till this evening.

HELMER: Very well. But now tell me, you extravagant little person, what would you like for yourself?

NORA: For myself? Oh, I am sure I don't want anything.

HELMER: Yes, but you must. Tell me something reasonable that you would particularly like to have.

NORA: No, I really can't think of anything—unless, Torvald—

HELMER: Well?

NORA: *(playing with his coat buttons, and without raising her eyes to his)* If you really want to give me something, you might—you might—

HELMER: Well, out with it!

NORA: *(speaking quickly)* You might give me money, Torvald. Only just as much as you can afford; and then one of these days I will buy something with it.

HELMER: But, Nora—

NORA: Oh, do! dear Torvald; please do! Then I will wrap it up in beautiful gilt paper and hang it on the Christmas tree. Wouldn't that be fun?

HELMER: What are little people called that are always wasting money?

NORA: Spendthrifts—I know. Let us do as you suggest, Torvald, and then I shall have time to think what I am most in want of. That is a very sensible plan, isn't it?

HELMER: *(smiling)* Indeed it is— that is to say, if you were really to save out of the money I give you, and then really buy something for your-self. But if you spend it all on the housekeeping and any number of unnecessary things, then I merely have to pay up again.

NORA: Oh, but, Torvald—

HELMER: You can't deny it, my dear little Nora. *(Puts his arm round her waist.)* It's a sweet little spendthrift, but she uses up a deal of money. One would hardly believe how expensive such little persons are!

NORA: It's a shame to say that. I do really save all I can.

HELMER: *(laughing)* That's very true,—all you can. But you can't save anything!

NORA: *(smiling quietly and happily)* You haven't any idea how many expenses we skylarks and squirrels have, Torvald.

HELMER: You are an odd little soul. Very like your father. You always find some new way of wheedling money out of me, and as soon as you have got it, it seems to melt in your hands. You never know where it has gone. Still, one must take you as you are. It's in the blood; for indeed it is true that you can inherit these things, Nora.

NORA: Ah, I wish I had inherited many of papa's qualities.

HELMER: And I would not wish you to be anything but just what you are, my sweet little skylark. But, do you know, it strikes me that you are looking rather—what shall I say— rather uneasy today?

NORA: Do I?

HELMER: You do, really. Look straight at me.

NORA: *(looks at him)* Well?

HELMER: *(wagging his finger at her)* Hasn't Miss Sweet-Tooth been breaking rules in town today?

NORA: No; what makes you think that?

HELMER: Hasn't she paid a visit to the confectioner's?

NORA: No, I assure you, Torvald—

HELMER: Not been nibbling sweets?

NORA: No, certainly not.

HELMER: Not even taken a bite at a macaroon or two?

NORA: No, Torvald, I assure you really—

HELMER: There, there, of course I was only joking.

NORA: (*going to the table on the right*) I should not think of going against your wishes.

HELMER: No, I am sure of that; besides, you gave me your word— (*Going up to her*) Keep your little Christmas secrets to yourself, my darling. They will all be revealed tonight when the Christmas Tree is lit, no doubt.

NORA: Did you remember to invite Doctor Rank?

HELMER: No. But there is no need; as a matter of course he will come to dinner with us. However, I will ask him when he comes in this morning. I have ordered some good wine. Nora, you can't think how I am looking forward to this evening.

NORA: So am I! And how the children will enjoy themselves, Torvald!

HELMER: It is splendid to feel that one has a perfectly safe appointment, and a big enough income. It's delightful to think of, isn't it?

NORA: It's wonderful!

HELMER: Do you remember last Christmas? For a full three weeks beforehand you shut yourself up every evening till long after midnight, making ornaments for the Christmas Tree and all the other fine things that were to be a surprise to us. It was the dullest three weeks I ever spent!

NORA: I didn't find it dull.

HELMER: (*smiling*) But there was precious little result, Nora.

NORA: Oh, you shouldn't tease me about that again. How could I help the cat's going in and tearing everything to pieces?

HELMER: Of course you couldn't, poor little girl. You had the best of intentions to please us all, and that's the main thing. But it is a good thing that our hard times are over.

NORA: Yes, it is really wonderful.

HELMER: This time I needn't sit here and be dull all alone, and you needn't ruin your dear eyes and your pretty little hands—

NORA: (*clapping her hands*) No, Torvald, I needn't any longer, need I! It's wonderfully lovely to hear you say so! (*Taking his arm.*) Now I will tell you how I have been thinking we ought to arrange things, Torvald. As soon as Christmas is over—(*A bell rings in the hall.*) There's the bell. (*She tidies the room a little.*) There's some one at the door. What a nuisance!

HELMER: If it is a caller, remember I am not at home.

[*Christine Linde enters. She and Nora had been schoolmates but Nora has not seen Christine for some years. The Lindes had lived in another town. Christine has been a widow for three years.*]

MRS. LINDE: (*smiling*) Nora, Nora, haven't you learnt sense yet? In our schooldays you were a great spendthrift.

NORA: (*laughing*) Yes, that is what Torvald says now. (*Wagging her finger at her.*) But "Nora, Nora," is not so silly as you think. We have not been in a position for me to waste money. We have both had to work.

MRS. LINDE: You too?

NORA: Yes; odds and ends, needle-work, crochet-work, embroidery, and that kind of thing. (*Dropping her voice.*) And other things as well. You know Torvald left his office when we were married? There was no prospect of promotion there and he had to try and earn more than before. But during the first year he overworked himself dreadfully. You see, he had to make money every way he could, and he worked early and late; but he couldn't stand it, and fell dreadfully ill, and the doctors said it was neces-sary for him to go south.

MRS. LINDE: You spent a whole year in Italy, didn't you?

NORA: Yes. It was no easy matter to get away, I can tell you. It was just after Ivar was born; but naturally we had to go. It was a wonderfully beautiful journey, and it saved Tor-vald's life. But it cost a tremendous lot of money, Christine.

MRS. LINDE: So I should think.

NORA: It cost about two hundred and fifty pounds. That's a lot isn't it?

MRS. LINDE: Yes, and in emergen-cies like that it is lucky to have the money.

NORA: I ought to tell you that we had it from papa.

MRS. LINDE: Oh, I see. It was just about that time that he died, wasn't it?

NORA: Yes; and, just think of it, I couldn't go and nurse him. I was ex-pecting little Ivar's birth every day and I had my sick Torvald to look after. My dear, kind father—I never saw him again, Christine. That was the saddest time I have known since our marriage.

MRS. LINDE: I know how fond you were of him. And then you went off to Italy?

NORA: Yes; you see we had money then, and the doctor insisted on our going, so we started a month later.

MRS. LINDE: And your husband came back quite well?

NORA: As sound as a bell!

* * *

NORA: You mustn't be angry with me. Tell me, is it really true that you did not love your husband? Why did you marry him?

MRS. LINDE: My mother was alive then, and was bedridden and help-less, and I had to provide for my two younger brothers; so I did not think I was justified in refusing his offer.

NORA: No, perhaps you were quite right. He was rich at that time, then?

MRS. LINDE: I believe he was quite well off. But his business was a pre-carious one; and, when he died, it all went to peces and there was nothing left.

NORA: And then?—

MRS. LINDE: Well, I had to turn my hand to anything I could do, first a small shop, then a small school, and so on. The last three years have seemed like one long workng-day, with no rest. Now it is at an end, Nora. My poor mother needs me no more, she is gone; and the boys do not need me either; they have situa-tions and can shift for themselves.

NORA: What a relief you must feel it—

MRS. LINDE: No, indeed; I only feel my life unspeakably empty. No one to live for any more. (*Gets up rest-lessly.*) That is why I could not stand the life in my little backwater any longer. I hope it may be easier here to find something which will pay me and occupy my thoughts. If only I could have the good luck to get some regular work—office work of some kind—

NORA: But, Christine, that is so frightfully tiring, and you are tired out now. You had far better go away to some watering-place.

MRS. LINDE: *(walking to the window)* I have no father to give me money for a journey, Nora.

NORA: *(rising)* Oh, don't be angry with me.

MRS. LINDE: *(going up to her)* It is you that must not be angry with me, dear. The worst of a position like mine is that it makes one so bitter. No one to work for, and yet obliged to be always on the look-out for chances. One must live, and so one become selfish. When you told me of the happy turn your fortunes had taken—you will hardly believe it—I was delighted not so much on your account as on my own.

NORA: How do you mean?—Oh, I understand. You mean perhaps Torvald could get you something to do.

MRS. LINDE: Yes, that was what I was thinking of.

NORA: He must, Christine. Just leave it to me; I will bring the subject very cleverly—I will think of something that will please him very much. It will make me so happy to be of use to you.

MRS. LINDE: How kind you are, Nora, to be so anxious to help me! It is doubly kind in you, for you know so little of the burdens and troubles of life.

NORA: I—? I know so little of them?

MRS. LINDE: *(smiling)* My dear! Small household cares and that sort of thing!—You are a child, Nora.

NORA: *(tosses her head and crosses the stage)* You ought not to be so superior.

MRS. LINDE: No?

NORA: You are just like the others. They all think that I am incapable of anything really serious—

MRS. LINDE: Come, come—

NORA: —that I have gone through nothing in this world of troubles.

MRS. LINDE: But, my dear Nora, you have just told me your troubles.

NORA: Pooh!—those were trifles. *(Lowering her voice.)* I have not told you the important thing.

MRS. LINDE: The important thing? What do you mean?

NORA: You look down upon me altogether, Christine—but you ought not to. You are proud, aren't you, of having worked so hard and so long for your mother?

MRS. LINDE: Indeed, I don't look down on any one. But I am both proud and glad to think that I was privileged to make the end of my mother's life almost free from care.

NORA: And you are proud to think of what you have done for your brothers.

MRS. LINDE: I think I have the right to be.

NORA: I think so, too. But now, listen to this; I too have something to be proud and glad of.

MRS. LINDE: I have no doubt you have. But what do you refer to?

NORA: Speak low. Suppose Torvald were to hear! He mustn't on any account—no one in the world must know, Christine, except you.

MRS. LINDE: But what is it?

NORA: Come here. *(pulls her down on the sofa beside her.)* Now I will show you that I too have something to be proud and glad. It was I who saved Torvald's life.

MRS. LINDE: "Saved"? How?

NORA: I told you about our trip to Italy. Torvald would never have recovered if he had not gone there—

MRS. LINDE: Yes, but your father gave you the necessary funds.

NORA: *(smiling)* Yes, that is what Torvald and all the others think, but—

MRS. LINDE: But—

NORA: Papa didn't give us a shilling. It was I who procured the money.

MRS. LINDE: You? All that large sum?

NORA: Two hundred and fifty pounds. What do you think of that?

MRS. LINDE: But, Nora, how could you possible do it? Did you win a prize in the Lottery?

NORA: *(contemptuously)* In the Lottery? There would have been no credit in that.

MRS. LINDE: But where did you get it from, then?

NORA: *(humming and smiling with an air of mystery)* Hm, aha!

MRS. LINDE: Because you couldn't have borrowed it.

NORA: Couldn't I? Why not?

MRS. LINDE: No, a wife cannot borrow without her husband's consent.

NORA: *(tossing her head)* Oh, if it is a wife who has any head for business—a wife who has the wit to be a little bit clever—

MRS. LINDE: I don't understand it at all, Nora.

NORA: There is no need you should. I never said I had borrowed the money. I may have got it some other way. *(Lies back on the sofa.)* Perhaps I got it from some other admirer. When any one is as attractive as I am—

MRS. LINDE: You are a mad creature.

NORA: Now, you know you're full of curiosity, Christine.

MRS. LINDE: Listen to me, Nora dear. Haven't you been a little bit imprudent?

NORA: *(sits up straight)* Is it imprudent to save your husband's life?

MRS. LINDE: It seems to me imprudent, without his knowledge, to—

NORA: But it was absolutely necessary that he should not know. My goodness, can't you understand that? It was necessary he should have no idea what a dangerous condition he was in. It was to me that the doctors came and said that his life was in danger, and that the only thing to save him was to live in the south. Do you suppose I didn't try, first of all, to get what I wanted as if it were for myself? I told him how much I should love to travel abroad like other young wives; I tried tears and entreaties with him; I told him that he ought to remember the condition I was in, and that he ought to be kind and indulgent to me; I even hinted that he might raise a loan. That nearly made him angry, Christine. He said I was thoughtless, and that it was his duty as my husband not to indulge me in my whims and caprices—as I believe he called them. Very well, I thought, you must be saved—and that was how I came to devise a way out of the difficulty—

MRS. LINDE: And did your husband never get to know from your father that the money had not come from him?

NORA: No, never. Papa died just at that time. I had meant to let him into the secret and beg him never to reveal it. But he was so ill then—alas, there never was any need to tell him.

MRS. LINDE: And since then have you never told your secret to your husband?

NORA: Good Heavens, no! How could you think so? A man who has such strong opinions about these things! And besides how painful and humiliating it would be for Torvald, with his manly independence, to know that he owed me anything! It would upset our mutual relations altogether; our beautiful happy home would no longer be what it is now.

MRS. LINDE: Do you mean never to tell him about it?

NORA: *(meditatively, and with a half smile)* Yes—some day, perhaps, after many years, when I am no longer as nice-looking as I am now. Don't laugh at me! I mean, of course, when Torvald is no longer as devoted to me as he is now; when my dancing and dressing-up and reciting have palled on him; then it may be a good thing to have something in reserve— *(Breaking off.)* What nonsense! That time will never come. Now, what do you think of my great secret, Christine? Do you still think I am of no use? I can tell you, too, that this affair has caused me a lot of worry. It has been by no means easy for me to meet my engagements punctually. I may tell you that there is something that is called, in business, quarterly interest, and another thing called payment in installments, and it is always so dreadfuly difficult to manage them. I have had to save a little here and there, where I could, you understand. I have not been able to put aside much from my house-keeping money, for Torvald must have a good table. I couldn't let my children be shabbily dressed; I have felt obliged to use up all he gave me for them, the sweet little darlings!

MRS. LINDE: So it has all had to come out of your own necessaries of life, poor Nora?

NORA: Of course. Besides, I was the one responsible for it. Whenever Torvald has given me money for new dresses and such things, I have never spent more than half of it; I have always bought the simplest and cheapest things. Thank Heaven, any clothes look well on me, and so Torvald has never noticed it. But it was often very hard on me. Christine—because it is delightful to be really well dressed, isn't it?

MRS. LINDE: Quite so.

NORA: Well, then I have found other ways of earning money. Last winter I was lucky enough to get a lot of copying to do; so I locked myself up and sat writing every evening until quite late at night. Many a time I was desperately tired; but all the time it was a tremendous pleasure to sit there working and earning money. It was like being a man.

MRS. LINDE: How much have you been able to pay off in that way?

NORA: I can't tell you exactly. You see, it is very difficult to keep an account of a business matter of that kind. I only know that I have paid every penny that I could scrape to-geher. Many a time I was at my wits' end. *(Smiles.)* Then I used to sit here and imagine that a rich old gentleman had fallen in love with me—

MRS. LINDE: What! Who was it?

NORA: Be quiet!—that he had died; and that when his will was opened it contained, written in big letters, the instruction: "The lovely Mrs. Nora Helmer is to have all I possess paid over to her at once in cash."

MRS. LINDE: But, my dear Nora—who could the man be?

NORA: Good gracious, can't you understand? There was no old gentleman at all; it was only something that I used to sit here and imagine, when I couldn't think of any way of procuring money. But it's all the same now; the tiresome old person can stay where he is, as far as I am concerned; I don't care about him or his will either, for I am free from care now. *(Jumps up)*

* * *

[*Krogstad, from whom Nora had borrowed the money, now threatens to reveal the secret to Torvald unless Nora uses her influence with her husband to make sure Krogstad is permitted to keep his post at the bank. When Torvald learns of Nora's forgery, he is furious and declares her unfit to rear the children. While he relents when Krogstad sends back the forged note and his position as head of the bank is no longer threatened, and he wishes to resume with Nora their usual relationship, Nora is no longer willing to do so.*]

ACT III

* * *

NORA: We have been married now eight years. Does it not occur to you that this is the first time we two, you and I, husband and wife, have had a serious conversation?

HELMER: What do you mean by serious?

NORA: In all these eight years—longer than that—from the very be-ginning of our acquaintance, we have never exchanged a word on any serious subject.

HELMER: Was it likely that I would be continually and for ever telling you about worries that you could not help me to bear?

NORA: I am not speaking about business matters. I say that we have never sat down in earnest together to try and get at the bottom of anything.

HELMER: But, dearest Nora, would it have been any good to you?

NORA: That is just it; you have never understood me. I have been greatly wronged, Torvald—first by papa and then by you.

HELMER: What! By us two—by us two, who have loved you better than any one else in the world?

NORA: *(shaking her head)* You have never loved me. You have only thought it pleasant to be in love with me.

HELMER: Nora, what do I hear you saying?

NORA: It is perfectly true, Torvald. When I was at home with papa, he told me his opinion about everything, and so I had the same opinions; and if I differed from him I concealed the fact, because he would not have liked it. He called me his doll-child, and he played with me just as I used to play with my dolls. And when I came to live with you—

HELMER: What sort of an expression is that to use about our marriage?

NORA: *(undisturbed)* I mean that I was simply transferred from papa's hands into yours. You arranged everything according to your own taste, and so I got the same tastes as you—or else I pretended to, I am really not quite sure which—I think sometimes the one and sometimes the other. When I look back on it, it seems to me as if I had been living here like a poor woman—just from hand to mouth. I have existed merely to perform tricks for you, Torvald. But you would have it so. You and papa have committed a great sin against me. It is your fault that I have made nothing of my life.

HELMER: How unreasonable and how ungrateful you are, Nora. Have you not been happy here?

NORA: No, I have never been happy. I thought I was, but it has never really been so.

HELMER: Not—not happy!

NORA: No, only merry. And you have always been so kind to me. But our home has been nothing but a playroom. I have been your doll-wife, just as at home I was papa's doll-child; and here the children have been my dolls. I thought it great fun when you played with me, just as they thought it great fun when I played with them. That is what our marriage has been, Torvald.

HELMER: There is some truth in what you say—exaggerated and strained as your view of it is. But for the future it shall be different. Playtime shall be over, and lesson-time shall begin.

NORA: Whose lessons? Mine, or the children's?

HELMER: Both yours and the children's, my darling Nora.

NORA: Alas, Torvald, you are not the man to educate me into being a proper wife for you.

HELMER: And you can say that!

NORA: And I—how am I fitted to bring up the children?

HELMER: Nora!

NORA: Didn't you say so yourself a little while ago—that you dare not trust me to bring them up?

HELMER: In a moment of anger! Why do you pay any heed to that?

NORA: Indeed, you were perfectly right. I am not fit for the task. There is another task I must undertake first. I must try and educate myself—you are not the man to help me in that. I must do that for myself. And that is why I am going to leave you now.

HELMER: *(springing up)* What do you say?

NORA: I must stand quite alone, if I am to understand myself and everything about me. It is for that reason that I cannot remain with you any longer.

HELMER: Nora, Nora!

NORA: I am going away from here now, at once. I am sure Christine will take me in for the night—

HELMER: You are out of your mind! I won't allow it! I forbid you!

NORA: It is no use forbidding me anything any longer. I will take with me what belongs to myself. I will take nothing from you, either now or later.

HELMER: What sort of madness is this!

NORA: Tomorrow I shall go home —I mean, to my old home. It will be easiest for me to find something to do there.

HELMER: You blind, foolish woman!

NORA: I must try and get some sense, Torvald.

HELMER: To desert your home, your husband and your children! And you don't consider what people will say!

NORA: I cannot consider that at

all. I only know that it is necessary for me.

HELMER: It's shocking. This is how you would neglect your most sacred duties.

NORA: What do you consider my most sacred duties?

HELMER: Do I need to tell you that? Are they not your duties to your husband and your children?

NORA: I have other duties just as sacred.

HELMER: That you have not. What duties could those be?

NORA: Duties to myself.

HELMER: Before all else, you are a wife and a mother.

NORA: I don't believe that any longer. I believe that before all else I am a reasonable human being, just as you are—or, at all events, that I must try and become one. I know quite well, Torvald, that most people would think you right, and that views of that kind are to be found in books; but I can no longer content myself with what most people say, or with what is found in books. I must think over things for myself and get to understand them.

HELMER: Can you not understand your place in your own home? Have you not a reliable guide in such matters as that?—have you no religion?

NORA: I am afraid, Torvald, I do not exactly know what religion is.

HELMER: What are you saying?

NORA: I know nothing but what the clergyman said, when I went to be confirmed. He told us that religion was this, and that, and the other. When I am away from all this, and am alone, I will look into that matter too. I will see if what the clergyman said is true, or at all events if it is true for me.

HELMER: This is unheard of in a girl of your age! But if religion cannot lead you aright, let me try and awaken your conscience. I suppose you have some moral sense? Or—answer me—am I to think you have none?

NORA: I assure you, Torvald, that is not an easy question to answer. I really don't know. The thing perplexes me altogether. I only know that you and I look at it in quite a different light. I am learning, too, that the law is quite another thing from what I supposed; but I find it impossible to convince myself that the law is right. According to it a woman has no right to spare her old dying father, or to save her husband's life. I can't believe that.

HELMER: You talk like a child. You don't understand the conditions of the world in which you live.

NORA: No, I don't. But now I am going to try. I am going to see if I can make out who is right, the world or I.

HELMER: You are ill, Nora; you are delirious; I almost think you are out of your mind.

NORA: I have never felt my mind so clear and certain as tonight.

HELMER: And is it with a clear and certain mind that you forsake your husband and your children?

NORA: Yes, it is.

HELMER: Then there is only one possible explanation.

NORA: What is that?

HELMER: You do not love me any more.

NORA: No, that is just it.

HELMER: Nora!—and you can say that?

NORA: It gives me great pain, Torvald, for you have always been so kind to me, but I cannot help it. I do not love you any more.

HELMER: *(regaining his composure).* Is that a clear and certain conviction too?

NORA: Yes, absolutely clear and certain. That is the reason why I will not stay here any longer.

HELMER: And can you tell me what I have done to forfeit your love?

NORA: Yes, indeed I can. It was tonight, when the wonderful thing did not happen; then I saw you were not the man I had thought you.

HELMER: Explain yourself better—I don't understand you.

NORA: I have waited so patiently for eight years; for, goodness knows, I knew very well that wonderful things don't happen every day. Then this horrible misfortune came upon me; and then I felt quite certain that the wonderful thing was going to happen at last. When Krogstad's letter was lying out there, never for a moment did I imagine that you would consent to accept this man's conditions. I was so absolutely certain that you would say to him: Publish the thing to the whole world. And when that was done—

HELMER: Yes, what then?—when I had exposed my wife to shame and disgrace?

NORA: When that was done, I was so absolutely certain you would come forward and take everything upon yourself, and say: I am the guilty one.

HELMER: Nora—!

NORA: You mean that I would never have accepted such a sacrifice on your part? No, of course not. But what would my assurances have been worth against yours? That was the wonderful thing which I hoped for and feared; and it was to prevent that, that I wanted to kill myself.

HELMER: I would gladly work night and day for you, Nora—bear sorrow and want for your sake. But no man would sacrifice his honor for the one he loves.

NORA: It is a thing hundreds of women have done.

HELMER: Oh, you think and talk like a heedless child.

NORA: Maybe. But you neither think nor talk like the man I could bind myself to. As soon as your fear was over—and it was not fear for what threatened me, but for what might happen to you—when the whole thing was past, as far as you were concerned it was exactly as if nothing at all had happened. Exactly as before, I was your little skylark, your doll, which you would in future treat with doubly gentle care, because it was so brittle and fragile. *(Getting up.)* Torvald—it was then it dawned upon me that for eight years I had been living here with a strange man, and had borne him three children—Oh, I can't bear to think of it! I could tear myself into little bits!

HELMER: *(sadly).* I see, I see. An abyss has opened between us—there is no denying it. But, Nora, would it not be possible to fill it up?

NORA: As I am now, I am no wife for you.

HELMER: I have it in me to become a different man.

NORA: Perhaps—if your doll is taken away from you.

HELMER: But to part!—to part from you! No, no, Nora, I can't understand that idea.

NORA: *(going out to the right)* That makes it all the more certain that it must be done. *(She comes back with her cloak and hat and a small bag which she puts on a chair by the table.)*

HELMER: Nora, Nora, not now! Wait till tomorrow.

NORA: *(putting on her cloak)* I cannot spend the night in a strange man's room.

HELMER: But can't we live here like brother and sister——?

NORA: *(putting on her hat)* You know very well that would not last long. *(Puts the shawl round her.)* Good-by, Torvald. I won't see the little ones. I know they are in better hands than mine. As I am now, I can be of no use to them.

HELMER: But some day, Nora— some day?

NORA: How can I tell? I have no idea what is going to become of me.

HELMER: But you are my wife, whatever becomes of you.

NORA: Listen, Torvald. I have heard that when a wife deserts her husband's house, as I am doing now, he is legally freed from all obligations towards her. In any case I set you free from all your obligations. You are not to feel yourself bound in the slightest way, any more than I shall. There must be perfect freedom on both sides. See, here is your ring back. Give me mine.

HELMER: That too?

NORA: That too.

HELMER: Here it is.

NORA: That's right. Now it is all over. I have put the keys here. The maids know all about everything in the house—better than I do. Tomorrow, after I have left her, Christine will come here and pack up my own things that I brought with me from home. I will have them sent after me.

HELMER: All over! All over!—Nora, shall you never think of me again?

NORA: I know I shall often think of you and the children and this house.

HELMER: May I write to you, Nora?

NORA: No—never. You must not do that.

HELMER: But at least let me send you—

NORA: Nothing—nothing—

HELMER: Let me help you if you are in want.

NORA: No. I can receive nothing from a stranger.

HELMER: Nora—can I never be anything more than a stranger to you?

NORA: *(taking her bag)* Ah, Torvald, the most wonderful thing of all would have to happen.

HELMER: Tell me what that would be!

NORA: Both you and I would have to be so changed that—Oh, Torvald, I don't believe any longer in wonderful things happening.

HELMER: But I will believe in it. Tell me? So changed that—?

NORA: That our life together would be a real wedlock. Good-by. *(She goes out through the hall.)*

HELMER: *(sinks down on a chair at the door and buries his face in his hands)* Nora! Nora! *(Looks round, and rises.)* Empty. She is gone. *(A hope flashes across his mind.)* The most wonderful thing of all—? *(The sound of a door shutting is heard from below.)*

I Remember Papa

Harry Dolan

The other night after attending a gratifying function which had been initiated to help the black man, specifically to help build a nursery for children of working mothers, and after seeing and hearing white people make speeches professing their understanding and desire to go to any length to help, I found myself suddenly cornered and forced to defend the fabled laziness of the black man.

What was especially surprising was the fact that I assumed this white acquaintance—since he had paid thirty dollars to attend this dinner held for the purpose of helping the black man—did, at least in part, have some sympathy with what his, the white people, had tried to accomplish.

As I stood there watching his eyes I became suspect of my own sincerity, for I stood attentively nodding my head and smiling. I lit a cigarette, raised an eyebrow, performed all of the white man's laws of etiquette, and all the while I knew if it had been others of my black brothers, they would have cursed him for his smugness and invited him outside to test his theory of black man's courage and laziness. Of course I did none of these things. I grinned as he indicated in no uncertain terms that as soon as the black man got off his lazy butt and took advantage of all the blessings that had been offered him for the last two hundred years, then he, the white man, would indeed be willing to help.

I could have answered him—and was tempted to, for he was obviously sincere. Instead, I found an excuse to slip away and let a white man fight my battle, a friend, even a close friend. I went to a far corner and blindly played a game of pool by myself as the voices of this man and my friend dissected me. I stacked the pool balls, leaned over the table, and remembered a black man I had known.

It was said of him later in his life that he had let his family down. He'd been lazy, no-account, a troublemaker. Maybe so, maybe so, but I can't help remembering nights of his pacing the squeaking floor muttering to himself, coming back across the floor, sitting down, his legs trembling as he listened to the woman plead for him not to do anything bad.

"I'll go to hell first before I'll let you and the children starve." God, how many times had I heard him say that! How many other men standing bunched in helpless stagnation have I heard vow to take a gun and get some food for their children! Yes, they were planning to commit a crime; yes, they were potential criminals. Then. They are usually black too—another crime, it seems.

I remember that man, but more I remember his woman, my mother. Curiously though, I never remember her dancing, running, playing; always lying down, the smell of disinfectant strong, the deep continuous coughing, the brown paper bag filled with the toilet paper red with bubbly spit and blood, lying half concealed under the bed.

I never remember her eating food such as bread, meat, potatoes; only apples and only Delicious apples. In those days five cents apiece. She was a small woman, barely five foot.

"Junior," she would say softly. She never spoke above a whisper, "Go to the store and get me an apple." The thin trembling hand would reverse itself and slide up and under the covers and under the pillow and then return as though of its own volition, the weight almost too much, and as I'd start out the door, she would always smile and say, "Hurry, Junior."

I'd nod, and always, always there seemed to be a need to hurry. Those trips were always made with a feeling of breathless fear. I didn't know why then, only that for some reason I must always come back as soon as possible.

I was returning with an especially large apple, walking along, tempted to bite just a tiny piece, when I turned the corner and saw the black police ambulance standing in front of my door. Suddenly I had to go to the bathroom so bad I couldn't move. I stood watching as two uniformed men came out with the stretcher, and then the sound of my mother's shrill voice hit me.

"Mama, Mama," she was screaming. I could see her twisting and swinging at the lady next door as she was held back. I stood there feeling the hot piss run down my trembling legs, feeling cold chills spatter through my body, causing frozen limbs to spasmodically begin to move. I forced myself toward the police wagon as the men opened the doors and slid the stretcher along the bare metal. I saw my mother's head bounce on the floor.

"Wait," I moaned, "don't hurt her." Then I was running, screaming, "Please don't hurt her."

I looked down at her pain-filled face, and she smiled, even then she smiled. I showed her the apple. The effort to nod seemed a terrible effort but she did, her eyes so very bright, so very shiny.

"You eat it, Junior, you and sis."

"What's wrong, Mama?" I asked softly. "You really, really sick now?"

She nodded.

"Your father will be home soon. Tell him I'm at the General Hospital. Tell him to—to hurry."

"I'll tell him, Mama," I promised. "I'll tell him to hurry, Mama." She nodded sadly and puckered her lips as she always did since we weren't allowed to kiss her.

That was the last time I saw my mother except at the grave. My father came to the funeral with two white men who stood on each side of him all the time. There were people crying all around us. My grandmother kept squeezing me and moaning. I saw my father try to cover his face but one of the men said something and he stood up stiffly after that. I didn't cry, because my mother seemed to look happier, more rested than I had ever seen her. For some reason, I was glad she was dead. I think maybe, except for us, she was too.

I was nine, my sister five. It was not until ten years later that I saw my father again.

We sat on opposite sides of a screen and talked into telephones. I had come there to tell him that in spite of my beginning, I had made it. I was nineteen, and a radioman in the U.S. Coast Guard, ready to fight and die for my country. There had been something mysterious about his smile.

"I'm proud of you, boy," he said. "You're a real man. You know I volunteered for the front lines too, but they turned me down."

We don't want you, I thought, we're not criminals, we're honest, strong. Then I looked again at this thief, this "Loaf-of-bread gunman"

as the papers had tagged him. He had taken five loaves of bread, along with twelve dollars. Suddenly I could not stay there condemning this man, my father. It seemed such a waste, this magnificently strong man sitting there, his tremendous chest barely moving, hands resting quietly. He seemed to have accepted his fate and yet I felt as though he were talking to me, his whole being showering torrents of words about me.

"Be careful, boy, there are so many ways to fail, the pitfall sometimes seems to be the easiest way out. Beware of my future, for you must continue, you must live. You must, for in you are all the dreams of my nights, all the ambitions of my days."

A bell rang and we stood up and a man pointed me toward a heavy door. I looked back, and saw him standing easy, hands at his side, so very calm, yet my mind filled to overflowing with the many things he had not said. It was to be ten years before he walked again as a free man, that is, as a physically free man.

I remember an earlier time, an earlier chapter of my growing up. I remember the first time my mother said we were taking lunch to my father's job. We had been down to the welfare line and I had stood with her, our feet burning against the hot pavement, and slowly moved forward in the sun. Years later I stood in chow lines over half of the world, but no desert, no burning deck was as hot as that day.

At last we reached the man sitting at the desk and my mother handed him the book of stamps. She smiled, a weak almost timid smile, as he checked her name and thumbed her to the food line.

As we headed home, my wagon

was loaded with cans of corned beef, powdered milk, powdered eggs, and white margarine that she would later color yellow to look like butter.

At home we made sandwiches and off we went to my father's job, to take him his lunch. I pulled my sister along in my wagon, a Red Flyer.

It was to be a picnic, a celebration really, my father's new job.

I remember the wagon did not have a tongue or handle but only a rope with which I pulled it wobbling along. We were excited, my sister and I, as we left our district of dirt streets and unpaved sidewalks and began to make our way along roads called boulevards and malls we had never had occasion to travel. The streets themselves were fascinating, so different. They were twice as wide, and there were exotic trees along the sidewalks and lo and behold trees down the center of the street as far as I could see and then we turned the corner and before us stretched an overwhelming sight. An overhead highway was being built. Columns rose to staggering heights, bulldozers thrust what seemed to me mountains of dirt before them, and hundreds, no thousands of men seemed to be crawling like ants hurrying from one point to another. Cranes lifted nets of steel and laid them in rows on the crushed rock.

I stared in awe at important-looking white men in metal hats, carrying rolls of papers which they intermittently studied, then pointing into space at what to me seemed only emptiness.

And then I saw my father. He sat among fifty other black men, all surrounded by great boulders marked with red paint. They all held steel chisels with which they cut along the marked lines. They would strike a certain point and the boulder would split into smaller pieces and as we approached there was a silence around them except for the pinging of the hammer against the chisel. In all the noise it was a lonely sound, futile, lost, oppressive. My father seemed to be concentrating, his tremendous arm whipping the air. He was stripped to the waist, black muscles popping sweat, goggled eyes for the metal and stone only. We stood there, the three of us, my mother, my sister, and I, and watched my father work for us, and as he conquered the huge boulder my chest filled with pride. Each stroke shouted for all the world to hear: This is my family and I love them! No one can tell me this was the act of a lazy man.

Suddenly a white man walked up and blew a whistle and the black men all looked up and stopped working. My father glanced over at me, grinned and winked. He was glistening with sweat, the smell strong and powerful. He dropped his big hand on my shoulder and guided me to a large boulder.

"Hey, boy, you see me beat that thing to bits? This one's next," he said, indicating the one that shaded us from the sun. "I'll pound it to gravel by nightfall." It was a challenge he expected, he welcomed. That was my lazy, shiftless father.

And then one day they brought him home, his thumb, index, and middle finger gone from his left hand. They sat him in the kitchen chair and mumbled something about carelessness. He sat there for two hours before he answered our pleadings.

"Chain broke, I—I was guiding boulder. I couldn't, I just couldn't get my hand out from under in time —I, goddam it, Jean, they took my fingers off. I layed right there, my hand under the rock, and they nipped them like butchering a hog. Look at my goddam hand."

My mother held him in her arms and talked to him. She spoke softly, so softly my sister and I, standing in the corner, couldn't hear the words, only the soothing softness of her voice.

"Joe, Joe, we can." And then he began to cry like—like I sometimes did when I was hurt deep inside and couldn't do anything about it.

After that there was a change in him. My father had been a fighter. He had feared no man white or black. I remember the time we were sitting on a streetcar and a woman had forgotten her fare—or maybe she never had any in the first place. Anyway, the driver slammed the doors on her and held her squeezed between them.

My father jumped up, snatched the driver out of the seat, and let the woman out. He and the driver had words that led to battle and Pop knocked the driver down just as a patrolman arrived. The patrolman didn't listen to any of the people that tried to explain what had happened. He just began to swing his night stick at my father's head. It was a mistake. My father hit him once and even today I can see all the people laughing at the funny look on the policeman's face as he staggered back all the way across the street and up against a building, slowly sagging down.

The police wagon arrived with four other policemen and one told him they were going to beat his brains in when they got him downtown.

My pop had laughed then and backed against the building.

"I guess ain't no sense me going peaceable then."

They knocked out all his upper front teeth that day, but as he said later, "Them four white boys will think of me every time they shave."

They finally overpowered him and dragged him, still struggling, to the wagon. One of them kept muttering, "He's one fighting son of a black bitch, he's a fighting son of a bitch."

All the time I hadn't said a word or cried or yelled as they stomped and kicked him. I had shut my eyes and held my lips tightly pressed together and I had done just as he'd always told me.

"You stay out of it, boy, stay real quiet, and when that wagon leaves, you run behind and keep it in sight. If they lose you, you ask someone where the closest police station is— that's where I'll be. You go home and tell your mother."

That's the way he had been before losing his left hand. Afterwards, well, it took a lot from him. He told me one day, laughing and shaking the nub as he called it, "If I'd only had the thumb, just the lousy thumb, I'd have it made."

Gradually he lost the ability to see humor in the nub. I think the whole thing came to a head the night I killed the kitten.

We hadn't had meat or potatoes for over two weeks. Even the grease drippings were gone and my mother was too sick to raise her head from the pillow. So I had gotten the skillet and put it in the open grate. We had two cups of flour so I mixed water with it and poured it into the greasy skillet. I can still recall the coldness of the room on my back and the warmth from the grate on my face as my sister and I knelt and hungrily watched the flour brown.

You know, today my wife marvels at how, no matter what she puts before me, I eat with relish. My children say that I eat very fast. I pray to God they never have to experience the causes of my obsession. But back to the story—the flour finally hardened and I broke a piece for my sister and a piece for my mother and left mine in the skillet on the table.

I took my mother's piece over to the bed and put it in her hand. She didn't move so I raised her hand to her mouth and she began to suck on it. Then I heard my sister scream, "Topsy is eating your food, Junior, Topsy's eating your food!" I turned around to see the cat tearing at my tiny piece of hard dough. I went wild. I leaped across the room and grabbed the kitten by the tail and began slamming her against the wall.

"That's my food," I kept yelling, "my food!" At last I heard my sister screaming, "She's bleeding, you're killing Topsy. Here, here, eat my bread. Please don't kill her."

I stopped, horrified, staring at the limp nothing I now held. It was two weeks later that they got me to speak and that same night my father left the house for the last time. I don't say that what he did was right. No, it most assuredly was wrong. But what I do ask is, what else could he have done? I need an answer quickly, now, today, for I tell you this, my children will not starve, not here, not in this time of millions to foreign countries and fountains to throw tons of water upward to the sky, and nothing to the hungry, thirsty multitudes a stone's throw away.

Diego Rivera. *Festival of All Souls in the City.*

Love and Like

HERBERT GOLD

He got to Cleveland rather late, telephoned his wife from the terminal, and asked if he could see their children early the next morning. She seemed easy and friendly and said sure thing, of course, why not? He sighed when they hung up and he stared at the telephone. Maybe his absence really could work to level things between them.

Why not? as she said. They had loved each other for ten years, or rather, for a part of those ten years impossible now to calculate. He remained sitting in the telephone booth. He was a young man with a thin, almost boyish body and a large head, heavily muscled at the jaw. On his face he wore the haunted, eaten look of a man whose accomplice has betrayed him. Whether his accomplice was his former wife or his conscience was not yet clear to him. Conscience was still talking to his wife: We didn't make it, kid! And he was remembering one of their last quarrels before they gave up, when he said, "We'll try! We'll try!" And paining badly with those fragments of the past, he heard again her quiet reviving words: "We'll try, darling."

If they had finally made out, it would all have been remembered as the progress and process of love; with failure it could seem all bad; he was determined to hold in retrospect to a mixed verdict—some pretty, some unpretty, and nevertheless the long Sunday afternoon habit of lovemaking spoke for a true intimacy. The hardest, most essential responsibility to a dead marriage is to preserve the ripe strength it once had. Must have had, must have! So without love now, if that's the way it was, why couldn't they simply like each other for the rest of time? Not just for the sake of the children. For the sake of themselves and what they had become through their marriage.

All this was a resolution. Done.

Then he fished for another dime in his pocket and called Sally. Okay, it was late, but wasn't she glad to hear his voice? She laughed sleepily and asked him how he could doubt it. She was sleepy was all. She had been thinking of him. All right then, dreaming. He could come right up—would it take twenty minutes? For splashing some cold water and getting good and waked and dressed.

"Don't bother," he said, and even in the telephone booth, under the falsification of wires and electricity, the laughter brought her back to him with all her soft, warm childlike sleepiness amid the fragrant bed-clothes. Sally's face was devoted to laughter, especially the very blue, waiting-to-smile eyes—those eyes of a blue which could give him what a strong awakening on a morning in the country gives, courage and appetite and great belief in the future. The morning he left, she had made breakfast for him, hotcakes with butter, sweet syrup, three kinds of jam —mostly he remembered the sweetness and much melted butter. With his mouth stuffed, he shook his finger to warn her that he had something to say as soon as he could swallow.

"What's the matter, darling?"

"Oh it's good! It's so good, Sally!"

And the lashes fell over the brave summer-sky eyes. She was timid despite her laughter's deep abandon. "Then why do you have to leave Cleveland?" she had asked. The question cost her heavily. Without me? her paleness said. The unsmiling blue eyes were trying to force his reply: Then why don't we . . . ?

Unspoken questions were, as al-ways, the ones he felt he had to answer and precisely could not. Therefore he had spoken in a tone of stubborn exasperation. "Because my wife lives here. Because it's too soon. Because I have to." He reached across the debris of eating to touch her more gently. If he could tell her, he would. "Now, with you here, Sally, it's all right. Right now. But I can't tell you about my wife and what she did to me. What she does! What she still does!" he cried out abruptly. His heart was pounding, and the good of the breakfast was gone. "And the children—I still can't imagine any other life than being their father——"

She was looking at his hand on hers with a curious withdrawn attention.

He said quietly, "Maybe I just haven't found the way to love them without loving their mother. Maybe that's it and there is a way."

His wife made him need Sally; if only she could free him entirely for Sally! This way, entangling him in her wrath, she hardly let him know who Sally was. He had to find her in a long history in which she had no part, past a furious, gyring woman, after outrage and love gripping each other as do alloyed metals under heat and pressure. Let Sally make him pure—as he never was—let Sally make herself felt!

But her eyes were answering nothing, were asking him: And why can't we? He realized that she had not heard him at all. Everyone is sometimes made deaf by intentions. He might as well have been brutal: And I just don't know about you, either, dollface. Seems as how I met you someplace before.

Now he shrugged and felt the

damp shirt moving on his back. Every time he said the word "wife" she should interrupt and say, *"Former* wife." Shaper to Shaper—*over,* he decided, ferociously willing it as if he were passing a message by radio, and the battlefield made communication difficult: *Over!* He had just finished writing a technical manual on the care and operation of a new, improved, long-range walkie-talkie. In Cleveland he had worked for General Electric on instructions for the installation of intercoms in language any mechanic could understand; in the identical off-white New York office to which he had arranged a transfer, under the same Armstrong Cork soundproofing, with thousands of little holes conking, stunning every whisper of distraction, he had labored with a Signal Corps semantics expert on further explanations of how to keep contact open under conditions of stress. "Limited contact under conditions of vital stress. . . ." He was making mass poetry for five-stripe sergeants who don't like the word *war.* It was a career for a careful explainer. But now over to Dan Shaper, please!

Deciding: Condition of stress not total chaos if receiver flashes emergency transistor filters out static toward coded meaning (see Fig. 3). Put into heart's English. Also see resources of regret, hope, and desire for possible decoding toward good conscience.

II

At seven-thirty the next morning he was running up the stairway toward his daughters. The buzzer had admitted him; the door to the upstairs flat was open.

"Paula, honey!"

"Hello, Daddy. Mommy and Cynthia are still sleeping. I knew you were coming. I got up and took my own breakfast. Wheaties and prunes and a pickle. Mommy says I can eat what's on the bottom shelf of the fridge and that's where the pickle was. Hello, Daddy."

Exactly what he wanted always, this time Paula was a six-year-old lady sitting on the couch. First she rang the buzzer to let him in, then she returned shyly to her place to wait for him.

"And I knew you would be coming back today because I counted the days," she said. "You said thirty. You promised." She was quivering with excitement beneath the formality with which she had vain hopes of hiding herself. A month is even longer in a child's life than in her father's. He took her in his arms and she said, "I'm *glad.*" Her eyelashes were wet when he kissed them.

"I'm glad to see you too, honey-bear."

They whispered together gravely to keep from waking the rest of the house. He felt strong in his month-long convalescence. He forgot that he had slept only the sleep of the recently divorced, that is, no sleep but that profitless, dream-jammed one of might-have-been and exhaustion. Paula gave him her news. Not only had she lost a tooth, but also the ragged edge of the new, grownup one had begun to appear. He looked and saw. At her command he appreciated it with his finger, too. They were interrupted by another child, running scared and barefoot to join the celebration, four-year-old Cynthia. Disappointed, Paula bravely permitted Cynthia to catch up with a minute of greeting.

Then he pulled them both onto the couch beside him, still bedwarm in their summer pajamas, and so were they three sitting when his wife came out. She had heard them, stayed a long time in the bathroom, and emerged only after satisfying herself that she was sweet and alert and nice to look at. He was very aware also of how he looked to her with the two little ones wriggling close to him.

She smiled and said, "Hello, Dan." Her hair had been cut in a new fashion and fluffed out to make a soft frame for the delicate bones of her face. Her brow, finely marked by frown lines, seemed wider now, peacefully expectant. What happens to a young woman when she divorces? Aren't her eyes puffy in the morning any more? Where does the blue lymph go? Is she no longer stringy and dismayed after the first glimpse of herself in the bathroom mirror? Not this one. Dark, slender, and cool, she was mobilized for love again.

Shaper was pleased by her greeting and—with only a twinge of self-judging jealousy—by her appearance. He put it down to his credit that she was not ruined for the hope of love. His moving out of town had done them both good. Sally last night had started the visit auspiciously. Paula was a marvel and Cynthia a wonder. And now his wife was the decent human being he had almost always believed in.

"Yes, thanks," he said, and they all four had breakfast together, just like a family. Since the first years of their marriage, his wife had barely eaten in the morning, drinking black coffee and nibbling at dry toast and then rushing to the scales, but today she took an extra piece of toast and covered it with butter in order to show him that she could now enjoy food. Well, let her have that pleasure, he thought with a fine expansive sense of tolerance. They both watched the girls munching greedily, proving something to their parents.

Later, while the children played with their new toys, they sat together in the front room and talked, balancing coffee cups on their knees. "It's all right now," she said. "You did right to go away for a while. There were good things, weren't there, Dan?"

"I'm glad you remember."

"I do. I do." Her narrow face with its wide brow wore a complex expression—frowning, tender, at peace. They had shared and not shared. They had been satisfied, unsatisfied. She had wanted, they had both wanted another child; but (stiff white mouth and tense, silent staring out the window) she had needed a guarantee that this time it would be a boy. She had found a doctor in Miami with fantastic notions about special diets, the power of raisins and lima beans and times of the month—litmus paper, fever charts, count the days—and a husband to be applied as an agent when the alchemical and astrological signs say yes. Desperate woman! Yet a man can grow to accept a partial madness when it is the condition of the rational universe he desires. Desperate husband! (Aren't many people superstitious? When she has her son, what difference by what illusion she was comforted?) But one day he faces the obvious question: All right, there's a fifty-fifty chance, say. And if it's not a boy? Kachoo—sneezing with boredom and despair.

And if it's *not* a boy?

As always, his besetting flaw was one of perverse hope and pride—the crazy patience of the man who needs love too much to take the necessary risk of losing it. When this danger is not challenged, love and like both are doomed. He had feared to compel his wife to follow his lead. He would not submit to being dragged along by her frantic jittering after happiness; she could not follow his bone-tired, patience-ended assertions of will; for them it had long been too late for all but knowledge, and perhaps in love it is always too late for that. The terrible guilt which he carried with him was a hope, not an understanding: Once she might have been made to want what you want, if you had pressed hard enough, and if you had said in time, No, not that way! Follow me and come to yourself!

The enduring of wrong amounts to an acquiescence in evil, a sin of active malignance which grew on the body of their marriage like a tumor and, like a tumor, sapped its strength to proliferate in this cruel organism, divorce. When the body is dead and all health irrelevant, only then will the cancer cease. In the meantime, every joy, every lively nerve, every vein of health must supply food for the fibrillar parasite. The dying creature curses its strong heart, its tuned body, because these prolong the agony. Each motive for happiness must be bled dry.

Shaper knew that he had not acted in time.

But finally he had said *no*.

NO!

Said no to lima beans and love by astrological suggestion—and maybe, kid, you're right about wanting to see Dr. Kasdan, he's supposed to be a good one—although he believed in that cure almost as little as he believed in the efficacy of lima beans to induce suckling boy babies instead of beautiful but incomplete girls.

A nightmare. Yet this was the intelligent, accomplished woman who could charm a stick into life. Seething and plausible. She took a cigarette from the box on the record cabinet, very sure of herself, very much the fresh young divorcee, and said, "Dan, I thought about you while you were away. We both lost . . ."

"Yes, despite everything we've lost a great deal."

"We lost *control*," she said firmly.

He felt rebuked, but it did not touch his reviving humor. Often she was bad, but sometimes she was steady and knew the way. She was right! He longed to regain this old sense of his wife's rightness. He needed it to go on from where he was.

"I don't blame you any more about that . . . that," she said. "It was wrong of me to be so . . . so . . ."

This time he did not make the mistake of supplying her words. Perhaps she was testing him.

"So angry, nervous with you," she said. *"Judgmental,"*—one of those damn labels again, he thought, but nodded encouragingly. Warily she went on with a speech that seemed almost rehearsed. He did not expect her to grant him a confession, but he thought maybe she was making it in her own way at last and that this could finally release both of them: I was jealous, and not merely jealous, also morbid and criminal and made furious by fantasies, until you had to give me justice for them because you, too, demanded someone totally committed to you, because jealousy itself is an infidelity which takes a wife from her husband. . . . But she said instead (and who would she be if she used his words? Her own were good enough): "I think I've learned something. We each learned, so it wasn't so bad. We probably couldn't have married anyone else, or anyone better, so why blame each other?"

"Right! We won't any more!"

She frowned and did not hear him. He felt her effort not to slip down and down into the despair which had driven him away. It had poisoned him—all that cleverness turned to making him out a monster—and it had poisoned her. Rage, like jealousy, is a fire which burns out the future: she apparently knew it, and the softness in her eyes, even this willed, desperate kindness at the door to the furnace, restored her to him as he needed her to be restored. Then they could both go on to love again, not merely beyond hatred, but also strong in a practical, difficult friendship. Perhaps that was too much to ask, his version of her romanticism, but at least it was a principle worth striving for.

"We had some good reasons together," he said.

"What? Oh! Of course! I know, I remember, Dan. It wasn't just our psychological set, we really liked each other—"

He grinned wryly. "So it seemed at the time."

She did not admit a poor effort at humor in his enforced grip on the past. "That's the way you remember it and maybe that's partly how it was. I suppose. Anyway,"—with great solemn effort for which he wanted to thank and thank and thank her —"whatever our motivations"—although he winced at the word—"we— yes—Dan—*we loved each other.*"

"Yes!"

And again she retreated. "But it was sick."

"Okay," he said.

And she folded her hands contentedly. "That's what I've learned."

Who's been teaching you now? he wanted to ask. He resisted successfully. He wanted her to understand about the passive faithlessness of the woman who expects her husband to be what he is not, what no man can be, who lies with him locked in her lonely dream of perfection, who sees her real marriage by the strange sunless radiance of a dream marriage which is unaltered by growth and event, by this particular day on earth or that one. Absolutist! Idealist! Romantic! These had been his epithets for her. And you can't make yourself into Eleanor of Acquitaine and me into a troubador by nagging me for talking with a girl at a goddamn party! All right, so I didn't, didn't, *didn't* light your cigarette! So neither of our children were boys!

He managed. He said nothing.

The admissions she had just made were an extravagant yielding. She felt loss, they meant; relief and giving up pain could help to teach her to praise life, and when she could bear the din of pleasure once more, she might even be grateful to him. Together they had exorcised childish ways—the lesson was nearly fatal;

they had also learned and practiced love. Once free of each other, they could go on with this lifelong music.

But how stiff she still was!

Nevertheless, in her own way, calmly, gently, and much sweeter than his secretly unleashed judgment, with but a slight stiffness of the mouth, she went on. He was impressed by her rapid peacemaking, so soon after the tears and threats and furious thrashing. There was a deep quiet note of loss despite the artificial speech. He listened: "It was inevitable. You were my mistake and I was yours. But neither of us is *bad,* and we could have married really bad people, real stinkers—so if we know it was inevitable. . . . Dan," she concluded rapidly, with a flush, "I said things, did things. Now I'm sorry."

She was in misery. He wanted to comfort her. Then she straightened up with a strong glance of resolution and touched her hair. "I'm sorry," she said.

"I'm sorry too, kid, for all I did wrong."

Apologies and pride and gratitude. Shaper felt a reviving flow of tenderness for her because she said *stinkers,* and maybe even thought stinkers, and not that cant of hers, *destructive personalities.* They sat very still, in a heated blush, both of them, their shame almost a courting shyness, and they listened to their children splashing in the sink and laughing. An odd music for courting. The children were his laughter and hers; her body and his were met in the children as they had been entangled in each other. All this willed and anxious sweetness between them might finally give him Sally, give her whomever she wanted, and not

merely send them raging after others. It could be the final creative act of their marriage—the two of them educated by their errors, not bound to them, freed for love and not greedily clutching at it. They would always have Paula and Cynthia together.

Once, on the day when their lawyers had first met, before this resilient stupor with which he shielded himself had settled in, he had cried out in this room, after they had both put the children to bed and he was facing her, ready to say goodbye, when he was at the door with all the unfinished business of their life meeting him in her huge hot eyes: "We didn't make it, kid."

"We tried. We tried."

"We tried, honey."

And they had fallen into an adulterous passion, still unshaken in their will on divorce, the adultery made painfully sweet. He remembered it with shame, his eye suddenly prickling and aching, and saw himself again covering her with kisses, forcing her mouth open while her body bent backwards, arched, received him. She had given herself without a word, moaning, and then he had fled like a thief.

Now he was merely quiet. The ache in his eye was an unheard sound, an unseen tremor of desire.

Still they would always have the children together.

He felt the day's heat already sweeping in and looked about for the fan. In his reticence in this familiar and very unfamiliar house, his and not his, he obeyed the stricture she had once made in anger—don't touch anything, nothing is yours any more! (Nothing but the children.) "Put your hands on me and I'll get an injunction," she had said the next day. It was her only reference to that strange, silent last lovemaking. "I'll slap a court order on you so fast it'll make you dizzy."

"Okay, kid, do you really think I want to touch you?" And that time his eye had not ached, not seemed invisibly swollen. It was the end. Almost comfortably he believed in playing out the rest of their career together without the danger of intimacy. Of course he was wrong, but something of the comfort remained in his waiting, his planning, his cautious observance of the new rules of tact and courtesy. It was only the children he needed.

She had rearranged all the furniture, painted, covered, hid, and replaced, so that the house could begin anew. He recognized it as he recognized her, from a distance, from a long time, from a dream of voyages. She had planned it that way. Fastidiously he had been plucked out of her skin and the scars dusted over, so that she could say, *Who? Who?* She wanted him to notice how much he was not at home. How managerial and intelligent of her to begin briskly anew! With what a rush she had retreated from him!

He smiled at her. "You look very well, kid."

She turned and pressed her lips together tightly. In pain she had a large mouth—childbirth, bereavement—he had been there with her; anger was a small one which sent him away. By the resenting white line at the corners of her mouth, he recognized his wife, the one who had first wanted him for pride in what he did—others wanted him, he was not like the herd of college boys; then wanted him to do nothing that did not include her, that might make

others covet him; and then when they were cozy in a common, including misery, rebuked and rebuked him for his failure to be her romantic ideal. O father she lost, O son I did not give her, how amply I was repaid for trying to replace you! What foolish pride that thought this abstract hungering of hers was desire for me! Impossible victories that I wanted!

Now thanks to that bloodless thin line and the stiff tremor of her response to his word for her, *kid*, he felt entirely at home. It was almost like the last years of their marriage again.

She reached for an envelope in her purse. "You forgot to pay my dental bill. It dates from before our settlement," she said with her marvelous cool telephone voice, "so of course you are responsible. Dr. Jonas' secretary called me about it and I said you had probably forgot. I said I'd give it to you personally when you got back to town."

He took the paper, put it in his pocket, and called like a drowning man, "Paula! Cynthia! Let's go outside, would you like that?"

They came screaming with pleasure into the room, and he believed that, yes, he could still keep his grip on the green feelings he had brought in from his month away.

III

He liked everyone, even his wife. Perhaps he loved no one but the children—he was so diminished by the belly-ripping, face-clawing final year of marriage—but he felt a reviving and undiscriminating benevolence toward the world. Plus desire for Sally, ah, that's important! It was recovery from an illness. When he returned to his hotel after leaving the children for their afternoon nap, his thoughts wheeled around to Sally, and they were imbedded within a silly glow of sympathy for everyone, irrelevantly, from the desk clerk with sinus trouble to the Japs at Hiroshima. Symptom of convalescence. "Key, please! Hot enough for you?"

He wanted to share this precarious perch on health. Sally had agreed to drive out to Chagrin Falls and go swimming with him. The first time he took a girl swimming was important to Shaper. It's a risk and rousing. You run the challenges of dirt, ants, and fatigue; he took chances on a cold noting of his skinny body in boxer trunks, the scatter of hair and the ropes of tendons and no strong rhythm of distraction. Enjoying the happy vanity of a lovely object, long used to triumphant waiting, Sally might have an eye for some tumbling beach athlete, not Dan Shaper with his glasses and his bony knees. Admire, admire me! her every movement had been saying to him for months. He had done it well. Was that a guarantee that she would love him?

No matter, he thought as he shaved in his hotel room; I don't want love now, I couldn't take it yet, what I need is just what Sally is giving me. Why worry about categories? He remembered that warning song: "Silkless silk and milkless milk! Love oh love oh loveless love. . . ."

Fully dressed, he lay on the bed, organizing himself to meet Sally, feet straight and toes up, watching the hand of the clock move down to

two. These few minutes of rest were necessary, despite his driving heart, after a morning spent with his children, seeing in their faces the intermingling of his wife and himself, feeling her breath with his gesture when they talked, his forehead with her dark, thick-lashed eyes when they laughed. Their eyes when he rolled on the floor to play bus or airplane as they had during the easy evenings of their babyhood (they remembered, they demanded it)—they wanted him to bathe them again, as he used to—and their mother's eyes, the tragic pouched eye of the beauty unsuccessful in love. And then the solemn politeness all around when he turned them back to her.

One minute to two! Think of Sally—full-breasted and full-hipped, blond, frequently amused—how lucky to be all those things that his wife was not! And Sally had her own secret sources, too! He remembered the yearning of her response to his first, long-delayed kiss, when in the renewal of timidity, the return to boyishness brought on by the long illness of his marriage, he had quite simply been afraid. No, it took wisdom to be full and warm like that. Her hand on his back was deep art, her weight on his shoulder was Plato and . . . and . . . Sally, not Plato! No romantic ideal! She did it for herself and for him.

Lord, ten after two already. He must have dozed. He threw swimshorts, towel, clean clothes into the blue Air France bag he usually carried for tennis. He took a taxi to Sally's apartment, where she was waiting for him at the entrance. She had no need to keep him waiting; she didn't play that game any more—

she felt his need for her and forgot to protect herself. She said: "Darling! It's such a beautiful day for us!"

That was intelligence, wasn't it? Her own marriage had been a revolving door, in and out with practically no noise or loss of heat, but she probably knew more about it than he suspected. Surely she knew more than she would tell. And she had her own automobile, a blue Ford convertible, to which she just automatically handed him the keys—that too made her clever and deep. Like all American boys, he had dreamed of girls like her with their blond heads thrown back against the leather seats of cars like these. . . . That it was her automobile only gave the joke a turn. Once he got over the divorce, lawyers' fees, travel costs, the rest of it, he, too, would have the open car which his wife had always thought too dangerous, bad for a dry skin, impractical.

"Why didn't I think of picking you up at your hotel?" Sally asked.

"Never mind, just move over and sit close to me."

"Are you a one-arm driver?"

"Like to be."

"Then I'll put a knob on the wheel for you."

Wryly he submitted to his retreat toward college-boy pleasures—the blonde, the convertible, the exaggerations of flesh in erotic gaming. Well, he would tolerate himself, he would come out of it. Let the submerged fantasies see the light, why not? Sally was there waiting, and more than merely *the blonde*. She was Sally. Her silent smiling said that she was with him part of the way at least.

And they teased and played while

he drove under the hot August sun through a part of the city, a part of the suburbs out to the private country lake he had chosen because neither of them had been there for years. They paid to enter. On a weekday like this, they had the short clean beach almost to themselves. When they met on the beach after dressing, he submitted to a moment of shocked, almost unpleasant awe at her perfection of body under the blaze of sunlight. Beauty is pitiless. She turned under his brooding with a model's pouting half-smile, her eyes blank, self-regarding, retreated into vanity. This must have been an old girlish habit. He hated the thought of her searching him for flaws, as he was now doing to her; but he needed to be looked at with pride, too. The different male and female ways about bodies would protect him, no? And hadn't they already begun the long study of each other in the close and secret dark?

Different, different.

She was a marvel, and that they both knew it somehow widened the space that had been closed by their first tentative lovemaking. A monument wants a pedestal.

Then she turned and ran, elbows pumping, toward the water. He caught her, and they went in together. They laughed and spluttered, and she began swimming, and he caught her again. "Let's go out to the dock," she said. She swam poorly, splashing and puffing; he swam well and easily. Sally's wisdom! When he helped her up onto the outer dock, she was no longer a monument. She leaned on him, gasping a little and laughing. He kissed her shoulder with its gleaming, running beads of water. She nipped his hand with her bared teeth.

They lay a long while under the descending late afternoon sun, stretching against each other, talking at intervals, listening to the few shrill voices back on shore. He spoke a little of his wife. He did not speak of the children. "Just smile," she said sleepily, her face close to the hot whitewashed boards. "I don't want you to worry about a thing."

He stopped worrying. He put his arm on her back and rubbed the strong articulations of her spine until his fingers ached. She smiled, saying, "More, more, more. Ah, that's nice."

She was very young, overproud of her body in the suburban way, just beginning to get past using it as a weapon instead of spending herself through the splendors of flesh. But she was learning, she was. And if her beauty alone had frightened him, could he ever have borne that rare loveliness which comes of beauty joined to a proper sense of its being? No, thank her for limitations. And she was learning to spend herself freely, too. And with him.

At sunset they sat up to watch. She leaned on him to share the radiating warmth of her body. They approved of the sky. He missed not seeing the children this evening, but he needed Sally now; he would go to them again tomorrow. With a fine sense of no hurry, no hurry for anything, they waited until the sun was gone.

But the swim back to shore made them furious with hunger. She suggested buying Chinese food and eating it at her apartment. They drank a great deal of tea and nothing else. The sunburn began to show on her face and arms. She put a stack of records on the machine. They made love slowly, patiently. "Why are you smiling?" he asked her, and she replied: "Why are you smiling?" He explored a generous body with its slowly stirring languors and bold risings to the touch. She explored whatever he meant to her. "Why *are* you smiling, now tell me!" His hand had discovered, with the sense of regaining a fine lost memory, the crisp, crinkly, blond hairs of her secret places.

They gave up talking. They loved each other until they both felt as light and pure as driftwood, and then all tumbled together they slept.

IV

Paula, who was six, said to her father, "Mommy says you don't love her any more."

Her father, who was thirty-two, replied, "No, but I like her."

"But, but," said Cynthia, who was just four. "But can I go out and find Gary?"

"Why don't you stay with me for a little while?" her father asked. Dismayed by his querulousness, he repeated the remark in another voice. "Stay here with me. I have to go back soon. Anyway," he added, "it's almost bedtime."

"Okay," said Cynthia, resigned. She was a very small child, pouting and serious, with overbusy limbs. She paced back and forth on the long, low, especially constructed, "contemporary" couch which her mother had bought partly because her father didn't want it.

"But *why* don't you love Mommy

any more?" Paula insisted. "You always told me you did."

"That was b-b-before." Her father stammered for an explanation, dulled by knowing that there could be no valid one for Paula. "I tried—we did—I wanted to. We just weren't happy together. You know how that is, Paula."

"No," she said flatly and firmly.

"We have to live separately. It's like when you and Cynthia are tired and quarrel. We put you in separate rooms until you feel better."

"When are you and Mommy going to feel better?"

"It's not exactly that way with grownups." The sly innocence of Paula's question brought his hand out to touch her pleadingly; he wiped away the smudge of dirt on her cheek. She always made herself up with a stroke of dust as soon as possible when her mother washed her face. Cynthia, humming to herself, was listening with a smudge of prying watchfulness across her eyes. With a premature false security, the two girls frowned for serious discussion. The children of the divorced are engaged too soon in love as a strategy. Joy recedes before strategy; these children are robbed of their childhood. The huge brooding of possibility which human beings have at their best comes of the long passionate carnival of childhood; no fear of cost down this midway, just another and another breathtaking joyride on the great rollercoaster! and another quiet gathering-in of food and rest—it should be. It should be a storing of unquestioned certainties for the infinite risks of being a person. But instead, instead. Heavily Shaper touched the two girls as if

to make them child animals again. It was not right that a father should feel this hopeless pity, and this need to enlist his daughters in the harried legions of rationality: "Here's how it is with grownups—"

"You mean Mommy and you?"

"Yes. Yes. Now listen. We feel better living in separate places. We're going to stay like that. But we like each other, Paula, and we love you and Cynthia. We both do."

"But, but, but, but," Cynthia sang, carefully wiping her feet on the pillows. "But heigh-ho, the derry-oh, the farmer takes a wife."

"Cynthia," said her father, "you shouldn't. Take off your shoes if you're going to play on the couch. It wasn't made for children."

Cynthia looked at him silently and, scraping the fabric, slid down beside him. Paula pulled between his knees, fighting to get closer than her sister. She began to suck her thumb. Her father pressed his lips together, resisted the temptation to remove her thumb from her mouth, and instead lit a cigarette. He decided that perhaps his silence would oblige her to remove the thumb and speak. It did not. At last he said, "I want you to understand. Mommy wants you to understand, too. Even though I'm not going to be Mommy's husband any more, I'll always be your father. I couldn't change that even if I wanted to, and besides, I would never want to. Don't you want always to be my daughters?"

Sucking busily, Paula said nothing.

Cynthia announced with a grin, "But I want a daddy who loves my mommy. I think maybe Uncle Carl, he loves Mommy——" The look on her father's face told her that he was not enjoying her joke. "But I *know* you're my *daddy* for real."

"I am. For real."

"Okay," she said, bored with the discussion.

Paula looked at her wet and slippery thumb, considered putting it back, had another idea. "Why doesn't Mommy say hardly anything to you no more?"

"*Any* more," her father said. "I already explained. Because we don't get along—just like we don't let you and Cynthia talk to each other when you don't get along——"

"But we do anyway! But that's only for a few minutes! But it's not, not, *not* the same thing, Daddy!"

"No," he said, "you're right. It's really not."

"Then *when?*"

"When what?"

"When are you coming to sleep here again?"

"I told you, I already explained. Mommy and I——"

"When you went away you said you'd come back to live here in a few days."

"Well, we thought maybe. I hoped. But it's worked out this way instead. Now listen to me, girls, it's not really so different. I see you very often. We go out together for milkshakes. We're just like before."

Silence from Cynthia. From Paula, coldly, suddenly with her mother's precise articulation: "It's not the same, and you know it."

"Okay, you're right, it's not." Her recognition of his hollow heartiness made him flush. She cut right through what he said. She remembered very well that he had been a part of the life of the house and she did not like her new sense of the house. He said, "I guess you're right, Paula, but that's how it is. That's all. We don't have to talk about it."

Silence. Then:

"So you really don't love Mommy any more." But she was a child again. The moment when she spoke with her mother's voice had passed. "Daddy," she said.

He resolved to go through it patiently once more. "No," he said, "and she doesn't love me. But we like each other, and we love and like you, both together, and we always will. You understand that, Cynthia?"

"Okay," said Cynthia.

Paula was sucking her thumb again. Her mouth was pulled around, working and bothering, as if she were trying to pull the skin off. She might be learning to bite the nail.

From the back of the house her mother walked toward the living room where the two children and their father were talking. She said hello, picked up a book, and returned to the bedroom. This meant that she would like him to notice that his time was up. A brisk, dark young woman, she was freshly showered and very pretty, although too thin. She wore a housecoat, but a girdle under it, stockings, and high-heeled shoes. Obviously she wanted to get the girls to bed early because she was going out.

He began to say goodbye to his daughters. He reminded them that he would come to see them at noon tomorrow. Cynthia threw her arms around his neck, laughing, and demanded: "Bring me something, maybe a surprise!"

"If you like," he said. He had a

sick lonely weakness in his stomach of something not yet done, not possible.

"Do you like me, Daddy?"

"I like you and love you, Cynthia kid."

Paula was rubbing her face against his hand, the thumb still in her mouth. He lifted her to kiss her saying, "And Paula too. Now goodbye until tomorrow."

As he started down the stairs, Paula stood with her swollen thumb dripping and shouted after him: "Oh, how I'm sick of those words love and like!"

V

The next day there were fresh flowers in a new vase on the coffee table. His wife was cool and abstracted, and the familiar house could not have looked more strange to him if he had returned to find it filled with angry, overheated growth, like an abandoned greenhouse. Even not considering the flowers, the entire room spoke to him with faint whispers of disarray about how his wife (ah no! *former* wife!) had been out late and importantly that evening. A wrinkle in the carpet informed on her. The piled pillows on the couch were his witness.

"The children are having lunch," she said. "Would you like something?"

"No thanks, kid, I had a late breakfast."

She hid a small yawn behind her hand. "Me too." The yawn was excessive. It was more than required. She was putting it on with a trowel. Well, still, this was very much better than the hysteria with which she had sent their marriage into darkness, like a couple trapped forever at an abandoned Luna Park in the spinning, jolting cars of the Bug. Of course, this had its own special touch of refined cruelty, which merely went to show that she still depended on his feelings for some of her satisfactions. "Sit down, please," she said.

"Make myself at home?"

She smiled tolerantly. He wondered if she had any imagination for how it disturbed him to visit the place which had been his home, which in some way still was, which was so mysterious, like a room dreamed of and then found and then you're suddenly unsure of whether you really dreamed of it or only now think you did. Of course she understood; she had worked the house out for herself, and had a right to; but any imagination past this redecorating, working-it-out variety requires sympathy. When he looked up at her, she was still dressed in that social smile. Excessive!

"Some coffee?"

"No thanks, really. How do you think the children are taking things? My being away. Visiting. I was wondering if maybe I shouldn't see them too much when I'm in town. You can't make up for the normal daily——"

"Yes, overstimulating them."

"Too exciting," he said, giving up his thought by agreeing with her.

"They don't show it, of course. It's funny," she remarked, smiling patiently, as if it really were funny, while she explained the joke. "At their age they can't express it, they're too well behaved, they don't have the vocabulary for discussions. . . ."

He gave up listening. He was trying to place himself in this room. Sometimes they had pulled the shades and made love on the floor. He heard her despite the noise of memory. Her mania for psychology had always annoyed him. Jargonizing. And yet, and yet, once when she could not sleep he had held her in his arms on the couch while they quietly talked all night about her father and brother—she could talk English when she wanted to—and she spoke of love and violated need and loneliness. Then they had talked about each other and their children and how different it would be. Then they had gone out on the porch to hold hands in the chill spring dawn and watch the lights go on in kitchens and bedrooms down the street. See, we didn't need that sleep, we've gained a night on everyone, he had said, and she had answered: Yes, yes, yes, we have something they don't have. I'll even make you a better breakfast!

"But," she was saying, "I'm afraid they will express what they're feeling about us now when they get to adolescence."

Another time, after a terrible quarrel—thinking that maybe with the guidance of their bodies which wanted to give to each other . . . — more than hopeful, prayerful, they had spread a quilt on the floor just below where she was now sitting. They had helped each other down as if crippled, slowly, slowly, and then safe on the floor, had flung all their strength into the cruel struggle to possess, a lurching grinding grasping assault on tenderness, and her head thrown back and her mouth open so that he could not see her teeth, only a dark place and pulsing groan issuing deeply from within it, and then fiercely she closed and bit his shoulder and the cry, *You did it—you did it—you did it.*

What?

Oh love, love, love.

Would the rock be there the next day for rolling up the mountain?

He blinked and straightened his shoulders in the heavy inert grip of sweated clothes. Now was now, and his former wife was talking. Now was also then, but his former wife was speaking. It was now ten summers for them. She had something to tell him about their children.

"—with sibling rivalry," she was saying. "They're beginning the latency period."

"Oh yes, yes." Didn't she remember how he disliked those words? What was she trying now? He went into the kitchen for a glass of water. The children made him rehearse what they would do together after lunch. The park and the swings and a milkshake. All the milkshakes they could drink.

When he returned she was sitting calmly in her chair, hands folded together, with a subdued half-smile on her face. He recognized her analyzed smile, the one she reminded herself to slip into when she talked psychology. That little smile cost me a year's pay. Okay, go easy, he told himself, it's only a year's pay. A convertible plus gas and insurance. A while in Europe. It's a funny nice little darling of a smile, really—all sympathy and comprehension and let's-be-mature-about-the-infantile-phase. No joy, no teeth either, but it's still better than the screams, much easier on the neighbors and a good deal easier on me.

"Did you visit Pete and Ellen last night?" she asked.

It gave him a little malicious pleasure in his turn to let her know that he had not been lonely, that he had not even gone to see his best friends in Cleveland. "No, I had something else to do. I'll see them in a day or so. I called them."

They were both silent. All the maneuvering and rivaling warned him. He was doing nicely, but better get the kids out soon, get out quickly. If she felt his strength of distance, if she sensed Sally and stopped being nourished by whatever it was in the room, the flowers, whatever it was, things might go poorly again. He followed her eyes. Three novels by Evelyn Waugh, new in their jackets, were piled on an end-table. A quick twitch of grin crossed his face. He had never liked Waugh, neither had she; ever since college they had taken literature passionately and together, even after she had begun her long crush on the bound volumes of *The Psychoanalytic Quarterly*. Now the little pile of Waugh was a roadmark meant to state: See, I've come under someone else's influence. I'm reading the books *he* recommends, doing the things he . . . It amused Shaper, but when he finally spoke his voice was husky and he had to swallow and it still didn't clear. "I talked with Pete," he said. "I suppose I should go out there."

"Yes, if you're not too busy. Really, Dan, you don't have to be afraid about our friends. I wouldn't spite you for the world—"

"I'm not worried. They know me. I mean to see them, but you know how it is, things come up, I'm only here a few days."

With a thrill of satisfaction, he felt the balance swing over: I'm giving her now! She's getting it! This while he knew how much she pained only by how she goaded him—so rusted together are the ways of untying man and wife. And this all the while that his eye began to ache, that he asked himself a fleeting, trespassing question about how he could expect to hold his job while committed to this continuous deep marital work, that he suddenly saw a tiny receding Sally fretting and scowling and making up her eyes and mouth three times over because she was jealous of his quarrel with his wife. Sally's moods were reflected in heavy ways with her lipstick. His own lip was sore inside where he had bit it. That was his wife's habit which he had borrowed. Divorce, divorce! he thought. Let us be divorced in the flesh as we are by law!

She was watching him shrewdly. She knew him. He could keep the bitten lip from her, there was no way she could palp it. It was a contest to hold secret the hurt eye with its invisible throbbing. And then Cynthia ran up and put her hands in his pockets to see if he had gum.

"Get out, Cynthia! Go wait for your daddy in the other room. Paula has enough sense. Can't you see I'm talking to him?"

Like successive waves of fever and chills, chills and fever, his tenderness for the children, who resembled both his wife and him, gave way to hatred, hatred, cold disdain for this woman who forgot so much, who destroyed so much. And then a sharp new ache in his eye, flicking all the way up into his brain, no, just the sinuses, no, cruelly into the brain: Maybe I should just have given in. Maybe I should have taken it on her terms.

No, no! He wanted to turn off one whole side of his head. The eye had a furious life of its own. He warmed it, comforted it with his palm. Bent to it for an instant, he looked up again strictly. "You don't have to sit here if you're busy," he said.

"Oh, I don't mind, Dan. And besides, there really are some little things we have to straighten out."

"Well, thank God most of that is over." And he added meaningfully, "We don't have to quarrel any more: We can concentrate on the children. We can be friends."

"Yes, of course, but turns out there are some other expenses—"

"Oh please, no, kid."

"Yes."

"What do you mean?"

"Some other little things. It's complicated raising children under these circumstances, all alone—well, never mind that. Let me see. Not just Dr. Jonas, but——"

"What? That again? I won't bother pointing out about that one bill, but as to any others, you know very well——"

"You know very well!"

"Please don't be sarcastic. Let me finish." But she was staring at him with her enormous eyes turning black—as with anger, as with love— the pupils dilating, and thick congested hate squirting like black arterial blood from a deep wound over the thin face. He struggled to be without memory. He thought again (how many times like this?) that if he could keep calm, keep easy, maybe she, too, would make the effort. He said in a low, forced, effortful voice, "You know very well that our agreement states that you meet all expenses out of the check I send you every month. There's no other way to do it, kid——"

"Don't call me kid! I never liked it! Your idea of an endearment!"

He went on stubbornly, quietly, "You know what my income is. I don't even— Well, I'm not complaining. But there's nothing more to discuss about money unless I fail to send you the check some month."

"Kid! Kid! You still want to call me kid, but if one of your children needs some special care, what's it to you——?"

"What's the matter?" he interrupted, frowning. *"What* special care?"

She mimicked him with ungainly, ferocious sarcasm. "It's not in the agreement. . . ."

"Are you going to tell me? Or is this all nonsense again?"

"I want to send Paula to the Bainbridge School."

"What for? We—*you* can't afford a private school like that. Anyway, what does she need it for?"

"A fatherless child!"

"Is that my fault? Did you want me? Did you *want* me?"—and he felt the harsh sting of self-pity like dust at his eyes and he shook it out angrily. "Didn't I fight to keep us together long after any other man would have run off or gone batty?"

She tapped her foot and did not answer. He had a suspicion that he could peel off her skin now and find a genuine, very satisfied smile, but with a forced calm—like his—she began patiently to explain. "I've discussed this with several people. The Bainbridge School——"

"I'll tell you just one thing: I don't have the money."

"—is oriented toward difficulties,

special problems, broken homes——"

"Oriented, oriented!" He called out the word as if it were a verdict brought down upon her. He stood up, shaking. He had a throbbing frontal headache just behind his eyes. "Please let me get out of here, I've got to get out," he said. "Let me take the children and get out for a while. The little thing we should orient ourselves to is orienting ourselves toward not talking like this when they're watching."

And they both turned and admitted the presence of the two, very quiet, very thoughtful little girls. "We were just talking," Shaper said weakly.

Shrunken and bent, all the pleasant civilizing of their time of separation scraped away, his wife managed to wipe the children's faces for their walk. At the door she turned to him and said severely, "When you come back, please leave them downstairs to play for a few minutes. I want to talk with you alone."

Blackmail! In order to see the children, in order to keep her from trying to rip him out of their lives as she had ripped him bleeding out of hers, he had to find some way of settling into a decent habit with her. The headaches that were already a tradition of their arguments made him almost blind in his left eye. He counted on her not interfering with the children—at least for this her psychology supported him—but he suspected that she wanted to think of him as an ideal monster, and then she could reason that for their own good they should *relate, transfer, orient* to some other man. Her words! That damn vocabulary! Lousy blackmail! He tried to soothe his eye by cupping the palm of his hand over it. Blackmail!

Within the dynamics of her romantic, absolutistic passion for the ideal, each failure between them had to be complete and each small difference total war. Symbols had become reality, reality had become a great abstract, timeless, and predetermined sign; her energy was bent furiously on making reality over in the image of her idea about it, even at the cost of destroying its value as reality. (She could not measure the cost. Cost is not a part of heavenly systems.) This way of life has a kind of internal logic, perfect and unaffected by experience; there was no opposing it. Once her husband was defined by her needs ("It happened to me when I was immature"), there was nothing he could do to become human again. Even yielding to her—that most human of acts—enraged her because it violated her idea about him. She hurt herself, and could not stop, and threw herself and the husband she once loved and anyone else in the way (including their children) under the clanking treads of the interlocking syllogisms by which she lived: Good and Evil are pure; the ideal exists only in heaven; my husband is here below with me, on earth; therefore he is evil. I must have perfection; I do not have perfection; then this is hell I live in and I am damned and I will destroy, destroy, destroy. . . . Still, someday, O someday I will find pure virtue in a man! He must exist because I can imagine him. I need him right now! I deserve him, I call him, I insist on him! I am pure, I have waited for him, this other creature has no value because see the way he shrivels, vanishes, like a bad idea! I will prove that he does not exist by making him recognize it himself.

And so silly, smart novels. And bouquets of roses in a cut-glass vase. Well, if he knew how she was getting at him, he should be able to get out of the way. And he didn't have to display Sally, either.

VI

Fortified by wry self-congratulation at his assent to whatever she now needed in order to do away with him, he could turn his back on her judgment and go to another. He did not believe in magic; her eyes could no longer make love to him, make harm for him, uncover into life that which was only recently buried but quite dead, and down the dust came sifting fast. It was done. There were the children left, for ever and ever the heirs. He would shower again and scrub hard.

But always there returned the tangled memories which dragged him off the straight road leading from life into the violent death of divorce and back into life again. Their children were not the only heirs, the only judges. It would be useful simply to hate her until he forgot her. It would be fine to despise her. Pride! But when he had almost succeeded (for example, in a sweet moment of fuel and money while having the oil checked in Sally's car), abruptly his greedy feeding on resentment came back to him, his joy and suspense before what she would do if she discovered that, for almost a year, there had been an-

other girl. She needed the lesson; he had almost told her himself to let her know how he, too, might be lonely. . . . But recognized his cruelty. And remembered that the night before she found him out, she had risen to his grief and guilt to comfort him in what she took for a passing depression; there had been a miraculous access of tenderness and gratitude, and she had said, "We have the best little kids in the world, I know it, Dan. I'm sorry about the boy, I know it's foolish. Be patient with me. Care for me."

And then the next day the gossip came round and she had confronted him and he had thought, Let her learn, let her burn awhile. We've tried everything else.

Divorce time, wake up! It was time to move past his malice and her answering sweetness, past his yearning and her vindictiveness, past their other swinging meetings and partings, and years of it, common efforts, successes, bitter ultimate failure. While she talked, talked, talked, harassed him, practiced meanness, he could still find the silent regret faraway—deep in her eyes—apology, helpless apology—and yet the two of them could not settle together into firm admission of it. Her regret and her ancient willingness to love were resin-soaked roots. She consumed them. She refused to look at them. So be it, she seemed to have decided, abandoning her will with relief, and threw the uncured wood on the fire; and the destruction of what she most needed to grow from gave a wild hissing edge to her sarcasm, "I do wish that you would please stop calling me *kid*. Customarily that's for the children of goats. I seem to recall that I have a name."

Arson. She still remembered something more than that from him. On her face—as the lover in the morning finds on his darling's face the marks of their excess together during the night—he saw the brand of her secret self-appraisal, secret wish that it might be otherwise. It was like that thickness and slight purpling of eyelids in the morning after the flesh has made fantastic avowals that moments do not die (even as the moment is dying), that what the intentions of lovers have brought together no God shall sunder (even as spirit and life notify us of matter and death). Who desires what they had now earned? She had wanted to want only him. To him she had cleaved for everything, rest, trust, energy, hope. "But I need a son, too," she had said. "Lots of women have sons. No one is perfect. I want *your* son, Dan, no one else's. Why can't I have the one thing I need?"

For everything, for too much.

As he recalled his chill judgment of her fever, the summer and the city spun crazily around him and he thought: It is as she says—all my fault, all! I was bored with her because I could not admit my need of her, because I feared that her weakness excluded me and she could do nothing for me. I took for boredom my dull despair of touching her without the qualification of a son, of receiving what a husband asks *without qualification,* so that I determined to need nothing at all from her and to build a life which did not depend on her. I sought to harden myself by freezing. Monstrous!

"You're a monster," she had shrieked when he admitted that his strength to indulge her had come from a little student at the Institute of Music.

"It didn't make any difference then. You were thinking about your skin, your waistline, your father, your analyst, your fantasies about a son—you didn't really care what I——"

The war again. Go! He leapt from the transport. His parachute was tangled. Other bodies hurtled free, mouths open, teeth bared, roaring with effort. They relished the free tumble into space; for him it was death. "Monster!" He plummeted. Didn't he have the same rights as others? No. He would be a sickness of gore on the earth below. "Monster, monster!" And then it came back to him, first with a belly-jerking violence, earth and life crashing up toward him as the parachute caught, opened, and he gasped, and then he lazily swung like a pendulum on a great clock: I am not a monster. Nor is she. But I am not, either.

He went to Sally that evening with something which he knew was not love, but she thought it was. Surely she was not really deceived; his need of her was what she needed, and love an irrelevance. Nevertheless she showed her doubt in ways that made it harder for him—her anxious trick of brushing and rebrushing her eyelashes with mascara, her rather heavy step into the kitchen, giving him a glimpse of the thickened waist to come, the overbright, slightly foolish, false glee of her smile, responding to uneasiness by an American piece of advice, *Keep smiling!* She came back with coffee, a glare of teeth, blond eyelashes on which he made out tiny beads of blacking.

He asked her to take off her make-up, and she did. His chill calculation shocked him. He wondered if his wife had so shot his nerves that all he could bear now would be a sweetly boring girl, worried only about pleasing him and getting the shine off her nose with the right shade of powder, a lover of musicals and of her beauty sleep—and of pleasing him! That quintessence of desirability! She watched, frowning. He grew tender and regretful. What could he ever do for her? She meant to be good to him. She studied him silently, unreproaching, with the pallor of the blond girl without make-up under bad light, and she suddenly seemed very beautiful, the most lovely woman in creation. He switched off the lamp and they moved toward each other in the dark.

When they lay down together, she said something which surprised him. It was trivial and familiar, like the breathless delicacies of undressing, and yet it was a trouble that she should cry out to him with words that seemed merely taken from a popular song: "Oh put yourself close to me baby."

How he wished for her silence then!

Afterwards he realized that the shock of vulgarity came not from her words but from his considering and judging her without interruption. He stroked her hair gently and felt sorry for her, for him. Someone was using someone. Somebody was somebody's weapon. There, there. "Sally, darling, I'm not good these days," he said. "I used to be, I want to be, but I'm in trouble."

"You are, you are, don't worry, you are good."

"I'm in bad trouble. I'm afraid. Nothing's clear any more."

"It will be, don't worry. You just need time, Dan. You make *me* happy."

"Do I?"

She put her finger on his mouth, shh. She wanted to sleep. He dozed too, asking himself, Why don't I ever think of her until I have to? That first day and the swimming was an accident, the false health of fever. Not right, not right, though I take my breath from her hair.

Then he slept.

In the middle of the night he woke with a start, with a sense of flapping inconsequential interruption, as if a cat had jumped onto his back. He lay staring into the murmurous dark and fearful of the dream he could almost remember which had awakened him. Oh yes, yes. Forget about it. He had been falling without a parachute for years and years. He did not crash to earth; he diminished. The starving man could not take food—even the thinnest broth was an agony to his constricted body.

The air conditioner threw its chill breeze into the room. Gradually, very slowly, reluctant to disturb the fine integrity of her sleep, he began to stretch against Sally for ease and new rest. She was warm, spreading, slightly curled as she lay. He pressed against the softly sleeping girl. He put his hand over her back and moved it down through slopes and valleys. She stirred, and their mouths were breathing and feeding into each other. In a few minutes he heard a hoarse voice, his own, saying, "Oh come to me baby." She gathered him in.

VII

The last day of the visit. Shaper wanted to store up memories of his children so that, like a camel, he could survive away from them through a season of drought. Yellow-eyed, lurching, he was still not a camel; he could not live for oases; he could not carry Paula and Cynthia in a hump. Still he had hope. Although it was already several days past the end for his wife and him, he believed that perhaps they could manage one more meeting for the sake of the children.

Sally gave him her automobile to drive out to the house. This time the mild luxury, moving through traffic in the sun with the top down, failed to reassure him. He found himself singing an old Louis Armstrong blues tune:

> *From milkless milk*
> *And silkless silk*
> *Love oh love oh loveless love*

On an impulse he stopped at Peter's, found him home, and asked him to come along. "My wife and I will both be on good behavior with a third party present." And he looked away. "That's the way it is, Pete. Sorry."

The children were shy; they knew he was going. They had no practice in goodbyes. For them he was already gone. Paula gave him a drawing of flowers and the sun to put on his wall—"where you live, Daddy." The sun was sticking out its tongue.

As the subdued and uneventful visit ended, it occurred to Shaper that Pete might find his insistence on having a witness somewhat melodramatic.

"Goodbye," he said to his wife.
"Goodbye," she said.
"I hope everything goes well."
"Have a good trip back."

They shook hands. Peter looked embarrassed. They went downstairs. Shaper glanced back from the driveway and at the same moment two things happened: The children came out to stand on the upstairs porch, and his wife came running down the stairs. She paused, breathing shallowly, before Shaper and Peter near the door of the automobile. She put out her hand with a slip of paper in it. "I enrolled Paula at that school," she said softly. "I forgot to mention it. Here's the bill."

He drew back and tried to look merely puzzled. "You know I'm not supposed to. . . . I can't pay these bills. We already decided about that."

She held to the low tone, but there was a stir and hiss in her voice. "Who decided? Who? Who decided?"

"Look"—and he was pleading although he knew this was the wrong way to move her. He had only found a steady right way when he was sick and needed her out of helplessness. It was the only way he could count on for sure. It was no way by which he could live. "Look, I sent a check for that dentist's bill, although you were wrong about the date. It came after our agreement. Now please try to understand, there's got to be a stop."

"A stop to what?"

"We have to start living the way things are. We have to adjust. You've said it yourself. Money shouldn't get in the way now. Our lawyers——" And he watched the pinched face and the black, swollen eyes replacing that bright young divorcee which was her pleasant role. He had hoped, but once more he misjudged her. Through those inky eyes, through the cloud of hate with which she blinded herself and poisoned him, she would not see what they had sometimes been for each other—she never saw that any more; she would not see the neighbors, she would not see Peter, she would not see the children. "Kid! Listen to me, please!" And he had no words to force her to remember how they had held hands and strolled in parks, and tenderly made up many quarrels, and congratulated each other that the girls were a fine combination of the best in both of them.

A stop? Her lips were saying, working within, white and diminishing.

"Listen, please, *please,*" he said, "you have to understand how things are with me now. I live in a room, don't you understand? One room. I'm not a college kid, I'm not used to it any more. One furnished room, do you hear me? I save my money so I can fly in to see the children——"

"Then don't do it."

"What did you say, kid?" He leaned forward, trying to see her. "I'm sorry, I didn't understand."

"I said stay there. Get yourself a kitchenette. Don't bother coming in."

He held his breath. He was peering into a night, and his eye was tired. It throbbed, and he wanted to comfort it with his palm. It seemed very important this time not to let her know about the ache in his eye. He straightened up and said, "Look at me. Look at us, kid." His eye wanted an eye to return its gaze.

"Just don't bother," she said, as if her mind were far away with household thoughts. "Stay away. I could spend the money. Who needs you here? Don't bother."

She drew closer to him with a little smile on her face.

"And I may as well give you one other bit of information. Why not? Since you're leaving again." She spoke this last phrase slowly, as if it were particularly important. He recognized the angry wenchy smell which her body gave up when she could no longer hold on. She said: "Now I feel like a woman. I've found a man who knows how."

In the first moment of almost prudish shock, he felt his face being fixed in a skeletal grin. "That had to come. The list is complete. Okay. All right." But the grin spread uncontrollably, attacking his bones, dissolving the sockets of his eyes.

"It was no good with you. Your fault. It was never any good."

"All right," he said, "congratulations." He turned to Peter. "You hear? My wife was a virgin all along."

"Yes," she said, "that's about it. I told Carl, too, how I was never touched. Since you're going away again, you might as well have this to take with you. Nothing. Never anything, though I tried to make you believe. . . ."

Their children were leaning over the railing to peer down on them. His hand flew toward his eye; he stopped it midway and put it down. He said: "All right, there's no point in discussing this either, is there? Our lawyers have it all settled now. We paid the lawyers to do this job for us."

"Lawyers!" And it was done.

Something new had been released, and in public, in the summertime, with people coming out on their porches or peering through the screens to listen and saying a *shame* and smiling to themselves. "You want lawyers to raise our children?" she shouted. "That's your idea of a father?"

"All right, all right," he said in a low hoarse voice. He held himself stiffly, shaking with fury and disappointment. "Can't you control it? All this nonsense. There are people. You're doing it on purpose, kid, don't you see?"

She screamed as if he had struck her. "Don't call me kid! Just get out of my life!"

"Okay, okay." He turned to climb into the car. Peter was already huddled miserably on the seat. As he moved behind the wheel, he felt an almost physical eruption, a brutal crack in his throat, and the word came out, "Fishwife!"

"Corrupt! Corrupt!" she was shrieking at him. "Get out of my driveway! Go! Corrupt! You never cared for me or your kids—*children*" —and desperately she sobbed the correction. "Or anyone but yourself! Get out of my life!"

That was how, standing silently together on the porch, the children saw him last. He was being chased by their mother's rage.

At the corner he pulled over to the curb and asked Peter to drive. "I'm not in control," he said, shivering, and went on as if this were the total explanation. "It was what broke us up. I had to stop, I couldn't be run by her temper. Once I saw Paula watching me back down like a fool just because of that look she gets in her eyes—you saw it, Pete—

when she can't be reached. When I saw my daughter judging me. . . . It didn't even help to slap her, I tried that once or twice. . . ."

Peter turned and drove through the long park along East Boulevard. He was explaining, comforting, merely talking. "Well, she has to feel better about what she's doing. She's not the kind of woman who can take ambiguity. You have to be all good or all bad, and you're *it,* man. Once you could do no wrong; now. . . . Well, women are like that. Don't think she's going to be nice for the sake of the past."

"What about the kids?"

"The past has got to be wiped out. Women have their feet on the ground—anyway they call it the ground, sometimes it's our faces— and one thing she thinks she can do for the next man—whoever he is, I don't know if she's telling the truth," he added hurriedly, "is just wipe you out. That way she gets to be a girl again. She said it plain enough. For her you've got to be pure mistake, friend, and nothing else. Evil Dan the Bad Young Man."

"No kindness at all? That's how I'd like to be. Let me tell you something: When we first decided to separate we both wept."

Peter shook his head. "She told Ellen *you* cried. She said you're maudlin. She said you held her hand until it ached. Make up your mind to take it. Women—" Did Peter believe what he was saying? No, he was tuckered out and embarrassed and his mouth was uttering for him the cynical clichés about women which men in club cars tell each other. It was what he thought Dan wanted to hear. He may even have believed it true at that moment, true because it sounded familiar, and with a deep breath of fatigue (he thought this was sympathy) he again spoke what he took to be comfort to his friend: "Women," he sighed.

"But doesn't she look all yellow, sick? Her skin that color and those enormous black eyes. Yellow, sick, and mean. She used to be so pretty. She still is. Oh she hates me!"

They both fell silent. When Peter felt that he had done his duty, he asked to be dropped at home. He needed to get away. He loved his own wife; he felt as disconnected, tired, and jittery as if he had been casually unfaithful to her. Naturally he resented his old friend for bringing him to this possibility, and also —because his wife was surely his best friend—he resented Dan for eliciting his jovial male cynicisms. And within, too, he felt a thrill of pride: not for him this failure! This is the pleasure we are said to feel at funerals. And at a still deeper, more solemn place, because he was a good and kind man, he felt regret for his friend and also knew something of what was happening to him. This sense came up through the barriers as discomfort, as a desire to squirm: I can't take it, I want to do my duty by you, Dan; I want to get away. He had his own wife and children to play with on a fine dry summer afternoon. Ellen had asked him to do some shopping.

VIII

I held her hand until it hurt her! And what a bore she was with her whining and her headaches and her suspicions! Suspicions!

Better. He used his old ritual for

remembering himself into gratitude for freedom: her nagging, her picking at her face in the mirror, her stiff jealous mouth with the white lines. Sure, finally he had given her cause for jealousy, why not? But first he had warned her that she was making him lonely when she stood before the mirror, plucking, squeezing, hating herself, dreaming of miraculous sources of happiness—her father fantastically restored to life, a baby boy instead of girls, an analyst who really understood and *said* something.

With dreamless dreams
And schemeless schemes
Love oh love oh——

It was fine to be driving Sally's car with the top down. She was waiting for him, and she would be all health and dazzling smiles, unskilled in moping, untrained for meanness.

But don't you have to take a woman from strength, not weakness, if you are to give her anything important and receive anything worth taking?

And after his ten-year marriage, wasn't he too distant from Sally to bring her anything but that cheating desire, the need for comfort? He had loved his wife despite everything, he knew he had, they had been young and unmarked together, he insisted on remembering—and he did not love Sally. He made too much of the hot pity of bodies. He had thought to find love by loving, but instead, at length, filled with crazy pride, he had discovered sex. The solution was as ineffective as the discovery was unoriginal. Like his wife, he made too much of things.

When what he needed to save his life was simplicity, a bare white room and sleep, restoring sleep, how long could greater and greater complication soothe him?

Didn't that mean only trouble to come?

Perhaps. He parked Sally's automobile on the street before the massive, teeming apartment building—luxury circa 1928—in which she lived with her closets full of clothes, her mirrors, her pink-feathered slippers, and her music-to-dream-by. But if he couldn't allow himself Sally, that beautiful and perhaps silly girl—yes, he should say it out, maybe it's true

—he might be unable to prevent doing what he had too often already considered. He had stood caught in the middle of his furnished room and thought it through. He had first been indulgent, then shocked by the persistence of the idea. Apparently it was one to reckon with.

He went up the long walk to the stuffy, overdecorated entrance, found her bell, and rang it. Could Sally stop his idea?

As he often did, he tried to think of what would happen to the children. That was a puzzler.

Diego Rivera. *Festival of All Souls in the Country.*

A Sense of Tribe

SARELLEN M. WUEST

Kids these days don't know what courage is. They've got calluses all over their bellies from lolling in front of the T.V., wallowing in somebody else's derring-do; but when it comes down to it they lack the personal guts to come straight home with a lousy report card.

Everybody is out of the house this afternoon except me and Miss Fumblemuch, and she's taking her relief. I'd rather be alone, I told the folks; but who listens to an old man? Then before I can put in my plates and give it, they all light out like spit off a cookstove. Ten to one the kids wind up ruining their eyesight in a three-D movie while life and death in their own family is being played in the flesh out at Greenwood.

I was just a little shaver of eight or nine the year my Grandma died. In those days when the old began to go, everybody who wasn't already living near tracked back to the farm; and when the big girls came down from Normal and the gifted brother even left his elocution coach, keep and tuition paid, we knew for sure she was a goner.

We little ones handled it like Christmas—up till all hours, piecing on sweets the whole day long, and backsliding on thumb-sucking and pants-mussing. We steered clear of the downstairs back bedroom where Grandma was doing her passing. All us kids and cousins believed, from having seen it happen three times in a row, that you never came out alive if you slept in that room. We thought the vines on that side died with the person within the farmhouse wall; and shooed away from the drawn blinds, ignored and neglected, we counted from a distance the final falling leaves. When the doctor's rig stood in front we moved out there, gawking like a flock of pullets with the gapes; and the longer he stayed inside, the worse we knew she was, it was exciting, and it would be a lie to say, we didn't relish it.

As a grandma, a baker of bread, and a giver of presents, she was gone for us already. Somebody else would sew on fresh sunbonnet strings next spring and lengthen our elastics when we were forced into long black stockings in the fall. This one never again would clamp us between her calico knees on the low rocker and explore our scalps for nits, or blere out in front of everybody to ask if our bowels had moved good. Sunken of cheek, out of her head, her best teeth in a glass of water, her hair in lank braid, she fought for life under a mountain of quilts in that darkened chamber of doom. A fine sheet of sermon paper lay ready if she should show the wish to write. Her keepsakes were locked safe from meddlers in their seashell casket. Puking her gruel, blind to the handsome orange a rich friend had sent, her work on earth was done.

She had borne fourteen and raised eleven, and the three little ones she had lost grew rosy-real in our minds as her life drew to a close. We saw them fondly beckoning her across a blue lagoon, holding out their baby hands like pink-faced angels in a seven-decker valentine, still wearing the little white gown in which they had died forty-some years before.

A dozen kinds of medicine reeked on the taboret, and in those days you knew what you had when you shook up the bottle—arnica, aconite, liniment, seena, and a compound for the bloat; not ten dollars for a tablespoonful of yellow swill you keep in the refrigerator and pay the doctor a day's wages for five minutes of that dead-pan look it takes him eight years to learn.

Fever or chills, coma or convulsions, the longer it lasted, the richer the pickins for us younguns. We counted fence pickets and boards in the barn to predict by odd or even whether she would pass in the night or in the daytime. Death to us was a cross between a bad Santa and a good witch, a vision that flew like a spear with a skull of charcoal, arms akimbo, pipestem legs in a scissors kick, and a long billowing costume of dirty curtains. For this apparition we combed the sky.

Upstairs in the best front bedroom the older girls lounged in their petticoats and corset covers, heating their curling irons over the lamp chimneys, rolling rose beads, copying coke receipts, and tracing curtain stencils, giggling and gabbing. Under the general gloom and sorrow that hung over us all, their whispering went on and on, trading T L's, love doting on their kissing cousins, reciting nasty memory gems, and making mink bait of the hired girls' reputations.

How could a lallygagging hussy who dried her drawers over the kitchen range get into the family way when the seeder crew came through in the spring and yet dance the two-step at the first fall husking be as flat as the pancakes she served bright and early next morning? A long-handled buttonhook was good for something besides the white kid tops to your french-heeled boots. And how come that crosseyed sass-box slouch to wear a ruby ring, and the boy's farm paid for, too? All cows are black at night.

One full-bodied beauty, the village pistol, I'll never forget. Slathered across her face, as if somebody had clouted her with a slice of salami, was a red-pebbled birthmark. She slaved for her betters with her pitiful disfigurement forever turned away until it gave you a wry neck yourself to be around her very long. But everybody knew she would do it under the bandstand, middle of town, broad daylight outside; and it was a matter of record she named her first little baby "Folly." That tiny mite was born with a sense of rhythm and grace of body that sent her dancing through great cities and across oceans to be the pet of kings and millionaires, and we who knew her "when" had a pretty good notion of where she got all that *oomph-ah!* The old bandstand burned down, but Folly, I hear, never cooled off.

There was a lot of vitality and enterprise in those hired girls. They were women of loins, and many a young buck might better have taken one of them to wife than the sisters' chums they chose among instead. But the wives and daughters of the better-off fought every way they knew to keep them from getting good husbands.

These wars, though, were not for children. Let us meechin' little snot-noses still in Ferris waists and draw-strings try to pick up what we could through the key holes, and we got slapped smartly downstairs in jig time by hands fragrant with Florida water.

Denied the secrets of the maidens

alone we tackled the food. Covered with fly-cloths, the rich desserts stood welcome all day long around the castor in the center of the dining table. For reasons beyond us, we were forbidden to pop corn or pull taffy, but we gloried in the lavish spread of cobbler and dip, fried cakes, Paris bread, and pickled watermelon, even store figs and dates. We kids ate high off the hog, and even the hog was not to be pitied. She was getting whole milk, the women folks being too busy to separate the cream.

The kitchen, the summer kitchen, the buttery and the pantry buzzed with their coming and going—washing and drying, shaking, brushing, boiling, baking, roasting and frying. Thick strong coffee, towards evening a regular stump-lifter, simmered all day on the back of the stove. The old maid daughter, a glaring misfit upstairs, now acting queen of the hive, was keeping off her feet in the kitchen rocker drawn close to the bedroom door, crocheting silencers for chamber-pot lids to quiet her nerves, and telling outsiders where things were.

"The big crock in the cool celler right behind the black-strap" or "In them lard tubs right under where the dill is up to dry" might be out cut short by a sharp *"Hark!"* which stopped the swishing skirts, the clattering pans and splashing water, and made the tired bright faces turn toward the narrow darkened door. They set a sight of store on last words and deathbed wishes in those days, and nobody wanted to miss hearing them if there were any. The moaning had stopped or the light shrieking had gone back to a mumble or the awful singing prayer was starting all over again. Whichever it was, in the lull they noticed us, pesky as pigweed, and shushed us on out the back door with orders to stay away from the cistern and not tantalize the geese.

There was more going on in the yard anyway. Grandma's old pet rooster was tied to the foot-scraper with a pink corset lace—in case she should ask for him. The men stayed close to the house the last day; only the newest hired hand would go as far as the woods or the pasture. Squatting in threes or fours or leaning again the house, they whittled sharp sticks to scratch lightning rods good and stout on the haybarn or somesuch, not rigging a second antenna for a separate T.V. for some overprivileged brats. The grave was already dug and lined with green boughs, and their shovels were propped out of sight around the corner of the shed. More and more neighbors and kin and even renters gathered. The hungry Bible salesman was back again; nobody knew where he slept. A sack of plug cut was passed around, and as the waiting got out longer something besides that was passing around for those who had their eyes peeled for it.

One young wag from the next town had us kids busting our gussets by getting the pigeons drunk back of the corncrib. He soaked some barley in stagger juice and they hogged it up as fast as he set it down. They'd run in circles with one wing in the dirt, loop the loop if they could get off the ground at all, then just like grown foals who had smelled the cork, they'd shimmie back and waltz around the lard lid begging for more, but darn it,

some killjoy came along and made him show where he had found the apple-jack jug stashed in the pie plant, and the fun was over. Not even a relative, he was ordered hangdog out to make himself useful by liming down the privies. Not too harsh a fate for this frisky Katzenjammer, we all thought. He was known to be the one who swiped the last two cards out of the Nights in Cairo stereopticon views, carried them on a mitten string around his neck for a nickel or ten soap wrappers per look, and none of us little fellers ever did get to see the Harem at the Bath.

Our jester liquidated, we sprouts were set at forced labor with flatirons over a peck of hickory nuts. A couple of hellers from the county seat were singing "Bring Me Nellie with the Novel Navel," a tearyanker ballad about a traveling man who came back to visit an old favorite only to find that she had died in harness. Anything about a horse in trouble always cut my quick, but the others said that wasn't it—to quit my blubbering so they could catch all the verses.

And so we were there when the dread moment came. Nobody remembered to search the sky for the dark flyer. Grandma's old mouser had a fit at the very moment of her passing and streaked off screeching and foaming to stiffen alone in the gooseberry briers. That's all there was to mark the point in time. We children looked only at the sad faces of our mothers, peering for some signal of what we were expected to feel. Those who had been cheerful, calm, and brisk turning out two days' work in one—the towers of strength—now went all to pieces.

And the ones who had sniveled worthlessly on the sidelines, failing in the little chores they did try to do, now turned into the bosses, directing the new rush of people and gifts, ordering the disposal of the bedding, the choosing of the laying-out clothes and ornaments, the very position of the saw-horses in the parlor and which sheets to use to drape them.

We children shuddered in fascinated ignorance around the burning which took place far down the yard in the afternoon, the bulky bundles which a hired man stoked and prodded in the coals until nothing at all was left. Old bedpads or bird's-eye dressing, I expect, but on that dim chilly day a village smart Alec of twelve whispered to us it was insides.

Grandma looked beautiful in peace, everybody said, in her purple Henrietta with ecru jabot and leg-of-mutton sleeves. Like iris in a florist box, it seemed to me. They had cleaned her hair with orris root and clasped her hands with the wrong thumb on top. I looked at the parched gray face, the wilted flesh on the diminished frame, the fancy dress tweaked up jauntily over the still body; and first grief welled up in me. No lap, I thought, never more no lap. I didn't take my second turn in line and went whimpering from the room with my first realization of death. Somehow I never was just a boy after that—no longer an island, as John Donne said, but from then on a part of the "maine."

The fellow they picked to hold the dogs back at the graveyard was simple. He was a fine hand with setting hens and would stir soft soap until he fell into the vat if somebody didn't tell him when to quit. But his wits didn't hold up under this job. He couldn't distinguish between a burying and a chivance, and he got confused and let them loose too soon. So the end was we each had to hold our own pup in the spring wagon, sort of half sitting on them to keep down the commotion, a captive audience if ever there was one, and we missed the part they had to do over.

And so, should the children be at the funeral? I say yes, and who has a better right, because mine is next. I've caught the young folks in three lies not counting evasions. All this is meadow-dressing about taking a rest and then calling in a new doctor. I wish they'd quit trying to spare me. I've got it comfortably worked out in my own mind and I hope I can slide away before I begin picking lint.

Fumble-much has "terminal case" written all over her face. She'll be back here in a minute, all relieved, with some brain-buster like, "How'd my boy-dolly like his nice back rub now?" She's smoked a couple of king size on the throne in the master bath, powdered her pussy with my daughter-in-law's prize stuff, and hitched up her falsies until coming at me against the light she looks like a harp.

Harp is a charged word around here now, but I know I'm going and I aim to be sweet about it. It's bad taste and poor timing to live too long. But lying here thinking, looking back and ahead, contriving my resignation, I'll have to admit to a recent hankering for the final flourish. Funerals are so different now; *services* no less! A chapel as pretty as the parlor of a swell hotel, mourners dressed up in party colors, statues smiling and canary birds singing, and electric organ piping you to the pearly gates, a nice ride in a big fat Cadillac, and all paid for later. Even tears are out of style.

Funerals are different, but children are not. How are they going to get a sense of *Tribe* if they're shielded from the primitives of life? Come my time, soon now, I want all the kids there. See that you don't go and deprive them of it.

For My Lover, Returning to His Wife

ANNE SEXTON

She is all there.
She was melted carefully down for you
and cast up from your childhood,
cast up from your one hundred favorite aggies.

She has always been there, my darling.
She is, in fact, exquisite.
Fireworks in the dull middle of February
and as real as a cast-iron pot.

Let's face it, I have been momentary.
A luxury. A bright red sloop in the harbor.
My hair rising like smoke from the car window.
Littleneck clams out of season.

She is more than that. She is your have to have,
has grown you your practical your tropical growth.
This is not an experiment. She is all harmony.
She sees to oars and oarlocks for the dinghy,

has placed wild flowers at the window at breakfast,
sat by the potter's wheel at midday,
set forth three children under the moon,
three cherubs drawn by Michelangelo,

done this with her legs spread out
in the terrible months in the chapel.
If you glance up, the children are there
like delicate balloons resting on the ceiling.

She has also carried each one down the hall
after supper, their heads privately bent,
two legs protesting, person to person,
her face flushed with a song and their little sleep.

I give you back your heart.
I give you permission—

for the fuse inside her, throbbing
angrily in the dirt, for the bitch in her
and the burying of her wound—
for the burying of her small red wound alive—

for the pale flickering flare under her ribs,
for the drunken sailor who waits in her left pulse,
for the mother's knee, for the stockings,
for the garter belt, for the call—

the curious call
when you will burrow in arms and breasts
and tug at the orange ribbon in her hair
and answer the call, the curious call.

She is so naked and singular.
She is the sum of yourself and your dream.
Climb her like a monument, step after step.
She is solid.

As for me, I am a watercolor.
I wash off.

Yonder Peasant, Who Is He?

MARY McCARTHY

Whenever we children came to stay at my grandmother's house, we were put to sleep in the sewing room, a bleak, shabby, utilitarian rectangle, more office than bedroom, more attic than office, that played to the hierarchy of chambers the role of a poor relation. It was a room seldom entered by the other members of the family, seldom swept by the maid, a room without pride; the old sewing machine, some cast-off chairs, a shadeless lamp, rolls of wrapping paper, piles of cardboard boxes that might someday come in handy, papers of pins, and remnants of material united with the iron folding cots put out for our use and the bare floor boards to give an impression of intense and ruthless temporality. Thin white spreads, of the kind used in hospitals and charity institutions, and naked blinds at the windows reminded us of our or-phaned condition and of the ephem-eral character of our visit; there was nothing here to encourage us to con-sider this our home.

Poor Roy's children, as commis-eration damply styled the four of us, could not afford illusions, in the family opinion. Our father had put us beyond the pale by dying sud-denly of influenza and taking our young mother with him, a defection that was remarked on with horror and grief commingled, as though our mother had been a pretty secre-tary with whom he had wantonly absconded into the irresponsible paradise of the hereafter. Our repu-tation was clouded by this misfor-tune. There was a prevailing sense, not only in the family but in store-keepers, servants, streetcar conduc-tors, and other satellites of our circle, that my grandfather, a rich man, had behaved with extraordinary

munifiicence in alloting a sum of
money for our support and installing
us with some disagreeable middle-
aged relations in a dingy house two
blocks distant from his own. What
alternative he had was not mentioned; presumably he could have
sent us to an orphan asylum and no
one would have thought the worse
of him. At any rate, it was felt, even
by those who sympathized with us,
that we led a privileged existence,
privileged because we had no rights,
and the very fact that at the yearly
Halloween or Christmas party given
at the home of an uncle we appeared
so dismal, ill clad, and unhealthy,
in contrast to our rosy, exquisite
cousins, confirmed the judgment that
had been made on us—clearly, it was
a generous impulse that kept us
in the family at all. Thus, the
meaner   our   circumstances,   the
greater seemed our grandfather's
condescension, a view in which we
ourselves shared, looking softly and
shyly on this old man—with his
rheumatism, his pink face and white
hair, set off by the rosebuds in his
Pierce-Arrow and in his buttonhole
—as the font of goodness and philan-
thropy, and the nickel he occasionally gave us to drop into the collection plate on Sunday (two cents was
our ordinary contribution) filled us
not with envy but with simple admiration for his potency; this indeed
was princely, *this* was the way to
give. It did not occur to us to judge
him for the disparity of our styles
of living. Whatever bitterness we
felt was kept for our actual guardians, who we believed, must be em-
bezzling the money set aside for
us since the standard of comfort
achieved in our grandparents' house
—the electric heaters, the gas logs,

the lap robes, the shawls wrapped
tenderly about the old knees, the
white meat of chicken and red meat
of beef, the silver, the white table-
cloths, the maids, and the solicitous
chauffeur—persuaded us that prunes
and rice pudding, peeling paint and
patched clothes were *hors concours*
with these persons and therefore
could not have been willed by them.
Wealth, in our minds, was equivalent to bounty, and poverty but a
sign of penuriousness of spirit.

Yet even if we had been convinced
of the honesty of our guardians, we
would still have clung to that bene-
ficent image of our grandfather that
the family myth proposed to us. We
were too poor, spiritually speaking,
to question his generosity, to ask
why he allowed us to live in oppressed chill and deprivation at a
long arm's length from himself and
hooded his genial blue eye with a
bluff, millionairish gray eyebrow
whenever the evidence of our suffering presented itself at his knee. The
official answer we knew: our benefactors were too old to put up with
four wild young children; our grandfather was preoccupied with business matters and with his rheumatism, to which he devoted himself
as though to a pious duty, taking it
with him on pilgrimages to Ste.
Anne de Beaupré and Miami, offering it with impartial reverence to
the miracle of the Northern Mother
and the Southern sun. This rheumatism hallowed my grandfather
with the mark of a special vocation;
he lived with it in the manner of an
artist or a grizzled Galahad; it set
him apart from all of us and even
from my grandmother, who, lacking
such an affliction, led a relatively un-
justified existence and showed, in re-

lation to us children, a sharper and more bellicose spirit. She felt, in spite of everything, that she was open to criticism, and, transposing this feeling with a practiced old hand, kept peering into our characters for symptoms of ingratitude.

We, as a matter of fact, were grateful to the point of servility. We made no demands, we had no hopes. We were content if we were permitted to enjoy the refracted rays of that solar prosperity and come sometimes in the summer afternoons to sit on the shady porch or idle through a winter morning on the wicker furniture of the sun parlor, to stare at the player piano in the music room and smell the odor of whiskey in the mahogany cabinet in the library, or to climb about the dark living room examining the glassed-in paintings in their huge gilt frames, the fruits of European travel: dusky Italian devotional groupings, heavy and lustrous as grapes, Neapolitan women carrying baskets to market, views of Venetian canals, and Tuscan harvest scenes— secular themes that, to the Irish-American mind, had become tinged with Catholic feeling by a regional infusion from the Pope. We asked no more from this house than the pride of being connected with it, and this was fortunate for us, since my grandmother, a great adherent of the give-them-an-inch-and-they'll-take-a-yard theory of hospitality, never, so far as I can remember, offered any caller the slightest refreshment, regarding her own conversation as sufficiently wholesome and sustaining. An ugly, severe old woman with a monstrous balcony of a bosom, she officiated over certain set topics in a colorless singsong,

like a priest intoning a Mass, topics to which repetition had lent a senseless solemnity: her audience with the Holy Father; how my own father had broken with family tradition and voted the Democratic ticket; a visit to Lourdes; the Sacred Stairs in Rome, bloodstained since the first Good Friday, which she had climbed on her knees; my crooked little fingers and how they meant I was a liar; a miracle-working bone; the importance of regular bowel movements; the wickedness of Protestants; the conversion of my mother to Catholicism; and the assertion that my Protestant grandmother must certainly dye her hair. The most trivial reminiscences (my aunt's having hysterics in a haystack) received from her delivery and from the piety of the context a strongly monitory flavor; they inspired fear and guilt, and one searched uncomfortably for the moral in them, as in a dark and riddling fable.

Luckily, I am writing a memoir and not a work of fiction, and therefore I do not have to account for my grandmother's unpleasing character and look for the Oedipal fixation or the traumatic experience which would give her that clinical authenticity that is nowadays so desirable in portraiture. I do not know how my grandmother got the way she was; I assume, from family photographs and from the inflexibility of her habits, that she was always the same, and it seems as idle to inquire into her childhood as to ask what was ailing Iago or look for the thumb-sucking prohibition that was responsible for Lady Macbeth. My grandmother's sexual history, bristling with infant mortality in the usual style of her period, was robust and decisive: three tall, handsome sons grew up, and one attentive daughter. Her husband treated her kindly. She had money, many grandchildren, and religion to sustain her. White hair, glasses, soft skin, wrinkles, needlework—all the paraphernalia of motherliness were hers; yet it was a cold, grudging, disputatious old woman who sat all day in her sunroom making tapestries from a pattern, scanning religious periodicals, and setting her iron jaw against any infraction of her ways.

Combativeness was, I suppose, the dominant trait in my grandmother's nature. An aggressive churchgoer, she was quite without Christian feeling; the mercy of the Lord Jesus had never entered her heart. Her piety was an act of war against the Protestant ascendancy. The religious magazines on her table furnished her not with food for meditation but with fresh pretexts for anger; articles attacking birth control, divorce, mixed marriages, Darwin, and secular education were her favorite reading. The teachings of the Church did not interest her, except as they were a rebuke to others; "Honor thy father and thy mother," a commandment she was no longer called upon to practice, was the one most frequently on her lips. The extermination of Protestantism, rather than spiritual perfection, was the boon she prayed for. Her mind was preoccupied with conversion, the capture of a soul for God much diverted her fancy—it made one less Protestant in the world. Foreign missions, with their overtones of good will and social service, appealed to her less strongly; it was not a *harvest* of souls that my grandmother had in mind.

This pugnacity of my grandmother's did not confine itself to sectarian enthusiasm. There was the defense of her furniture and her house against the imagined encroachments of visitors. With her, this was not the gentle and tremulous protectiveness endemic in old ladies, who fear for the safety of their possessions with a truly touching anxiety, inferring the fragility of all things from the brittleness of their old bones and hearing the crash of mortality in the perilous tinkling of a teacup. My grandmother's sentiment was more autocratic: she hated having her chairs sat in or her lawns stepped on or the water turned on in her basins, for no reason at all except pure officiousness; she even grudged the mailman his daily promenade up her sidewalk. Her home was a center of power, and she would not allow it to be derogated by easy or democratic usage. Under her jealous eye, its social properties had atrophied, and it functioned in the family structure simply as a political headquarters. Family conferences were held there, consultations with the doctor and the clergy; refractory children were brought there for a lecture or an interval of thought-taking; wills were read and loans negotiated and emissaries from the Protestant faction on state occasions received. The family had no friends, and entertaining was held to be a foolish and unnecessary courtesy as between blood relations.

Yet on one terrible occasion my grandmother had kept open house. She had accommodated us all during those fatal weeks of the influenza epidemic, when no hospital beds were to be had and people went about with masks or stayed shut up in their houses, and the awful fear of contagion paralyzed all services and made each man an enemy to his neighbor. One by one, we had been carried off the train—four children and two adults, coming from distant Puget Sound to make a new home in Minneapolis. Waving goodbye in the Seattle depot, we had not known that we had brought the flu with us into our drawing rooms, along with the presents and the flowers, but, one after another, we had been struck down as the train proceeded eastward. We children did not understand whether the chattering of our teeth and Mama's lying torpid in the berth were not somehow a part of the trip (until then serious illness, in our minds, had been associated with innovations—it had always brought home a new baby), and we began to suspect that it was all an adventure when we saw our father draw a revolver on the conductor who, in a burst of sanitary precaution, was trying to put us off the train at a small wooden station in the middle of the North Dakota prairie. On the platform at Minneapolis, there were stretchers, a wheelchair, redcaps, distraught officials, and, beyond them, in the crowd, my grandfather's rosy face, cigar, and cane, my grandmother's feathered hat, imparting an air of festivity to this strange and confused picture, making us children certain that our illness was the beginning of a delightful holiday.

We awoke to reality in the sewing room several weeks later, to an atmosphere of castor oil, rectal thermometers, cross nurses, and efficiency, and though we were shut out from the knowledge of what had happened so close to us, just out of our hearing—a scandal of the gravest character, a coming and going of priests and undertakers and coffins (Mama and Daddy, they assured us, had gone to get well in the hospital)—we became aware, even as we woke from our fevers, that everything, including ourselves, was different. We had shrunk, as it were, and faded, like the flannel pajamas we wore, which during these few weeks had grown, doubtless from the disinfectant they were washed in, wretchedly thin and shabby. The behavior of the people around us, abrupt, careless, and preoccupied, apprised us without any ceremony of our diminished importance. Our value had paled, and a new image of ourselves—the image, if we had guessed it, of the orphan—was already forming in our minds. We had not known we were spoiled, but now this word, entering our vocabulary for the first time, served to define the change for us and to herald the new order. Before we got sick, we were spoiled; that was what was the matter now, and everything we could not understand, everything unfamiliar and displeasing took on a certain plausibility when related to this fresh concept. We had not known what it was to have trays dumped summarily on our beds and no sugar and cream for our cereal, to take medicine in a gulp because someone could not be bothered to wait for us, to have our arms jerked into our sleeves and a comb ripped through our hair, to be bathed impatiently, to be told to sit up or lie down quick and no nonsense about it, to find our questions unanswered and our requests unheeded, to lie for hours alone and wait for the doctor's visit, but this, so it seemed, was an oversight in our training, and my grandmother and her household applied themselves with a will to remedying the deficiency.

Their motives were, no doubt, good; it was time indeed that we learned that the world was no longer our oyster. The happy life we had had—the May baskets and the valentines, the picnics in the yard, and the elaborate snowmen—was a poor preparation, in truth, for the future that now opened up to us. Our new instructors could hardly be blamed for a certain impatience with our parents, who had been so lacking in foresight. It was to everyone's interest, decidedly, that we should forget the past—the quicker, the better—and a steady disparagement of our habits ("Tea and chocolate, can you imagine, and all those frosted cakes—no wonder poor Tess was always after the doctor"), praise that was rigorously comparative ("You have absolutely no idea of the improvement in these children") flattered the feelings of the speakers and prepared us to accept a loss that was, in any case, irreparable. Like all children, we wished to conform, and the notion that our former ways had been somehow ridiculous and unsuitable made the memory of them falter a little, like a child's recitation to strangers. We no longer demanded our due, and the wish to see our parents insensibly weakened. Soon we ceased to speak of it, and thus, without tears or tantrums, we came to know they were dead.

Why no one, least of all our grandmother, to whose repertory the subject seems so congenial, took the trouble to tell us, it is impossible now to know. It is easy to imagine

her "breaking" the news to those of us who were old enough to listen in one of those official interviews in which her nature periodically tumefied, becoming heavy and turgid, like her portentous bosom, like peonies, her favorite flower, or like the dressmaker's dummy, that bombastic image of herself that lent a museum-like solemnity to the humble sewing room and made us tremble in our beds. The mind's ear frames her sentences, but in reality she did not speak, whether from a clumsy sense of delicacy or from a mistaken kindness, it is difficult to guess. Perhaps she feared our tears, which might rain on her like reproaches, since the family policy at the time was predicated on the axiom of our virtual insentience, an assumption that allowed them to proceed with us as if with pieces of furniture. Without explanations or coddling, as soon as they could safely get up, my three brothers were dispatched to the other house; they were much too young to "feel" it, I heard the grownups murmur, and would never know the difference "if Myers and Margaret were careful." In my case, however, a doubt must have been experienced. I was six—old enough to "remember"—and this entitled me, in the family's eyes, to greater consideration, as if this memory of mine were a lawyer who represented me in court. In deference, therefore, to my age and my supposed powers of criticism and comparison, I was kept on for a time, to roam palely about my grandmother's living rooms, a dangling, transitional creature, a frog becoming a tadpole, while my brothers, poor little polyps, were already well embedded in the structure of the new life. I did not wonder what had become of them. I believe I thought they were dead, but their fate did not greatly concern me; my heart had grown numb. I considered myself clever to have guessed the truth about my parents, like a child who proudly discovers that there is no Santa Claus, but I would not speak of that knowledge or even react to it privately, for I wished to have nothing to do with it; I would not cooperate in this loss. Those weeks in my grandmother's house come back to me very obscurely, surrounded by blackness, like a mourning card: the dark well of the staircase, where I seem to have been endlessly loitering, waiting to see Mama when she would come home from the hospital, and then simply loitering with no purpose whatever; the winter-dim first-grade classroom of the strange academy I was sent to; the drab treatment room of the doctor's office, where every Saturday I screamed and begged on a table while electric shocks were sent through me, for what purpose I cannot conjecture. But this preferential treatment could not be accorded me forever; it was time that I found my niche. "There is someone here to see you"—the maid met me one afternoon with this announcement and a smile of superior knowledge. My heart bounded; I felt almost sick (who else, after all, could it be?), and she had to push me forward. But the man and woman surveying me in the sun parlor with my grandmother were strangers, two unprepossessing middle-aged people—a great-aunt and her husband, so it seemed—to whom I was now commanded to give a hand and a smile, for, as my grandmother remarked, Myers and Margaret had come to

take me home that very afternoon to live with them, and I must not make a bad impression.

Once the new household was running, our parents' death was officially conceded and sentiment given its due. Concrete references to the lost ones, to their beauty, gaiety, and good manners, were naturally not welcomed by our guardians, who possessed none of these qualities themselves, but the veneration of our parents' *memory* was considered an admirable exercise. Our evening prayers were lengthened to include one for our parents' souls, and we were thought to make a pretty picture, all four of us in our pajamas with feet in them, kneeling in a neat line, our hands clasped before us, reciting the prayer for the dead. "Eternal rest grant unto them, O Lord, and let the perpetual light shine upon them," our thin little voices cried, but this remembrancing, so pleasurable to our guardians, was only a chore to us. We connected it with lights out, washing, all the bedtime coercions, and particularly with the adhesive tape that, to prevent mouth-breathing, was clapped upon our lips the moment the prayer was finished, sealing us up for the night, and that was removed, very painfully, with the help of ether, in the morning. It embarrassed us to be reminded of our parents by these persons who had superseded them and who seemed to evoke their wraiths in an almost proprietary manner, as though death, the great leveller, had brought them within their province. In the same spirit, we were taken to the cemetery to view our parents' graves; this, in fact, being free of charge, was a regular Sunday pastime with us, which we grew to hate as we did all recreation enforced by our guardians—department-store demonstrations, band concerts, parades, trips to the Old Soldiers' Home, to the Botanical Gardens, to Minnehaha Park, where we watched other children ride on the ponies, to the Zoo, to the water tower—diversions that cost nothing, involved long streetcar trips or endless walking or waiting, and that had the peculiarly fatigued, dusty, proletarianized character of American municipal entertainment. The two mounds that now were our parents associated themselves in our minds with Civil War cannon balls and monuments to the doughboy dead; we contemplated them stolidly, waiting for a sensation, but these twin grass beds, with their junior-executive headstones, elicited nothing whatever; tired of this interminable staring, we would beg to be allowed to go play in some collateral mausoleum, where the dead at least were buried in drawers and offered some stimulus to fancy.

For my grandmother, the recollection of the dead became a mode of civility that she thought proper to exercise toward us whenever, for any reason, one of us came to stay at her house. The reason was almost always the same. We (that is, my brother Kevin or I) had run away from home. Independently of each other, this oldest of my brothers and I had evolved an identical project. The purpose dearest to our hearts was to get ourselves placed in an orphan asylum, for we interpreted the word "asylum" in the old Greek sense and looked upon a certain red brick building, seen once from a streetcar near the Mississippi River, as a sanctuary for the helpless and a refuge from persecution. So, from time to time, when our lives became too painful, one of us would set forth, determined to find the red brick building and to press what we imagined was our legal claim to its shelter and protection. But sometimes we lost our way, and sometimes our courage, and after spending a day hanging about the streets peering into strange yards, trying to assess the kindheartedness of the owner (for we also thought of adoption), or after a cold night spent hiding in a church confessional box or behind some statuary in the Art Institute, we would be brought by the police, by some well-meaning householder, or simply by fear and hunger, to my grandmother's door. There we would be silently received, and a family conclave would be summoned. We would be put to sleep in the sewing room for a night, or sometimes more, until our feelings had subsided and we could be sent back, grateful, at any rate, for the promise that no reprisals would be taken and that the life we had run away from would go on "as if nothing had happened."

Since we were usually running away to escape some anticipated punishment, these flights at least gained us something, but in spite of the taunts of our guardians, who congratulated us bitterly on our "cleverness," we ourselves could not feel that we came home in triumph so long as we came home at all. Our failure to run away successfully put us, so we thought, at the absolute mercy of our guardians; our last weapon was gone, for it was plain to be seen that they could always bring us back, however far we travelled, or that we would bring ourselves back, too soft to stand cold and hunger, too cowardly to steal or run away from a policeman; we never understood why they did not take advantage of this situation to thrash us, as they used to put it, within an inch of our lives. What intervened to save us, we could not guess—a miracle, perhaps; we were not acquainted with any *human* motive that would prompt Omnipotence to desist. We did not suspect that these escapades brought consternation to the family circle, which had acted, so it conceived, only in our best interests, and now saw itself in danger of unmerited obloquy. What would be the Protestant reaction if something still more dreadful were to happen? Child suicides were not unknown, and quiet, asthmatic little Kevin had been caught with matches under the house. The family would not acknowledge error, but it conceded a certain mismanagement on Myers' and Margaret's part. Clearly, we might become altogether intractable if our homecoming on these occasions were not mitigated with leniency. Consequently, my grandmother kept us in a kind of neutral detention. She declined to be aware of our grievance and offered no words of comfort, but the comforts of her household acted upon us soothingly, like an automatic mother's hand. We ate and drank contentedly; with all her harsh views, my grandmother was a practical woman and would not have thought it worth while to unsettle her whole schedule, teach her cook to make a lumpy mush and watery boiled potatoes, and market for turnips and parsnips and all the other vegetables we hated, in order to approximate the conditions she considered suitable for our charac-

ters. Humble pie could be costly, especially when cooked to order.

Doubtless she did not guess how delightful these visits seemed to us once the fear of punishment had abated. Her knowledge of our own way of living was luxuriously remote. She did not visit our ménage or inquire into its practices, and though hypersensitive to a squint or a dental irregularity (for she was liberal indeed with glasses and braces for the teeth, disfiguring appliances that remained the sole token of our bourgeois origin and set us off from our parochial-school mates like the caste marks of some primitive tribe), she appeared not to notice the darns and patches of our clothing, our raw hands and scarecrow arms, our silence and our elderly faces. She imagined us as surrounded by certain playthings she had once bestowed on us—a sandbox, a wooden swing, a wagon, and a toy fire engine. In my grandmother's consciousness, these objects remained always in pristine condition; years after the sand had spilled out of it and the roof had rotted away, she continued to ask tenderly after our lovely sand pile and to manifest displeasure if we declined to join in its praises. Like many egoistic people (I have noticed this trait in myself), she was capable of making a handsome outlay, but the act affected her so powerfully that her generosity was still lively in her memory when its practical effects had long vanished. In the case of a brown beaver hat, which she watched me wear for four years, she was clearly blinded to its matted nap, its shapeless brim, and ragged ribbon by the vision of the price tag it had worn when new. Yet, however her mind embroidered the bare tapestry of our lives, she could not fail to perceive that we felt, during these short stays with her, *some* difference between the two establishments, and to take our wonder and pleasure as a compliment to herself.

She smiled on us quite kindly when we exclaimed over the food and the nice, warm bathrooms, with their rugs and electric heaters. What funny little creatures, to be so impressed by things that were, after all, only the ordinary amenities of life! Seeing us content in her house, her emulative spirit warmed slowly to our admiration; she compared herself to our guardians, and though for expedient reasons she could not afford to depreciate them ("You children have been very ungrateful for all Myers and Margaret have done for you"), a sense of her own finer magnanimity disposed her subtly in our favor. In the flush of these emotions, a tenderness sprang up between us. She seemed half reluctant to part with whichever of us she had in her custody, almost as if she were experiencing a genuine pang of conscience. "Try and be good," she would advise us when the moment for leave-taking came, "and don't provoke your aunt and uncle. We might have made different arrangements if there had been only one of you to consider." These manifestations of concern, these tacit admissions of our true situation, did not make us, as one might have thought, bitter against our grandparents, for whom ignorance of the facts might have served as a justification, but, on the contrary, filled us with love for them and even a kind of sympathy—our sufferings were less terrible if someone acknowledged their existence, if someone were suffering for us, for whom we, in our turn, could suffer, and thereby absolve of guilt.

During these respites, the recollection of our parents formed a bond between us and our grandmother that deepened our mutual regard. Unlike our guardians or the whispering ladies who sometimes came to call on us, inspired, it seemed, by a pornographic curiosity as to the exact details of our feelings ("Do you suppose they remember their parents?" "Do they ever say anything?"), our grandmother was quite uninterested in arousing an emotion of grief in us. "She doesn't feel it at all," I used to hear her confide to visitors, but contentedly, without censure, as if I had been a spayed cat that, in her superior foresight, she had had "attended to." For my grandmother, the death of my parents had become, in retrospect, an eventful occasion upon which she looked back with pleasure and a certain self-satisfaction. Whenever we stayed with her, we were allowed, as a special treat, to look into the rooms they had died in, for the fact that, as she phrased it, "they died in separate rooms" had for her a significance both romantic and somehow self-gratulatory, as though the separation in death of two who had loved each other in life were beautiful in itself and also reflected credit on the chatelaine of the house, who had been able to furnish two master bedrooms for the greater facility of decease. The housekeeping details of the tragedy, in fact, were to her of paramount interest. "I turned my house into a hospital," she used to say, particularly when visitors were present. "Nurses were as scarce as hen's teeth, and *high*—you can hardly imagine what those girls were charging an hour." The trays and the special cooking, the laundry and the disinfectants recalled themselves fondly to her thoughts, like items on the menu of some long-ago buffet supper, the memory of which recurred to her with a strong, possessive nostalgia.

My parents had, it seemed, by dying on her premises, become in a lively sense her property, and she dispensed them to us now, little by little, with a genuine sense of bounty, just as, later on, when I returned to her a grown-up young lady, she conceded me a diamond lavaliere of my mother's as if this trinket were an inheritance to which she had the prior claim. But her generosity with her memories appeared to us, as children, an act of the greatest indulgence. We begged her for more of these mortuary reminiscences as we might have begged for candy, and since ordinarily we not only had no candy but were permitted no friendships, no movies, and little reading beyond what our teachers prescribed for us, and were kept in quarantine, like carriers of social contagion, among the rhubarb plants of our neglected yard, these memories doled out by our grandmother became our secret treasures; we never spoke of them to each other but hoarded them, each against the rest, in the miserly fastnesses of our hearts. We returned, therefore, from our grandparents' house replenished in all our faculties; these crumbs from the rich man's table were a banquet indeed to us. We did not even mind going back to our guardians, for we now felt superior to them, and besides, as we well knew, we had no choice. It was only by accepting our situation as a just and unalterable arrangement that we could be allowed to transcend it and feel ourselves united to our grandparents in a love that was the more miraculous for breeding no practical results.

In this manner, our household was kept together, and my grandparents were spared the necessity of arriving at a fresh decision about it. Naturally, from time to time a new scandal would break out (for our guardians did not grow kinder in response to being run away from), yet we had come, at bottom, to despair of making any real change in our circumstances, and ran away hopelessly, merely to postpone punishment. And when, after five years, our Protestant grandfather, informed at last of the facts, intervened to save us, his indignation at the family surprised us nearly as much as his action. We thought it only natural that grandparents should know and do nothing, for did not God in the mansions of Heaven look down upon human suffering and allow it to take its course?

part **5**

Innovating...

An Attempt at Reform

AUGUST STRINDBERG

She had noticed with repugnance how girls were brought up to serve as housekeepers for prospective husbands. And so she had taken up a profession which could provide for her under no matter what conditions of life. She had learned how to make artificial flowers.

He, on the other hand, had been distressed to see how girls sat waiting to be taken care of by their husbands-to-be; for he was looking for a free independent woman to marry —a woman who could earn her own living, someone he could look upon as an equal and who would be a companion for life, and not a mere housekeeper.

And then the inevitable happened: they met. He was an artist—a painter, and she made artificial flowers, as mentioned before. And it was in Paris where they had been introduced to these new-fangled ideas.

There was style in their marriage. The two of them lived in three rooms in Passy. He had a room of his own at one end of the flat and she had hers at the other end; and the room in the middle served as their studio. This did away with the common bedroom and the double bed, that abomination which has no counterpart in nature and is responsible for so many excesses. It also did away with having to dress and undress together in the same room. Phew! No, it was far better for each to have a separate room—and then a neutral common meeting place, the studio. No maid, for they were both to share the cooking chores; they would only employ an old charwoman part time.

The whole thing was well figured out and it had logic in it.

—"But when you get children— what then?" demurred those who

371

had their doubts about such an arrangement.

—"We are not going to have any children!"

Fine! They were not to have children!

Everything went along beautifully! He went to the market in the morning and did the buying. Then he made the coffee. She swept, made the beds, and did the dusting and straightening up. After that they started to work.

When tired of working, they chatted for a while, made suggestions to each other, and had a thoroughly enjoyable time. And when dinnertime approached, he started a fire in the stove, and she washed and cleaned the vegetables. He kept an eye on the beef broth while she ran down to the grocer's; and after that she set the table, while he dished up the meal.

Yet they did not live like brother and sister. They bid goodnight to each other at night, and when that was done, they went to their respective rooms. But then there would come a knock at her door, and she would call out: "Come in!" The bed was narrow, however, and they each woke up in their own bed in the morning. And then there would come a knock on the wall.

—"Good morning, my little girl! How do you feel this morning?"

—"Oh, I feel fine! How do you feel?"

It was always a new experience when they met in the mornings; their encounter at breakfast never grew stale.

Evenings they would often go out together and meet with some of their countrymen at Syrach's. And she never had any objection to the smell of tobacco smoke, and she never in any way embarrassed him.

It was an ideal marriage, everybody thought, and they had never seen a couple that was so congenial and happy.

But the girl had parents, and they lived far away. And they kept writing, always inquiring whether Lisen would not soon be expecting; for they were yearning to have a grandchild. Lisen should remember, they said, that marriage was made for the sake of the children and not for the sake of the parents. But Lisen thought this an oldfashioned view. Retorting to this, Mamma asked whether it was the purpose of these newfangled ideas to root out mankind. This was a thought that had not occurred to Lisen, but she did not let it perturb her. She was happy, and so was her husband; and the world was envious because it had at long last seen a happy marriage.

And they certainly showed their happiness. Neither one lorded it over the other one; and expenses were paid out of their common purse. One time he earned more, another time, she; and that equalized matters.

And when they celebrated a birthday—then their old charwoman would come in and wake her with a bouquet of flowers and a little note with flowers painted on it, and the note would read: "Congratulations to Madame Flowerbud from her dabble-daub husband, who wishes her many happy returns of the day and requests her presence at a light but luscious breakfast in his room— right now!" And with this there came a knock on the wall, and she hastily put on her morning wrapper, knocked at her husband's door and was greeted with a warm "Come in!" And then they breakfasted sitting on his bed, and that day the old woman stayed with them the whole forenoon to do all the work. It was delightful!

And this kind of life went on for a couple of years, and they never tired of it. And all the prophesying of the prognosticators turned out to be wrong. For theirs was the ideal marriage.

But then it came about that the wife was suddenly taken sick. *She* thought the wallpaper had something to do with her illness, and *he* suspected it came from microbes. Yes—it must be caused by microbes!

But something was radically wrong, definitely wrong. She was not herself at all. They wondered whether it could be a cold. But then Mrs. Flowerbud began to grow decidedly stout. Could it be that she had developed a growth of the sort which they had read so much about lately. Yes, it must be a growth. And so she went to a physician. And when she came home, she broke into tears. Yes, it was indeed a growth—but the kind that eventually would seek daylight and blossom and, in time, bear fruit, that also.

The husband did not weep, however. He thought it was just fine; and the rascal even went about boasting of it whenever he visited

Syrach's. But his wife wept more tears. What would become of their position in relation to each other? From now on she would not be able to earn anything, and she would have to eat his bread. And then they would have to have a maid. Oh, those maids!

All their painstaking efforts, all their precautions and prudence and foresight had been wrecked by force of the inevitable.

But her mother-in-law wrote enthusiastic letters of congratulation, repeating over and over again that marriage was a divine institution, made for the purpose of having children and that the gratification of the parents was only of secondary importance.

Hugo, the husband, swore it would not matter in the least whether she earned anything or not. Wasn't the help she would give him by slaving for his child sufficient; and wasn't that worthy of payment also? She contributed her share. But for a long time the thought of being supported by her husband rankled in her. But when the little one arrived, she forgot all about that. And she remained his wife and companion as in the past, but more than that, she became his child's mother—and that, he thought, was the very best of all.

Jim/Ann/Victoria and Fran/Greg/Phil

GERHARD NEUBECK

Jim had closed the door and was leaving Ann behind in her house. They had talked and walked, gone to the movies and to bed, cleaned up the house and various bills, listened to music and themselves, and now he was leaving and was going into a new world in which Ann had no place for the time being. As good a time as they had had, he was now free of her and her ways. But it had been nice—notwithstanding the blowup Tuesday night over the bills—it had been nice. To Ann he could always be strong and let her rest and cuddle against him. And she had learned not to ask questions. Must be tough on her, but that's the way it is. And now he had closed the door.

Victoria, of course, expected him. In fact, she was out in the yard. The mail had just come and she waved at him with letters in her hand as he came out of the car. They hugged and she rubbed her hands over his arms and shoulders, and it felt to him as if she were working her fingers into the material, almost into the material of him, so strong was the impact. They went inside. The wonder of her, the way she looked at you. He felt that she read his mind the way she looked at him. She said, "You think we should talk first or get into bed first?" "Right," he said. "When I come to you I am always torn. I don't want to treat you just as a thing, but God, I'm turned on so. So strongly it's hard to keep my hands off you." He went over and while she muttered—"But you know I feel the same," they fell over each other. No, they never made it to the bed. He came on so strongly that it was over with before it had a chance to be consummated. He was so chagrined, but she only laughed and that helped. "Later," she said as she scrambled up. He came back out of the bathroom and she had prepared some food, sardines on toast. "Oh, she knows me well," he thought. "She, Vicky." And Ann flashed into his mind. "What is Ann doing now? On the phone for her boss no doubt." And then he obliterated Ann. Ann disappeared into a telephone wire on the horizon, then into a fluffy cloud that curled into the sky . . . away . . . gone.

The smell of sardines was overpowering. "I've been thinking," said Vicky, "that Norwegians have been exporting sardines for years. Are they in the Common Market?" The talk drifted to economics, the damage of a windstorm to the house, the population problem in India, and before long their feelings of affection had overpowered them again. Though this time they paced the tempo of their engines and it all came off well. "Was this love?" he thought. That word brought Ann back in and he knew then that he had said sentences with love in it to Ann not too long ago. Vicky had her back to him and as he contemplated that picture of her back he repeated to himself, "Love, love, love—but I feel what I feel and this is what I feel . . ." "Jim," Vicky brought him back. "Rub my back." And so he rubbed her back. She purred, a good sound making gooseflesh. Geese flying. He wrote big letters into the sky . . . V-I-C-T-O-R-I-A. "Jim," she said. "Jim."

They slept—they shopped, they ate dinner, they went to a concert, they went to bed, they made love again, and they slept.

Then it was Monday and he went off to the office. He called her at noon—the first day after he was back with Vicky he had trouble dialing her number—Ann's was too fresh in his mind—but Vicky's voice clear and direct said, "James, you'll be here tonight, do you know what that means?" "Yes, it means you are cooking dinner for two," he said. "Right," she said, "for two. I won't have to be alone. Do you know how lonely it is during the stretch when you are not here?" (How they avoided saying openly, "when you are with her." When they had done that in the beginning it had spoiled things so.) "Do you realize what it is to be by oneself? Of course, you don't," she said. "She probably wants to call me a bastard," he thought, "but she doesn't." "Vicky," he said, "no, I probably don't really know. Look now, don't let it get you down now. Five hours and I'll be there."

He brought her a candy bar, a ten cent bar, a Hershey. She tore the wrapper off impatiently—there it was again—Ann would have probably taken the candy out of the paper and thrown the wrapper into the wastepaper basket, offered him a bite and then taken one herself. All of a sudden he disliked Vicky, her directness and coarseness. She must have noticed something because she poked him and said. "Don't you like your little piggy?" He almost slapped her in the face then, but caught himself and patted her on the bottom instead, converting the anger into contact with her femaleness and it brought him back to reality. A reality that contained Vicky and him in this cozy kitchen with Monk playing a hard piano on the phonograph; liver, which he loved, smelling up the air; and Vicky after all saying to him, "Go, goose, read the mail. Dinner will be ready in fifteen minutes."

He went up to the john and there was a note pinned to his towel. The note read, "Towels are like skin—you remember?" Yes, he remembered how he had toweled her off after their first shower together and how she had revelled in the touch of the towel stroking her bottom and daring it for the first time—his cheek rubbing the same spot and her even greater delight at that. Towels. And then he saw rushing out of the darkness of his imagination Ann's bathroom. The dainty curtains and pink washcloths. He shook his head as if to brush that picture aside, but Ann seemed to come right out of her shower toward him and he hurriedly left the bathroom sniffing the air to be reminded of Vicky. And there she was in slacks and sweater, cigarette in her mouth, and a glass of sherry in her hand, looking at *Newsweek* magazine. She glanced at the kitchen timer. "Baked potatoes will be done in seven minutes. Give me a bit more sherry. Did you know Bernstein's retiring from the New York Philharmonic? He wants to compose full time. You like Jews?" And then they laughed over the old "Some of my best friends are. . . ." It never made any difference. It could have. They had talked that out at the time. The residue of some of those feelings. I suppose one never gets completely over them. It was good that Ann liked Jews. There it was again, Ann. Ann Lillegren, blonde and Scandinavian. Go away Ann, go away. He looked at Vicky and she blew smoke in his face. (She must know some-

times what was going on in his mind.) His concern must have shown. Vicky said, "Don't fret so, let's eat."

She brought up the point of loneliness after dinner. He had his head on her lap—they were on the couch. "When you called me and we talked about how hard these times without you are I didn't want to bug you, and really, I survive quite well, but you really ought to now what this place is like without you. A bore, really, and I get tempted to try the system for me, too. Why shouldn't I? I don't want to, that's why. Maybe in the future but not now. I suppose I could love others, but at this time in my life, I really want no one else. How does Ann feel?"

The surprise of Ann's name hanging there in the air was complete. He sat up and wanted to say, "You shouldn't have." The unbroken rule had been broken, but all he managed to say was, "Ann does have another. She can't stand to be alone."

So that was that. Silence now. Vicky and Jim both retreating from this delicate territory back to safety. His head back in her lap and she sliding her hand over his face and hair—the liver smell was too penetrating and they opened the windows widely. "Let's take a walk." "Yes, let's. It might rain so let's take raincoats." "Better close the windows." "Yes." They went for a walk.

Before going to sleep that night and Vicky asleep already, Jim was thinking, "So different they are—the way I wanted it—Ann, I think of her here—almost like something seeping in from the outside. Funny, seeping in, what a word, never gets in though, really. Vicky is there, all the way there, such a lot of Vicky. Vicky,

Victoria." He turned to her and curled around her and fell asleep.

FRAN

The car disappeared out of the driveway and Fran followed it all the way. "I have to go shopping," she thought, "I am out of practically everything. I better wait for Greg though. He likes to go to the market and tell me what to get. I am in for soups. God that man loves soups day and night. Night. I better do a bit of cleaning too. Phil sure doesn't care. Phil, good boy. Phil gone away for a month now. Go away from my window Phil," she hummed, "go away from my door."

She went upstairs and scrubbed the bathroom tiles, threw linen and towels in the hamper and brought out fresh things and folded the towels neatly. "That's Greg for you," she thought, "everything has to be in place. Crazy. Maybe Phil is too much of a slob, but what a comfortable good-natured slob." Her musing was interrupted by a bang on the door and then a loud voice. "Where are you honey?" And there was Greg, 6′ 2″, a big brute and she ran downstairs and into his bearhug. Greg had come in with two huge suitcases which were now standing in the middle of the hall. He viewed them, then swiftly took one in each hand and carried them upstairs to get them out of the way. Fran had cleared plenty of drawers for him and space in the closet. She knew he would be restless until he had put things away and changed from his business suit into khakies and polo shirt. So she went upstairs with him because she knew that's the way he

wanted it. "Run me a tub sweetie," he said and she knew what that meant too. "Sorry bud, you gotta wait another day Greg." "Oh, no." "Oh, yes," she said. He grabbed her and she wanted him too, but he would have to wait until the next day. So he took a tub and came down spic and span. "A well scrubbed guy," Fran thought. "It's nice to be seen with him. I like to be in his company, with people-wondering about us." And she wouldn't have to back the car out of the garage. Damn that driveway anyway. But Greg would handle that car like a toy. That big man had such a gentle touch. He slapped her on the rump. "Fran," he said, "I brought you something." With that he came up with a package from behind his back. Inwardly Fran groaned, "The usual ten pair of hose." To her surprise, however, he had taken the last discussion about gifts to heart and she unwrapped a pair of very pretty earrings. She was stunned if not shocked, but almost immediately panicked as her hands went to her ears. Had she taken Phil's earrings off? Yes, by God, she had. Greg looked at her as her hands were dropping down again. Had he guessed? Perhaps he had. Why shouldn't he? After all, reality. She remembered when she and Phil had gone to that party and this social psychologist had talked about gregariousness. The word had made her look up only to see Phil smile at her. He knew and he understood.

"Greg, you didn't. . . ." How corny that sounded. "You did." So that was good and he had also remembered to pay the insurance premium and he had brought a picture album to put her photos in, and he inquired about her kidneys and wondered if she still had to go so much. "Don't worry, silly. I'm OK. I really am. Let's go out and get some groceries." "You waited. Hmmm, good." They drove off and took the long way by the ravine where by now small crocuses could be seen, something that always thrilled her. But Greg paid no attention to the flowers. "I got tickets for the Sunday game," he told her and she could feel how much he was looking forward to that.

So they talked about the team's chances for the season. Fran knew her baseball all right. That was one side that he had brought out in her and she had learned to yell with the noisiest of fans. "Might be cool on Sunday," she said, "we'll need sweaters maybe. Good that I didn't put them away yet."

They loaded up on foodstuffs and it was fun to go to the store with a man. Why wouldn't Phil ever go? Well. Back home they stuffed the groceries away and then Greg went to fix the gutter on the back roof. She heard the clatter over the Grand Canyon Suite she had put on the record player and the music sounded just right. Greg, that big bear, out there fixing . . . he came back in. "Anything else to fix, hon?" "Fix me and you a drink, that's what you can do." He made her an old fashioned and had a beer for himself.

"Fran," he said, "I want you to read something." Greg went and pulled a magazine out of his briefcase. No, of course, it wasn't poetry, that was not Greg. It was an article on the gathering of wild rice in the marshes of Northern Minnesota. Greg had some Indian blood in him and there was some identification

with Indians which he held up to Fran almost as if to say, "See, kid, I'm really a wild Indian. Can you take it? Can you take me?" She took it fine and laughed over the whole rice business. She promised to read the article but was surprised that Greg had taken to reading the *New Yorker*. He guessed what she was thinking and said, "You know I don't read this kind of stuff usually, but somebody pointed it out to me." They talked a bit about this and that, one thing leading to another. Wild rice led to the film, *Bitter Rice,* to the starving people in the south to Greg's work which kept him informed about economic conditions all over the world. This was always fascinating to Fran. With Greg she felt engaged with the whole globe. So that brought out atlases and maps and she learned how the lira stood and how banking started in Italy. They had a book on Florence and when they paged through it they found a picture of a guy who looked a lot like Phil to her. She quickly went on to another page and Greg never noticed. "We ought to go to Florence. We must, we must."

"Careful now," Fran said as he kissed her and he was. And then there was morning, and then there were other mornings. On one of them, she woke with a fever. Greg took over, cancelled appointments at the office, called the doctor, took her temperature, changed the sheets, fed her pills, read to her, closed the curtains, wiped the sweat from her forehead and other places, and heard her breathe easier again when she got better. He scolded her for not having a tidier medicine closet. He couldn't find a damn thing. Fran promised to reform, only to mutter to herself, "What's the difference?" When she was better she asked if she had been talking in her fever, afraid of having said things that perhaps would—"You talked about filter cigarettes, you dummy," Greg said. Fran smiled.

She was well again. They discussed her taking a job. She had been a crack private secretary and their latest discussions convinced her that she was getting stagnant, rotting on the vine, and she longed for involvement with real moving matters again. Greg was against it. He didn't want her to complicate her life. "But dear, there is no complication. I am drying up here. I need to feel in touch with things going on. I'll be the better for it at home, believe me." In the end he gave in reluctantly. Then he went to work on it.

Made some contacts for her. Drove her to the interview and finally was proud of her; she had latched onto a good job with an electronics firm.

They celebrated by going out to eat enormous steaks and he persuaded her to try beer once more. Once more she tried it and once more she disliked it. Greg flirted with the waitress, but she didn't mind. But she tried to get out of the restaurant quickly because all of a sudden she couldn't stand Greg being so hearty and good humored. She wanted to get away and she got away to a fantasy that had Phil sitting on the porch whittling away at a piece of ash—no, not that, and she blushed. So she snuggled against Greg in the car, hungry now for getting him to her—yes, to and into. Greg was there. That's what counted.

from

Proposition 31

Robert H. Rimmer

"Your book will be quite shocking to middle-class suburbanites," I told her. "They play on the fringes of unconventional sexual behavior, but retreat into monogamy or divorce when a crisis rears its head." I was needling Nancy purposely. "Already one California newspaper has speculated that the Albertis didn't contribute five million dollars to Future Families of America out of altruism. Despite everything you've said, you'll be linked to the swingers and wife-swappers. Many people will be certain that Proposition Thirty-One is little more than a license for lust and sex gone amok."

Nancy shrugged. "The world is composed of people who would rather stagnate than embrace a new idea. Proposition Thirty-One unsettles their minds. Don't forget we have thousands of enthusiastic supporters and we have more than enough signatures to put Proposition Thirty-One on the state ballot this fall. Now it will be up to the voters. I've a feeling there is still considerable pioneering spirit in California. In a sense we're offering the only real adventure left in the world for men and women. If Proposition Thirty-One becomes law, a great many couples are going to experiment with corporate marriage. Even if they fail, the failure will do little harm to the participants, certainly nowhere near equivalent to divorce in a monogamous environment."

I had been casually examining the naked beauty of Tanya and Nancy, one a natural blonde, the other a dark-skinned brunette. An interesting contrast to my own untanned brown nudity. Even in their late thirties, their faces reflect a lovely ingenuousness that many females lose early in life. It occurred to me to test the depth of their conviction.

"Doesn't it bother you that my presence here on occasional weekends is a disrupting influence, at least for Horace?"

Tanya smiled. "In a typical monogamous situation, I simply couldn't tolerate you. But in that situation you wouldn't be here in the first place. Obviously, if Horace or David wanted to have an 'affair' with you (Such an ugly word!) you'd have to meet in your apartment. Your very existence would undermine the strength of a monogamous home. In this environment it is different. Because you are our friend, if you sleep with Horace or David occasionally, *here in our compound,* you're accepting an involvement with our total family. What's more, if you do, knowing you, you would *want* to be involved. Of course, I'm speaking for myself. Why don't you ask Nancy?"

"Sylvia!" Nancy grinned at me. "Neither Tanya nor I are sex-starved. If David or Horace wants to go to bed with you, it will be because they can come to you the way they come to Tanya or me, not for seduction or possession, but for the joy of complete surrender to a person who warmly understands them and can surrender to them. I'm sure that all of us, in the past year, have had opportunities to fuck with another person—I use the word fuck advisedly, because it sums up for me a sexual act devoid of meaning—yet I'm quite sure none of us has."

* * *

"It's ridiculous," David told Nancy, as he reluctantly dressed for their first California evening out together. "Where are we going that's so damned important that you had to pirate the Shea's babysitter?"

"Out!" Nancy said irritably. "Just out. We are going out! I've been cooped up in our dream house for six weeks, and I'm bored; I hate sunny California. You can at least take me out to eat. If you have no better idea, after we eat you can take me to a drive-in movie."

"A drive-in movie?" David exploded. "For God's sake, Nancy, are you regressing? Drive-in movies go on forever. Have you forgotten those interminable intermissions while the kids glut themselves on hot dogs and pizzas and play feelies with each other when there's nothing left to eat?"

"You used to think it was romantic."

David sighed as he patted Nancy's cheek. "That was fifteen years ago. We've grown up!"

"No, we haven't. We've grown down! Down to boredom." Nancy gestured out the bedroom window to the Shea house. "If you could take *her* to a drive-in, I bet you wouldn't be so damned grown up."

David smiled. He knew his silence would irritate Nancy, but it was probably true. After a certain number of years of marriage, the grass in the sack next door did look a little more virginal. In recent years Nancy had become somewhat methodical. Making love to her was like putting the same old Tchaikovsky symphony on the turntable. He could anticipate the climaxes and denouements. There never seemed to be a new phrase or motif. Maybe, if the truth was said, Nancy felt the same way about him. Maybe Nancy would be exciting and mysterious in a drive-in movie with Horace Shea, who didn't know her so well. Tanya was a natural blonde. Grinning at himself in the mirror as he tied his necktie, he had a sharp visual impression of a naked female stomach. Milky-white, it curved into the soft blond hairs of a female triangle; the cleft below it, darker and inviting, was lost in a love-tangle, waiting to be pierced by the touch of his fingers. It was the lure of the unknown. It occurred to David that his knowledge of females was somewhat limited. Prior to marrying Nancy, nearly fifteen years ago, he had known only two other women, both of them brunette. He chuckled; so he owned a brunette love nest, or it owned him. The truth was that copulation was only a minor motif in the symphony of marriage. Most marriage counsellors insisted this was true, and sensible people knew it, too. What was more, not all daydreams could become realities.

David suddenly realized that Nancy, fully dressed, was staring at him with tears in her eyes. Had she read his mind?

"Didn't you hear the telephone?" Nancy sobbed. "It was Franny's mother. Franny is going to the Sheas tonight. Franny's father is a friend of Horace Shea. He told her it wasn't proper for her to disappoint the Sheas."

"There must be other baby-sitters," David said, somewhat relieved that mind-reading wasn't one of Nancy's accomplishments.

"Well, you just try and find them!" Nancy shouted. She knew her hysteria was not justified by the circumstances, but she couldn't help herself. "It was *your* idea to move to California. The teen-agers out here

are all too busy trying to make babies. They haven't time to mind them."

"Then let's have a romantic evening at home," David said as he put his arm around her. "I'll call Rizzo's and order veal cutlet Parmesan to take out. I'll make you a drink, and I'll open a bottle of Chianti. We'll eat after the kids are in bed; then we can look at television."

Nancy scowled. "Maybe Jimmy'll lend you one of his cap guns. Then you can have man-fun together while you watch the Saturday night cowboys and help him shoot the bad guys. As for me, I'm going to bed alone."

* * *

Staying friends, as millions of families have discovered, is even more difficult than staying married. Since both marriage and friendship presuppose a mutuality of interests and common achievement goals usually lacking in a society where male specialization reduces communication to the level of communally shared bridge, television, golf, or amorphous Saturday nights drinking and eating out, it is easy to predict that friends will eventually tread on each other's toes. When this happens the proper spouses will return to the bastions of their married existence, and from this vantage point are likely to excoriate "those so-called friends of ours" with newer friends who, at the moment, may be more sympathetic.

Thus far the Herndons and the Sheas had become deeply enough involved to grin, or to laugh a little at each other's peccadillos, or sophisticated enough to see, in the superior quality of their friendship, something unique and thus quite satisfying. Unlike most friends they were unafraid to analyze themselves. "We must not become victims of the Saturday Night Syndrome," Horace, the self-appointed philosopher of the group, told them. "As the song says 'everybody loves Saturday night.' But people get as mad as hell if Saturday night doesn't love them. Ultimately, it means that most people expect a Constitutional guarantee that man is entitled to pursue happiness even if he is unhappy doing it."

Nancy wasn't wholly convinced. "In Boston, David always took me out to dinner on Saturday night; Mother sat for us. Since it's so difficult to find baby-sitters out here, why couldn't we sit for each other on alternate Saturdays?" Nancy pointed out that Tanya got out of her house every day to her job at Bayberry's. "The rest of you can afford to feel superior; I'm the only one who's cooped up here all week. I need Saturday night to love me."

Though they tried Nancy's plan, later on they agreed it was really more fun when they were all home together. Then they could visit back and forth, and when they were together the conversation was brighter than when they weren't.

"David can sit opposite me, stare at the other diners, be pleasant as hell with the waitresses, and not offer one interesting bit of conversation through an entire dinner," Nancy told Tanya.

"Years ago Horace talked incessantly," Tanya said. "He would spend whole evenings trying to explain how he felt about the tiniest thing. Nowadays when we dine out we often end up just grunting at

each other like a couple of old sows."

The men agreed with the women that years of marital togetherness had put a damper on extended conversation. But they all discovered an interesting phenomenon; together, as a foursome, some of their premarital curiosity was restored. Since their proximity was limited to a few hours on Saturday night, with an occasional Wednesday or Thursday thrown in to break up the week, the process of getting acquainted was a little slower than in ordinary courtship. And while David and Horace could talk directly to each other, as Nancy and Tanya did, David could only learn about Tanya or Horace come to know Nancy, somewhat casually, through their general conversation.

To save running back and forth between houses David wired an intercom from the children's bedrooms to the two living rooms. When the project was finished they had become such good friends that it was even possible for David to suggest, in jest, that he could also wire up the adults' bedrooms. "Then we could spend Saturday nights in bed and converse with each other," he said. He winked over his highball glass at Tanya.

"Or we could listen to each other," Tanya laughed. "Who knows, it might be stimulating for Horace."

Nancy glared at David. "Mostly you would hear David snoring," she said. He embarrassed her. More often than not lately his conversation seemed to ramble into bedroom subjects.

As she came to know the Sheas better, Nancy discovered that she and Horace were the steady members of their respective families. Unless he had a few drinks, Horace was somewhat introverted, just as she was. He lacked the carefree and effervescent temperament of Tanya. And she had noticed that, though David was often taciturn when they weren't with the Sheas, he seemed to be recharged whenever Tanya was around. At such times he assumed a bantering ease and debonair attitude that reminded Nancy of the days when he was courting her.

Once Nancy tried to explain this to Horace and received a rather surprising reply. "I think Tanya and David enjoy flirting in front of us," he told her. "Our apparent stodginess becomes an exciting foil for them." Nancy was not quite certain what Horace meant. She was aware that their marriages had been guided by the theory that likes attract but opposites marry, but she was confident that she and Tanya were reasonably good middle-class wives and mothers. Perhaps their marriages were nothing to write a book about. But really, except for those bedhopping actresses over in Hollywood, whose normal marriage was an interesting subject for fiction, or even for a factual book, for that matter? Nancy was at least sure that the dull life of predictability was preferable to passionate love followed by drugs, tears of penitence, and an overdose of sleeping pills.

* * *

Nancy, too, was somewhat of an intellectual disappointment to Horace. She admitted that her destiny in life was to be both a sex machine for David and a reasonably good mother for her children. "When I first married I used to fret about it," she told

Horace. "After four years of Smith I wasn't settling for suburbia. I was going to be famous, do something in the world. But after a day with the kids it takes more energy than I have left, and anyway, what does it all add up to? Before long we all die, and in the interim a woman's job is to devote her life to her children and to her husband's ego."

Though Horace guessed Nancy didn't really believe that, the truth was that she did, or at least had until they moved next door to the Sheas. From the moment that Tanya Shea had appeared in her living room, smiling so sweetly at David while she apologized for breaking in on them, Nancy had begun to feel somewhat frowzy. Somehow Tanya always managed to appear both casual and feminine, as well as crisp and self-assured. Yet Tanya was no brain. Nancy knew that, given the opportunity, she herself could really converse intelligently with Horace, even read his sociological stuff and understand it too. After all, she had majored in psychology at Smith. But all Tanya had ever done was graduate from high school; then she had studied art and fashion designing at some unknown college in San Francisco. Yet, though Tanya might be a flit-brain, she somehow managed to put her small talents to work. She of all people had successfully combined the joys of motherhood with a satisfying career.

She tried to avoid thinking too much about Tanya. In the morning, if she happened to be gardening in her tiny front yard, Tanya would gaily wave at her as she drove off with Mitch and Sam in the front seat of her car. Nancy knew that Tanya left them at the Andrew Jackson Grammar School, and then she could forget her maternal responsibilities until midafternoon. From nine to three Tanya enjoyed the professional world of fashion as artist and advertising director for Bayberry's in Long Beach. What's more, the seventy-five hundred a year she earned made her not dependent on Horace. Nancy stressed the *not dependent* rather than independent. She believed a female should give her male a feeling that she depended on him, though she wished she wasn't quite so dependent on David. If she ever had any possibility of being her own woman, it was lost forever. For security she had traded the only thing most females have to offer. Instant, hot-off-the-griddle sex. Of course, she loved David and the kids, but lately she had begun to worry. He didn't seem to mind that she served him a more hurried, cooler version of . . . what did David call it when they were first married? Nookie! "His little nookie." Ugh! Well, it wasn't the female's role to be the aggressor. Maybe Tanya Shea didn't believe that, but Nancy Herndon did. And it was too late to change now.

Nancy knew she must have some exceptional abilities; the trouble was she hadn't been able to define them. Once in college she thought she might become a writer, but after her first short story came back with a printed rejection slip she felt ill for a week, and that was the end of her literary career. Well, she might not have a career, but at least she was a better cook than Tanya. And kept a tidier, more organized house! Tanya seemed constitutionally unable to hang up her clothes, or put the dishes in a dishwasher until the last clean plate or glass was used. Nancy was sure that David, who was so finicky neat, couldn't live under the same roof with a woman like Tanya. She wondered if David appreciated her firmer grasp on the realities of homemaking. She knew Horace did.

* * *

As they became closer friends Nancy had been somewhat concerned that their Saturday evenings, even while they were playing bridge, seemed to creep with increasing detail and hilarity into the subject of sex. Not just general sex and dirty stories, though David and Tanya seemed to pick up in their business contacts an endless fund of stories. Neither David nor Tanya seemed embarrassed to say any of the dirty words that made Nancy shudder when she heard them. But even worse, the stories often led to blushingly frank personal observations on the sexual idiosyncrasies of their respective partners.

"Horace doesn't waste time," Tanya chuckled. She patted Horace's cheek. "Each year it gets faster and faster. Pretty soon I guess all we'll do is talk about it."

Horace wasn't angry. "The trouble with Tanya is I have a feeling right at the crucial moment she's planning a new advertising campaign for Tom Bayberry." He stroked her arm. "Or maybe she is pretending I'm Tom Bayberry."

* * *

"I like to speculate on what is going on in the lives of other people."

"Because your own life is so dull?" There was a tremor of anger in Nancy's words.

David just grinned. "You and I are the salt of the earth. A dull husband, a dull wife, and two dull children. There's millions like us." He drank his orange juice. "We did, presumably, move to the land of fresh oranges."

"If that's sarcasm, I don't appreciate it. I'll be glad to buy you fresh oranges."

David shrugged. "You seem a little edgy this morning. Why didn't you stay in bed for another hour and get your beauty sleep."

"I most always get your breakfast." Nancy angrily brushed the tear from her eye. Why was she so tense today? She couldn't help it. "I suppose you would deprive me of getting your breakfast."

"Now what in hell does that mean?"

"Oh, I don't know." It really was impossible to explain. "Here, put this maple syrup from good old Vermont on your waffle. It's snowing there. I wish it would snow here."

"Overshoes, snow shovels, snow tires, slush, runny noses." David smiled. "It's better thinking about it than being there."

"If it snowed here, maybe all the millions of automobiles on the freeways would skid together and crack up all at once instead of piecemeal. David, why do you have to go to work so early? Joe Casey, across the street, doesn't even own his own business. He works in downtown Los Angeles. He never leaves until seven thirty."

David munched on his waffle.

"David, did you hear me? Why do you . . . ?"

"I heard you, Sea. For the fortieth time, in the three years since we've lived here, I'll tell you. There are

three reasons. All my life I woke up early. My New England heritage. I can't sleep after five thirty. The early bird catches the worm. Look around you. This is a pretty expensive worm I've caught. The average income in these United States is about six thousand dollars. I make nearly four times that much. I'm thirty-six years old, and I believe that if I keep getting up early, someday I may make four or five times what I now earn. Since you obviously enjoy some of the luxuries of living, like that housecoat you're wearing—it probably cost at least fifty dollars—you can't possibly object to that. The second reason is: A man who runs a business *has* to set a good example. Arrive before his employees. Give them the evil eye when they are late. Convince them that the American dream can be theirs, too! Convince them that the way to wealth and happiness is through their own efforts. They, too, can catch the worm."

Nancy was aware that David was baiting her. She just stared at him, her eyes round and wide, an indication that she was in her story-listening mood. David finished his coffee. "As I've often said the third and most important reason is that between six and seven on weekday mornings only worm-catchers like me are on the freeway. This means that I can drive to the plant in forty-five minutes instead of an hour and forty-five minutes." David kissed Nancy's nose. "There's a tear in your eye. Don't cry, Nancy; I'm not going on a world tour. I'll be home tonight by seven."

"David, can't I go to work, too? Tanya does. Couldn't you use another secretary? Once I was a very good typist." Even as she said the words Nancy was angry at herself. Why was she pleading? If she were getting her period it would make sense, but that was at least two weeks ahead.

David walked toward the front door. "The trouble with you, Sea, is that you've been reading too many of those damned books. Some women haven't a thing to do except irritate other women into a frenzy because the modern American female supposedly isn't fulfilling herself. If you *had* to join that rat race, if you *had* to go to work every day, you'd be sick of it in a month. You're not aggressive like Tanya. What good would it do anyway? If you earned a hundred dollars a week, you would have to pay a woman seventy-five to take care of the kids." At the door David kissed her feebly on the lips. "I'll see you tonight, hon."

Nancy watched him drive off in his Pontiac toward the freeway. She brushed the tears from her eyes. Damn him anyway. Not so aggressive as Tanya! Someday she'd show him just how aggressive she could be.

*   *   *

Tanya had snuggled in his arms and murmured into his chest. "Oh, I don't know, David. If Horace and Nancy were as involved as deeply as you and I, then maybe they wouldn't be shocked at us." There were tears in her eyes as they made love. "Sometimes I'm shocked at myself. Why can't I give us up? Why can't we just go back to being friends? But I love you, too." She smiled at him sadly. "It's hopeless."

David had no answers. For two years now he had succumbed to that necessity of the American middle-class male, the other woman. No, that wasn't essentially true. Tanya was more than just another woman, much more than a sexual spree. He had acquired a deep need for the brightness and vivacity of her. She offered a necessary counterpart to his New England stodginess.

*   *   *

David felt as if he had been kicked in the stomach. Had Nancy discovered the truth? Why didn't he have the nerve to admit the whole business? But what was the truth anyway? Did he want to divorce Nancy and marry Tanya? What would that accomplish? He wasn't even sure that Tanya wanted to marry him. The truth, which he might not admit to Nancy or Tanya, was that he wanted to have his cake and eat it too. But society demanded a choice. Either he and Tanya would have to stop seeing each other, at least in bed, or they must divorce their present mates and marry each other. In the United States, in the twentieth century, there was no middle ground.

As David eased his car into the lane for the Santa Monica Boulevard exit he knew he had been ignoring the root of his problems. There was no longer any doubt that Tanya was pregnant; she had passed the time of her second period. "I'm certain it's our child, David," she had said, "And I can't say I'm really sorry. I hope it's a girl." But despite Tanya's determination to accept fate, their stolen Wednesdays had now taken on a quality of subtle recrimination.

"I thought you always took pills," David said. "Anyway, I was careful."

"Not too careful." Tanya grinned. "It happened the day we went swimming near Mendocino. You weren't too careful that day!"

David remembered that he had flown up to San Francisco in August, for a week. The story he told Nancy concerned an important contract in the offing for a new chain of jewelry stores. At the same time Horace went off to a conference of sociology professors and to teach summer school in Chicago; Tanya had left the week before, presumably to visit her mother in New Mexico. Actually, she met David at the International Airport, and they rented a car and drove north. For four glorious days they stayed at Mountain House. Tanya wanted him to see the art colony at Mendocino, because it reminded her of New England.

As they paused in their wandering through the town of weathered clapboard houses to examine a painting, Tanya expressed what David was thinking. "Oh, it's too bad," she said. "Nancy would love this place. Do you feel guilty, David?"

He shook his head. "I love you, Tanya. These few days are a very tiny but important part of our lives. I'm simply closing off the moralizing part of my brain. We aren't really hurting either Horace or Nancy."

*   *   *

I'm not being flippant or brave, David. I'm scared, too. Everything we've been saying skips the main point. Why, when we have perfectly good mates—nice families—why didn't we resist? Why did we let us go this far? Why wasn't one man enough for me? What was missing, that I could respond to you? Why do I love you and still care very much for Horace? Do you love Nancy? Do you love me enough to divorce Nancy and marry me? If your answer is yes, then what happens to our kids? Oh, David, we've a thousand questions to ask ourselves, haven't we?"

*   *   *

If he had never moved his business to California, he would never have known Tanya. But if he had stayed in Boston, might he not have met some other woman, another Tanya, and become just as involved? Was it his environment or his marriage to Nancy that was at fault? Weren't he and Tanya merely creatures of their times? They lived in a world that sanctified monogamy but was humorously permissive in its attitude toward adultery. Of course, what was comedy when it concerned someone else became tragedy when you yourself were involved. Could there be any doubt, furthermore, that the seeming lack of purpose or meaning in the average man's life made absolute fidelity unrealistic? Didn't many marriages collapse because a lifetime of monogamy was too big a burden for any human to bear in a society where the danger of starvation was only a quaint fact of history and people were conditioned to indulge their appetites?

Most lives were narrowly circumscribed by the daily task of making a living, paying taxes, saving money for the children's education, and scrimping to buy insurance to protect the family. Whatever time was left over was devoted to becoming good citizens, good spending-buying-consuming units.

*   *   *

There was one other dangerous adventure left in a world where everything was cut and dried and prepackaged and pronounced good for human consumption. The adventure of discovering another human being was a great deal more than the urge to pursue the female that had first impelled him to ask Tanya to lunch. The adventure was both sexual and mental. Who was Tanya, really? How did she think? What was important to her? In the long run would her flesh be more compelling to him, more of a refuge, than the flesh of Nancy? The first adventure a male and female had with each other was supposed to culminate in marriage, but did the first adventure also have to be the last?

With Nancy he was one person, with Tanya another, and yet both persons were essentially himself. He believed the adventure with Tanya actually opened the possibility of a different adventure in his relationship with Nancy, if only Nancy would understand, which she would never do. Once she knew about Tanya, she would insist that his behavior toward herself was less than was required of marriage. Comparison and criticism, the dread spoilers of man's relationship to his fellow man, would set in. Without his saying a word, or no matter what he said, Nancy would be convinced that Tanya was a better partner than she was. But that wasn't the explanation. They were very different women. Though they both had the same female equipment, the minds that set it in motion made each of them unique.

*   *   *

Amazingly, when he had implied his indecision to Sylvia, she had said, "Oh, Horace, you're *so* rooted in middle class concepts. Sure, we've enjoyed each other sexually, but that doesn't mean I have any designs on your life. It has simply been a pleasant experience; and don't you agree it has cleared the air between us, and made it possible for us to understand each other on a deeper plane?"

But Horace had not been able to absorb the idea of free love so easily. To his own surprise his attitude toward Sylvia had acquired a subtle tinge of possessiveness. He smiled at his thought. Of course, monogamy was a male concept. It was a one-way street of fidelity imposed by the male on one female or, if society permitted it, a whole harem.

In the twelve years of marriage, up until these two episodes with Sylvia, he had been faithful to Tanya. There it was, the old cliché again. Even he couldn't escape it. What did it really mean to be faithful or unfaithful to another person? Joining his sexual organs with those of Sylvia was no more an act of faithlessness than the pursuit of his career was. In neither of them could Tanya share whatever made him the man he was. For Tanya, sociology was equivalent to social work and hence in her words "a socially approved form of meddling in other people's lives." During the early years Horace had often wondered whether he would have been more deeply involved in marriage with a woman who could identify with his intellectual interests. But he had made his choice of Tanya consciously. A man needed a woman who loved him enthusiastically, boisterously, warmly, affectionately, proudly. Tanya had given him this kind of love. Wasn't that sufficient for any marriage? Whether Tanya read the books he read or thought the thoughts he thought, or whether he was vitally interested in Tanya's field of art and fashion advertising was of little relevance. And, of course, if he had married a woman like Sylvia the day-to-day challenge might have been too great. They might have spent their lives trying to excel each other.

*   *   *

It was simple logistics. Tanya loved him; Tanya was his wife. Nancy loved David; Nancy was David's wife. A man's own wife should be sufficient for any man. Even if she weren't, men and women made commitments. For its own protection, society created the laws and the environment that forced compliance.

Grinning a little at his thoughts, Horace turned into the driveway of his house. In an hour or so Tanya would be home and the little womb of domesticity that he and she had created in an alien world would bathe him in its warmth. Why did men even consider relationships that would destroy the solidity and permanence that were a vital need of the psyche? Had the Greek philosopher Plotinus summed it up in the words, "Out of conflict comes harmony?" But pure, continuous harmony would be as intolerable as permanent conflict.

*   *   *

As the time approached when she must leave the motel Tanya knew,

much as she had tried to prevent it, that her own feeling of hopelessness had communicated itself to David. Their affair was over; all that remained was a renunciation of each other. Lying together, their passion spent, they watched the afternoon shadows gather in the room. Words between them had dwindled, yet in the pressure of his hand on her back, the brush of his lips on her neck and breasts, Tanya knew that David was denying the possibility that they might ever have to give each other up.

She crushed her body hard against him and tried to prevent him from seeing her tears. I love you, David, she thought. I love Horace. If love for one man is good, can it be evil when it is multiplied? Why had she taken a chance on David? Her marriage was good. As a sociologist, Horace himself had defined it as beautifully average. She rarely fought with Horace over anything fundamental. They loved their boys. While they didn't identify with their house in do-it-yourself projects the way a lot of families did, they enjoyed the comfort of their home. Evenings when they didn't see the Herndons or other friends they pursued their own different interests. Separate, but comfortably together, she at a drawing board, working on new fashion layouts for Bayberry's, while Horace prepared a lecture or was deep in some book on sociology or psychology. Maybe neither of them was too much interested in the other's occupation, but how many husbands and wives were? In bed they were still able to release many of their deeper thoughts without embarrassment. Maybe their lovemaking had settled into inevitable routines. But was that sufficient excuse for a childish pursuit of romance? Why had she needed David? She and Horace really leaned on each other. With David she had to fall back on her own strengths; he gave her no feeling that he was protecting her from the world, as Horace did. He accepted her as very much his equal. Figuratively, marriage with Horace meant walking one step behind the warm shelter of his masculinity. With David, you fought the demons together and marched side by side. Tanya's thoughts amused her, though only briefly. She knew she lacked no freedom in her marriage with Horace. The truth was that both David and Horace had become necessary to her; in combination they extended the boundaries of her life.

To avert the suspicions that might be aroused if they both arrived home at the same time, Tanya left the motel an hour ahead of David. As she drove toward home, David's words, "I love you, Tanya," were echoing in her mind; she could still feel his hungry, fearful, bruising kisses on her lips. She felt both a compulsive need to apologize to Horace and a dreadful longing for David. Could she ever explain to either of them separately, or both together, the fact that she was a passionate female? Not that she would lie down with a total stranger, but that, so long as she was secure in their love, she was perfectly capable of wholeheartedly loving both Horace and David?

What would David think if he knew? Perhaps he had guessed. Tonight, if Horace felt affectionate with her, she would accept his love warmly and hungrily. Any feeling of guilt would come only from her inability to explain to either of her men how this was possible. And could a society that insisted a female love one man exclusively ever accept the possibility that a woman could love two men as loyally as one?

*   *   *

In less than three hours she would be home. It would be five in the morning in Boston, and Dad, fretting a little because he would be late for work at his office, would be waiting with Mom at Logan Airport, worried by the suddenness of her return after a three-year absence. Home. But in fact it was no longer her home; it was only her father's and mother's home. After fifteen years of not living there she thought of herself as being reduced to little more than a picture leaning on the mantelpiece. "That's my daughter with her husband and two children," her mother might tell a guest. "She lives in California, you know."

Momentarily, when Nancy arrived home, her mother would be just as ecstatic as she had been over the telephone. "Oh, darling, you've answered my prayers. I was just telling your father last night, if only you and David and the children would come home for Christmas, it would be all I could ask."

But would Mother be so enthusiastic when she discovered that David wasn't with her? What would she think when she finally realized that her daughter might become a permanent guest? But that was impossible; the house wasn't big enough for her to move in with Jimmy and Susan. Any arrangements would have to be temporary. Susan could sleep with her in the spare room, and Jimmy could bunk on the convertible sofa in Dad's den. Had she been coming home normally, David would have insisted that they stay in a motel, because he couldn't stand being cooped up with strangers. Her mother and father were strangers, now; they loved her for the past, for the child she had once been. And, of course, they were proud that she had given them two grandchildren to talk about. Still, they wouldn't relish a dependent daughter with youngsters being thrust upon them. Today females might have more freedom, more chances to function as full human beings, but once they had children they had no choice but to swap independence for security. Wise women accepted the inevitable; there was no double standard for them. They must be faithful.

Would David follow her to Boston? What if he decided that he just didn't care? What if he decided that he had had his fill, that she had given him the perfect way out? In the past few years she hadn't been the easiest person to live with; David had a right to expect more warmth and affection. Sick with fear at the thoughts cascading through her mind, Nancy could only pick at the dinner the stewardess put before her. Oh, God, please, please let him need me, she thought. Let him come after me, not with anger nor with recriminations, but just with the words, "I love you, Nancy." With the right words she could release the flood-gates of her misery. She needed to love him with a torrent of words and passion, and in his response he could convince her that he understood her infidelity. She could never beg for forgiveness. And could he ever be-

lieve that she had never actually made love with Peter Alberti? What difference would it make if she had? Would she be any the less Nancy who loved David and would spend the rest of her life quietly proving it to him?

If he didn't pursue her to Boston, then she must expect to spend the rest of her life trying desperately to make ends meet. Even with alimony, she would be a drain on Mother and Dad. Someone would have to mind the children while she went back to work. It was all so stupid, really. I love you, David. The words trembled on her lips. But no matter what, she would never telephone him and beg his forgiveness.

*   *   *

How did any man, alone, sustain himself? Remove the props of marriage from those who were married, and the very structure of their existence collapsed. Yet marriage with one person for a lifetime was a challenge that more and more people seemed unable to cope with.

Had Tanya needed David to give a meaning to her life that he, with his academic background, had been unable to provide? If they had had more friends, would she have been less likely to take up with David? Except for the Herndons, they had few friends. There were his colleagues at the Institute, but Horace had never made any effort to become involved socially with them. The reason he had made no effort had something to do with Tanya's lack of academic background, her too-purposeful ingenuity in social gatherings, and the fact that after a day of teaching, Horace preferred the escape to home or to the different world of the Herndons, that scarcely acknowledged a need for teachers.

Horace was driving toward Santa Monica and Sylvia Mai. Finding refuge with her was scarcely an adult way to face the problem of Tanya, but he couldn't go back. Accepting that your wife had made love with another man was one thing; adultery was a fact of life. But to watch your wife grow fat with another man's child in her belly while she told you she loved you was impossible. How could Tanya have been so careless? There was only one answer. Her love for David had been so great that she didn't give a damn.

Sylvia, on pills, never would take chances. Three times they had enjoyed each other sexually as an outgrowth of their day-to-day companionship. He had told Sylvia that she would like Tanya, and she had told him that, despite appearances, she was not promiscuous. She said she found it "kind of sweet" to discuss common interests, without tension, with a man she admired intellectually. Their love-making was a pleasant counterpart to their working relationship. Someday she might marry, especially if she found a man who would understand that an occasional afternoon of dalliance with a colleague didn't threaten their marriage. Was Sylvia's reaction so different from Tanya's? Did all of them, Nancy, David, Tanya, Sylvia and himself, in their reaching out for a special love relationship, point to a larger driving force, to man's deep need to find identification, to merge his own mind with the mind of all mankind? Was the blending of flesh simply a necessary point of departure?

As he pushed the floor number in the automatic elevator at Sylvia's apartment, Horace had a sudden ridiculous glimpse of himself. It was a vision of the cuckolded husband, disheveled and miserable, running to another woman. What did he expect from Sylvia, balm for his punctured ego? Right now she might be in bed with another man; he really knew very little about her private life, and had not asked about it, because he sensed that his lack of commitment to her gave him no right to probe. But he knew she had once been in love, and wanted to marry. "But not at the expense of my individuality," she had said. "In a way, I suppose I'm like your wife. I want children, a home, but not as ends in themselves. The reason Tanya is so valuable to you is that she hasn't tried to become a carbon copy of her husband."

In his marriage with Tanya and in this brief affair with Sylvia both women had given him a feeling of deep involvement. Here was their strength; it grew out of a seemingly defenseless giving of themselves as individuals. Yet it was now painfully apparent that Tanya had been able to maintain a segment of her emotional life completely independent and apart from him. It was the shock of this discovery, perhaps more than the physical aspects of her sexual relations with David, that dismayed him. Yet, hadn't he duplicated Tanya's performance? Not quite. He had had intercourse with Sylvia only three times, and after each time he had felt guilty toward Tanya. How many times had Tanya made love with David in two years? And was his child stirring in Sylvia's womb? It was not!

* * *

Thinking about it now, Horace could only feel that his attitude represented the greater logic. He was also being a damned sight more permissive than most men would be toward their wives. He could accept the human need for Tanya and David to want to experience each other sexually. But to blithely accept a continuation of marriage, with David's child in his house as a constant reminder of her infidelity, to live next door and accept David both as his friend and his wife's lover was impossible.

* * *

"Oh, God, Horace, I don't love you less. Just seeing you like this, knowing that I'm responsible, makes me want to hug you and tell you how sorry I am."

Horace scowled. "But you're not so repentant that you resent David. He's at least partially responsible for the jam you're in. What does he plan to do?"

"I told him to go to Boston and bring Nancy home. She *is* his wife. He doesn't have to tell her about me, and I won't tell her." Tanya shrugged. "Since he has strayed from the paths of virtue himself he can afford to be magnanimous and forgive her minor sin."

Horace shook his head grimly. "For Christ's sake, you're balmy. The truth is that you're counting on me to accept this mess. After I get over my indigestion, you're hoping I'll adjust to necessity. Do you really expect me to maintain the illusion the we have a happy marriage, even

to pass out the cigars when the new child is born?"

"No, God damn it, I'm not expecting anything. I don't need to live the rest of my life with a martyr for a husband. I'm perfectly capable of supporting my children."

"How does David feel about that?"

Tanya shook her head. "Men are so nice and fawning when they need a woman, and so damned pompous when they think a woman needs them. David refuses to go after Nancy or even telephone her. He feels that when she knows the whole story, she'll divorce him. According to David, Nancy could do penance for her own sins but, with her New England upbringing, could never live with him knowing that he had had sexual intercourse with me."

"It's not Nancy's background," Horace said stiffly. "Damned few men or women would tolerate it. Poor Nancy! She's the type that will take sleeping pills when she knows the whole damned story. Then you and David will have that cross to bear."

Tanya was crying. "I'm not so crass as you may think. I wish I could talk with Nancy face to face. I know she's not very tough."

* * *

Exactly ten days ago the Sheas and the Herndons had been, at least on the surface, reasonably contented middle-class citizens, with reasonably happy marriages, nice children, and comfortable homes. They were simply good neighbors who had become good friends. What had happened? He wondered if the trouble was that man no longer paid heed to an authoritarian religion or to a God who helped him control his natural instincts. Or was what they had all done symptomatic of a deeper, more basic risk-taking that led some people to climb mountains or dive deep in the sea or explore space, and others, perhaps even braver, to seek their adventure in the wonder and diversity of other human beings? He could ask himself whether the need to have intercourse with Sylvia was because of Tanya's inability to satisfy him sexually, but he knew the answer. Tanya was as capable as any female in the erotics of sex. While Tanya and Sylvia were different, the difference that reflected itself in bed was not in the physical act, but rather in their widely varied capitulation. And that *really* was the adventure. In the surrender of the mind, with their defenses down, each couple set the limits of their physical merger and with it the potential of expanding the horizons of each other's lives. In a mechanized society the only true unknown was the never-to-be-discovered mystery of another human being. Monogamous marriage, because it limited the opportunity to know even one other person or have one other friend, came a cropper on the natural yearning of the human soul.

* * *

"The truth is," he said, "that for the past two years David and Tanya have been lovers." Conscious for a moment of her stupified silence and of its contrast with the commotion of the crowded breakfast room, Horace watched the impact of his words register on Nancy's face.

* * *

But now she couldn't hold back the tears, "Oh God, my God, Horace! Why have you done this to me? Did you have to come three thousand miles to humble me even more?"

Horace's fingers encircled her wrist. "Nancy, we didn't make love casually. I care for you. Maybe that means I love you." He grinned feebly. "Who can really define love? Strangely, after the initial shock, I'm not angry with David. Perhaps he and Tanya found something in each other that you and I were unable to provide." He shrugged. "On the other hand, Tanya and I have lived together fifteen years. The fact that she and David have made love doesn't erase the lives we've both shared. I'm beginning to believe that Tanya is quite honest when she says she still loves me, and I'm certain that David still loves you."

Angrily, Nancy pulled out of his grasp. "*You* can have that kind of love. In my family we get married to one person, and we don't go around screwing the whole world even if we feel like it!"

"You're not being honest. You enjoyed last night. With a new lover, perhaps you even released a Nancy you've always held in check with David."

"I was a whore," Nancy said scornfully. "It was a cheap business. You tumbled into bed with me after an orgy at the Albertis. What kind of person are you, anyway? Go back to Tanya, who loves you, too. You and David can share her." She shuddered. "My God, what I did with Peter Alberti was harmless by comparison."

* * *

Horace smiled at our attentiveness as he reached for his drink. "I know this relatively calm appraisal may seem like ducking the issue, but if you think about it, the fact that we all see eye to eye on these three aspects of living probably makes us members of a fairly small minority group. Actually I suspect it's lack of conflict in these areas as much as our living next to each other that has made us more comfortable with each other than with other friends. Since David has committed himself, I'll join him and say to Nancy and Tanya that I like you both. In addition," and here Horace chuckled and glanced at Tanya and me, "you're both very different and enjoyable bed companions."

*   *   *

"Maybe that's not even it," David mused. "Maybe nobody can ever wholly release himself to all the aspects of another person."

"Meaning that you are David with Nancy, and David with Tanya, and neither Nancy or Tanya could ever know the complete David."

"It sounds complicated," David said, "and I'm not sure I completely understand my own motivations. All I know is that, whether or not the four of us ever define love, I have my own understanding of it. Because I feel completely protective toward both you and Tanya, I must love you both. That's not playing the game according to the old rules, but it's true."

"Why aren't you jealous? Right this minute Tanya is probably making love with Horace as violently as I have with you."

David smiled, "One thing I am

sure of. The act of love is a gift, not a possession. I enjoy the gift both you and Tanya have given me. As a male, I accept the trust of the surrender. It's the most wonderful commitment a female can give a male. It's the reason why prostitutes and the casual affair are always inadequate. Yet, it would be no good if the male couldn't accept the gifts without an equal commitment." David sighed. " My God, I'm getting as involved in words as Horace. All I'm sure of is, that a trade-off won't solve anything."

Snuggled with my behind curved against David's stomach, his hand cupping my breast, I thought about Horace. Somewhere I had read that the inability of a male and female to make a final commitment was a sign of immaturity, and I wondered if it were true. Could the four of us expand our lives together? I liked Horace; I enjoyed his searching approach to life. I knew, now, that we could both surrender to each other in the act of love. But tomorrow, after he had been with Tanya, could I casually go to bed with him? If I could, I'd still have to love Tanya, and that seemed impossible. That damned word love again!

*   *   *

At quarter of seven in the morning the four children, still in their pajamas, tiptoed into our room, but when David invited them into bed with us their silence turned into an uproar.

"We woke Uncle Horace and Tanya," Susan said as David tousled them. "They went to bed without any clothes on, too!"

"If it's Uncle Horace," I said, trying to divert her from our nudity, "it should be Aunt Tanya."

Mitch thought that was funny. "That sounds silly. Tanya doesn't go with aunt."

"Neither does Nancy," said Jimmy. "Anyway, Tanya's not my aunt."

Horace, dressed in a sweatshirt and chino pants, appeared in the doorway with Tanya. "Maybe they should call us all Mommy and Daddy," he said, and grinned, obviously enjoying my disheveled appearance.

Jimmy shook his head. "Daddy said I could call him Dave. I know a kid in school named Horace. They call him Ace."

"Ace it is," Horace laughed. "I like it. Enhances my ego. Makes me feel like a Las Vegas gambler."

"Daddy calls Mommy Sea." Susan said. "I like that."

"Poor me," Tanya sighed, "I'm stuck with my Hollywood name."

"Tan's a good name," Sam volunteered.

David chuckled. "Sea and Tan. It's settled. Sounds like a lotion to soothe Ace and Dave."

"Why don't you get in bed with us?" Susan asked Horace and Tanya, who were hovering near the foot of the bed.

"With four kids, plus Sea and Dave, I don't think there's room."

"Dave and Sea have no clothes on," Jimmy announced. "Ace and Tan can't see them without clothes on."

"Why?" Sam demanded.

"Because they're not married, stupid," Mitch said.

"I think you all better leave," I told them. "We'll get dressed."

At breakfast Susan speculated on a new idea. If the four of us slept in the other bedroom where there were four more bunks, we could talk in bed instead of staying up all night arguing. Tanya smiled at me. "Did you listen to us?" she asked Susan.

Susan shrugged. "For a while, but it sounded kind of noisy and dull. We were listening to Sam tell ghost stories. They were scary. Maybe Ace and Daddy, I mean Dave, could tell you and Sea some good stories."

* * *

I suggested we all go to bed together and we did. We tumbled into one of the big double beds, Tanya and I were squeezed into the middle, and Horace flopped down beside me, insisting that he still had three nights left with his new wife.

"What do we do now?" David laughed. "Make love and watch each other?"

"Not on your life!" Tanya shrieked.

I murmured my agreement. "I think that would be sleazy. Even animals don't copulate in groups. Making love is a subjective experience. If four of us made love together, it would become objective. We'd be voyeurs; it would be ugly."

Horace squeezed my hand. "So we've got one problem solved. Personally, I feel very comfortable. It's snowing again. We don't have to have intercourse every night. Why not just enjoy the warm cocoon the four of us have made together? We could just sleep."

"Before we do, I'll offer a challenge." David, leaning on his elbows, smiled across the bed at us. "We've tried to define love. Now I'll ask a question about sex that may be just as hard to handle. What do you feel at the height of intercourse, and what do you feel afterward?" We pondered on that silently for a while. Outside, the snow snapped against the window.

"Are there any words for it, really?" Tanya asked. "The moment is too short to hold. Maybe the real delight is in the surrender, and the wonder that you can surrender and want to."

* * *

Sometime before morning, David sleepily took my hands and we tottered off to the other bedroom, because we still didn't have any explanation for the children. I knew this was a reflection of the larger problem we would have to face if we tried to convince the world that we weren't entirely mad.

We spent the days with the children in a warm little island of our own making, but while we slept together according to Horace's plan, sex was not the keynote of our existence. Rather, as in an oil painting whose undercoating gives the final colors their vibrancy, the delight in our sexuality became a warm undercurrent between us. We had plunged into a new world where the intimacy of our private surrender to each other pierced the smog that people ordinarily exude in their daily contacts; we had colored all of our actions with a warm, understanding laughter.

I suddenly realized that I no longer felt jealous of Tanya or possessive of David. I had discovered myself to be a hungrily responsive person quite willing to share the easy affection that had become quite natural between all of us. The children, sensing the unity we had created, responded in an undifferentiated way that embraced Tanya and me as friends and equally valid mothers. Susan and Sam were entranced with Tanya's art lessons and spent hours drawing pictures and unrecognizable portraits of all of us. Mitch and Jimmy built an igloo with Horace. David, who had become intrigued with a book by Ralph Borsodi called *Education and Living,* occasionally went outdoors and supervised, but mostly he lay on the sofa reading excerpts to anyone who would listen. After one of our daily saunas, which we all took together naked, all eight of us, shivering, crawled into the kid's snow house. Mitch and Jimmy crumbled the roof on top of us. Thrashing and laughing, we disappeared in a huge pile of snow.

* * *

Horace grinned. "I don't think we'll find the specific answers in any book. If we dared to try communal marriage, the trick would be to succeed within our present environment, using all the artifacts of our culture, bending them to our needs, and not vice-versa. We would only reject values which didn't contribute to our basic need as a group family . . ."

"Corporate family," David said. "A corporate marriage seems more plausible and more contemporary than a group marriage."

While we continued to discuss the idea, we agreed with Horace that we were not trying to predicate Utopias as a solution for our personal problems. But, of course, the truth was that four adults, attempting to work together toward a common goal for themselves and their children, were delving deep into the basic reason that large-scale Utopias had never worked—*People!*

"There's another reason why Utopias fail," Tanya pointed out, "Any perfect society, with no problems, would be damned boring. If the four of us could merge our families, I don't think we could act like saints. Maybe it's necessary in the contained environment of the Chesley lodge, but back in San Pedro, Nancy and I would inevitably be in much closer contact with each other than either of you. If we needed to have a slam-bang argument or to hate each other for a while, I think we should."

"Forget the neighbors for a minute," David picked up the thread of his thoughts. "Since we've arrived at a point where we can discuss and solve most of our problems, except how to explain our sleeping arrangements to the kids, I'd like to propose an idea that's been flitting around in my mind for the past few weeks. It might be the step, if we dared to take it, that Nancy feels is so necessary. It would be a larger commitment to the belief that what we're doing is sound. I propose that we sell these two houses, pool all our resources, and incorporate as a joint family. If we did, we could build a home that made more sense for all of us. In the process we'd have to explain to the kids exactly what we hope to achieve."

Horace, stretched out on an air cushion, stared into the star-pricked sky. "How do you propose to do it?" he asked. He leaned on his elbow to examine my face and breasts which were right-angled to his belly. "Nancy looks startled."

David shrugged. "Not with the idea. It's just that she can't believe the rugged Republican individualist she married could accept communal property sharing. It's an interesting fact that most middle class families, at least among themselves, have a greater reluctance to discuss their financial strength or lack of it than they do their sexual relations."

"Or lack of sex," Tanya grinned. "It's one problem we've overcome."

Horace trickled his fingers across my shoulders. "I'm not so rich that I care."

"I've known you three years," David said. "I could only guess at your earnings."

"Seventeen-five from Cal Institute." Horace laughed. "I'm in the ranks of the poorly paid pedagogues. Tanya has been earning seventy-five hundred. My father left an estate of seventy thousand, mostly in blue-chip stocks, which earn about thirty-five hundred in dividends. All told, discounting Tanya's salary for the long haul, about twenty-two to twenty-three thousand."

We spent the rest of the evening in an excited money discussion. David was earning twenty-four thousand from Herndon Showcase. If the company had a good year he could take a few thousand more, but more likely he should plow the money back into inventory and new machinery.

"My God," Tanya marveled. "Together we're filthy rich. We have a total income of nearly fifty thousand dollars."

As we explored in detail our total assets, we discovered that our combined savings totaled twenty-six thousand dollars, the equity in our houses came to thirty thousand dollars, and the paid-up value of our insurance policies was about eighteen thousand.

David estimated the Herndon Showcase Corporation was worth about two hundred fifty thousand dollars, and Horace pointed out that the Herndon assets were greater than his inheritance. David didn't think that was important. "We can't measure this merger by money values. We're four people in reasonably good health. The question is, do we have the nerve to love and cherish each other until death do us part?" David explained in detail Sub-Chapter S of the Internal Revenue Code, under which he thought we could legally form the Herndon-Shea Corporation. All of our assets would be transferred to the corporation. Our combined income would be funneled through the corporation without double taxation, with one corporate tax on our earnings. Moreover, we would have the numerous advantages of corporate deductions and provisions such as the pension plans permitted to corporations. "You have to understand this is a special type of partnership corporation. It's permitted under the law as long as there are no more than ten stockholders." David laughed. "We're under the wire with nine of us; one to go!"

Tanya and I were slightly bewildered with the details. "Do we understand you right? All of us including the kids would be equal stockholders?"

"It's the only practical way, because it would minimize taxes. If we issued one thousand shares, each

of us and each child would immediately be given one hundred shares while a hundred unissued shares would be left for contingencies."

"If you're grinning at me, forget it," I said. "No contingencies. I take my pills religiously. Nine under one roof is enough."

We tossed the idea around for several days. The potential for us as a united family was fascinating and intriguing to contemplate. Our corporation would have sufficient income and assets to swing a mortgage on a unique house designed for a unified family. The corporation could not only be planned for so that we would have ample financial security for the four of us, but it could become a continuing source of income for the children as they grew older. If we designed our new home correctly, it could become a living thing, expanding to meet the needs of the entire family. Later, if any or all of the kids wanted to live with us, our corporate home could become a true family compound.

Tanya immediately got us involved in drawing plans for the kind of home we would need. Horace wanted to gamble. Could we immediately create what he called a "three-generation home?" He grinned at David. "You and I are without living parents, but Sea and Tanya each have a mother and father who are alive."

"You mean build a house large enough so that my mother and father could move in?" Tanya was stunned. "Jesus, NO!"

I agreed. "My mother and father would drive us insane. But you don't have to worry, the immorality of the whole business would give mother a fatal heart attack. She'd never come in with us."

Horace wanted to know what Tanya and I would do if our father or mother died. Put the survivor in a senior citizen center? I shrugged. It was something I didn't like to think about.

"I agree with Horace," David admitted. "After all, do we or don't we believe in what we're doing? If we don't, we should split up. If we do, then the purpose is to create stability for an entire family. If you think ahead, in thirty years, if we all live, we're going to be outcasts. The kids will have to decide what to do with us."

Tanya suggested we might begin to dislike each other after just a few years. What if we couldn't tolerate each other? Jokingly, Horace suggested that we design the house with hate-each-other apartments. David's intriguing contribution was, if someone should die or we couldn't get along, to have a prior agreement to sell the individual shares in our corporation to an outsider of the same sex. "That way," he said, "if Horace and I got bored, we could sell out to some impotent old man who might get a charge out of having two young wives."

Tanya and I, hysterical with laughter, vetoed the idea. If Horace and David got bored servicing two women, the only alternative would be for them to sleep together.

"What if one of us dies?" I asked.

"That's just the point," Horace said. "Let's say I die first. You and Tanya and the kids would have the kind of security with David that he and I are both trying to achieve for you independently."

"David would be a bigamist." Tanya wiped the tears of laughter from her eyes. "That would be illegal."

"Somehow, I think we're illegal right now." I was convinced of it. "I wish one of us dared ask a lawyer. Could we be put in jail for immoral conduct?"

"The hell with the legality of our private, married lives." David frowned. "We aren't hurting anyone. What we do together is nobody's damned business but our own!"

* * *

David and Horace, hell-bent to unite our activities as a family, began searching for an auxiliary sloop so that we could sail together and perhaps spend weekends on Catalina. Sundays, while they explored boatyards and marinas with the kids, Tanya and I caught up on our housework, which had become an increasingly co-operative effort.

While Tanya and Horace were enthusiastic over David's plan of incorporating as a joint family, and we finally agreed that David should ask his lawyer to form our corporation, I still had lingering doubts.

"It isn't only that you and I may one day come to blows over systems and methods of housekeeping," I told Tanya, as we tried to organize the accumulated week's confusion of the Shea house. "But there are other things that bother me. David's assumption that we can make a commitment by pooling our income and resources is all right, so far as it goes, but, damn it, we still lack social legitimacy."

I tried to find the words to express the daydream shaping in my mind.

"If we're going to survive the social upheaval we're creating, there should be some way we could get married in a civil ceremony as two couples." Tanya was piling clothes in the washing machine as I continued. "If there were such a form of marriage, approved by the state, I think a lot of people would find it a solution for their problems. At least, those who didn't agree could only say we were insane, not immoral or illegal."

Tanya whistled. "You've been reading too many of Horace's books. Most females wouldn't take the risk of sharing their husbands."

"You and I have."

"Maybe we're crazy. I've noticed, though, in the past four weeks since we've been eating meals together, that you're getting a little sloppier and I'm getting a little neater. That's a marriage I never thought we'd achieve!"

I laughed. "Since we aren't in competition for the males, we're becoming happier females. While I don't see eye to eye with you on many things, and I know I'm more of a worrier than you are, somehow I actually love you." I blushed. "I suppose that's a dumb thing to say to another female."

Tanya's eyes glistened. "Oh, Nancy!" She hugged me quickly. "I need you, too. It's strange. Sometimes I feel like an older sister, protective toward you, and yet I know it's mutual. I'm the dopey one who leans on you."

"Still," I persisted, "if there was such a legal marriage arrangement, it would give us protection against the Sylvia Mais and Betty Vinsons. Betty is still working for David. How do you know Horace isn't continuing his afternoons with Sylvia?"

"My God!" Tanya beamed at me. "Maybe, but I doubt it. I suppose as a last recourse we could compare notes. But my hunch is neither David nor Horace has enough stamina to keep three women satisfied, even if they wanted to try. Anyway, they're so ecstatically happy with the whole adventure of their new family that I don't think it's a problem. At the moment, I think the children should be our only major concern. It's obvious that Sam and Susan, if not Jimmy and Mitch, know that the four of us act differently toward each other than other families do toward their friends." Tanya shrugged. "Since I doubt if the four of us will live to see the day when we're legal, I think we should face the problem of the kids head on. We have to tell them; they have to understand that the details of our family, even playing around naked together, are not for publication."

Of course Tanya was right, though the challenge was not only to make them understand, but also to create an environment that would give them a strength and security, a way of life they could believe in should we ever be exposed to public controversy, as I was sure we must be. Tanya and I agreed that spontaneous affection expressed with bodily contacts would help in the reconditioning process. All four of us had been raised in an atmosphere where overt emotions were rarely revealed. Just plain happy hugging, face rubbing, or tears of joy in our eyes because we liked each other were considered embarrassing or unfashionable.

At first the children thought we were silly when we snuggled them good-bye in the morning. They watched curiously when we embraced Horace and David, urging them to hurry home because we would be lonesome. Tanya and I pecked them all on the lips and cheeks, while we ranted with any joyous nonsense that crossed our minds. Within a few weeks even Susan, who at first starchily withdrew from the confusion of good-byes and hellos, joined the fray, hugging Tanya and Horace as cheerfully as she did David and me. The men complained that it was taking as much as ten minutes to leave and even longer to be welcomed home, yet they had to admit that our first constrained attempts at affection had passed into a spontaneous nuttiness they looked forward to. While the warm feeling of belonging may not have been intellectualized by the children, it was obvious they were responding enthusiastically to the communal atmosphere.

In the crazy-quilt pattern of the life we were stitching together, Tanya's pregnancy and the coming baby became a continuing subject of dinner-time discussion. Tanya had reluctantly gone to a local doctor for prenatal care, but he had tried to make her give up her determination to have a natural childbirth. One night late in February when David arrived home, he seemed unusually effervescent. Having kissed me good-bye first in the morning, he reversed the order—hugged Tanya, felt her belly and pulled us both down on the floor while he tumbled with the kids. I smelled his breath and told Tanya he hadn't been drinking. Horace arrived in the middle of the confusion and joined us, remarking to David that he must have sold a new account.

"Nope, business was just its dull self," David said as we sat down to eat. "I've a surprise for Tanya, though, but I'll tell you all about that later." He sipped a glass of tomato juice and grinned at the kids. "I was wondering on the way home if you all knew where Tan's baby really came from."

Tanya and I looked at each other in shocked silence. Was David going to tackle the impossible without consulting with us? He glanced at us indulgently, as if to say the time had come to let a business tycoon handle the problem.

"Sure," Sam grinned back at him. "Ace planted a seed in her belly."

I couldn't help laughing. Let the big boss wiggle out of that one.

"Do you know how a female gets pregnant?"

Susan smiled knowingly at Jimmy, but didn't say anything. Then to my surprise, Jimmy answered. "Sure, when people get married they go to bed and make love."

Tanya gasped at the next question. "Do you know *how* a man and woman make love?"

"They kiss each other," Mitch suggested.

"That's not all," Susan said, showing her superiority. "A man puts his penis inside the woman. He has seed there, and a woman has an egg inside her." She smiled at me. "Sea told me all about it. It happens when they get married, and that's why people get married."

"They don't have to get married to do that," Sam said, "but I guess they should."

* * *

"What we're involved in definitely

isn't monogamy. Any extracurricular sexual activity affects three other people, not just one. I think it would devalue us all." She smiled at Horace and David, who were listening closely. "You've both traded philandering for the absolute security and certain delights of two different bed-mates. Any more are out!"

Gradually, after David's opening gambit with the children, we had arranged a Sunday-through-Saturday alternation of bed partners. Tanya and I were the queen bees of our own bedrooms. While this created some inconvenience for the men, who once a week shifted their clothing from house to house, mostly they complained about our combined washing program, which often ended up with their discovering they were wearing each other's underwear or socks. The children took naturally to the changing bed scene. Saturdays and Sundays, when we often slept later than they did, they ran courier service between the houses, indiscriminately, and without waiting for invitations, bursting into our bedrooms and tumbling in bed with us in groups of two or four. On weekends, all eight of us reunited for a naked plunge in the Shea pool.

*   *   *

I took the opportunity to harass Horace with my doubts.

His patience was formidable. "Sea, we've taken a giant step. David knows it, I know it. You and Tanya have nothing to fear. While the four of us, interacting on each other, have severed the chains of monogamous marriage, I think we're all constantly aware of each other, not as owned husbands, and wives, but as individuals. The we-are-alone atmosphere of two married people trying to forge a life together in this madly confused world has vanished. I think the idea of the four of us is giving us a unique strength and confidence, something we've never had before. What's more, I have no doubt, because every day we *dare* a little more to be the individuals we really are, that David and Tanya and you feel exactly the same way. In the process, sex and love have blended, into a new component. You love me and I love you, but our surrender is enlarged and magnified in the larger cohesive mirror of the four of us united as a meaningful family. No outsider could possibly give David or me the sexual satisfaction we have with you and Tanya." Horace kissed me, silently tracing with his fingers the curves of my breasts, back, and behind. He looked at me with tears in his eyes. "Sea, are you aware how you've changed? How Tanya has changed? Not competitively, but each realizing yourselves? Yet you have one wondrous thing in common—a lovely, warm, laughing affection. David and I've discussed it. We're not such Utopians that we'd be willing to blend our economic lives if we didn't believe we were doubly lucky in having two women who love us. Most men scarcely have one."